Lightning Engineering: Physics, Computer-based Test-bed, Protection of Ground and Airborne Systems

Paul Hoole · Samuel Hoole

Lightning Engineering: Physics, Computer-based Test-bed, Protection of Ground and Airborne Systems

Springer

Paul Hoole
Wessex Institute of Technology
Southhampton, UK

Samuel Hoole
Electrical and Computer Engineering
Michigan State University
Michigan, MI, USA

ISBN 978-3-030-94727-9   ISBN 978-3-030-94728-6   (eBook)
https://doi.org/10.1007/978-3-030-94728-6

This Springer imprint is published by the registered company Springer Nature Switzerland AG
The registered company address is: Gewerbestrasse 11, 6330 Cham, Switzerland

In loving memory of our amazing parents of seven children

Richard and Jeevamany Hoole

Treasured gifts of God's grace

And out of the throne proceeded lightning and thunder ….

(The Book of Revelation)

Preface

Lightning is a naturally occurring phenomenon of the global atmospheric electric circuit that evokes a sense of beauty, wonder, and grandeur. In its ability to cause electrical disruptions and damage, the engineer has to plan and design for lightning protection. The scientist seeks to explore the physics behind lightning and the details of its interactions with earth systems such as electric power systems, and airborne systems such as aircraft navigation and safety systems, and the important part lightning plays in the ecosystem, climate change, and severe storms. A lightning flash is a sudden flow of electrical charges within a cloud, from one cloud to another, from cloud to air, from a cloud to ground, or from ground to cloud. The direct and indirect effects of the lightning flash can adversely sever the operations of grounded installations and structures as well as ungrounded structures as in the case of aircraft inflight systems. Man has been adapting to these enigmas and threats in devising protection measures to alleviate the severity of these effects on structures and electric systems. These are what this book is about.

An Outline of the Contents of the Book

This book gives a contemporary and comprehensive overview of the physics of lightning and of lightning protection systems. It is based on close to 40 years of research, teaching, and consultancy work by the authors, in UK, USA, Asia, and the Pacific. The book is organized into various chapters to give the readers a comprehensive view of the physics of lightning and lightning protection measures. Each chapter of this book is designed to be a standalone chapter. As such, the introductory chapter gives a broad and we believe a helpful preview of the thunderstorm, the different components of the lightning flash, and the lightning protection principles based on the basic science of the lightning flash. Chapter 2 gives an overview of the climatology of lightning and electric storms. Further, it provides an overview of the lightning discharge, beginning with the preliminary discharges or processes such as corona, stepped leader, streamers, and inter-stroke process, moving to the dart (or dart stepped) leader, first and subsequent return strokes, including the important sub-microsecond threats, and continuous current.

Chapter 2 also delves into the very important aspect of pre-lightning electrostatics, which is critical in building structures (whether aircraft or a smart city) and systems that are less susceptible to both lightning induced Electrostatic Discharges (ESD) and lightning strikes. Chapter 2 also provides in-depth insights into pre-lightning strike and electrostatics. It discusses the subject in the context of the complex lightning-aircraft electrodynamics. The chapters also provide a topical section on the three-dimensional software modeling of dipoles to determine the electric field buildup on the surface of any structure under the electric thundercloud. Using the knowledge of both the electric charges and the electric fields, it is possible to form zones over the aircraft body to indicate areas of high risk to lightning strike. Further, this powerful software tool can be used to design electric substations, ground structures, and aircraft to minimize the lightning strike threat. These are the areas in which microelectronic, navigation, and instrumentation equipment will be subject to high electrostatic stress, possibly leading to electrostatic discharges. Although the technique for determining the pre-lightning, thundercloud electric charge generated electrostatic fields, electric charges, and voltages may be applied to ground structures and systems, as well as to airborne structures and systems, the application of the technique is illustrated for an aircraft, considered as a floating electrode, entering into the severe electric field environment of a thundercloud.

Chapter 3 builds on the initial chapters to present basic and effective measures to protect against lightning threat to ground and airborne structures and systems, protection measures to be used in high voltage to low voltage computer and communication systems, as well as in commercial and domestic buildings. The chapter helps the reader to gather lightning protection know-how in an understandable way by focusing on lightning protection of domestic houses, including the air-termination system, grounding system and internal protection of electrical and electronic equipment from lightning surges. Chapter 3 also helps the reader to move into broader thinking and appreciation of lightning protection principles by looking at the lightning protection of boats, historic buildings, Photovoltaic (PV) systems, and the wind turbine.

Chapter 4 moves from Chap. 3 into other associated protection techniques and is devoted to protection detailed measures in key installations and infrastructures that are the lifeblood of any nation's economy. These installations and infrastructures include energy, communication, and transport systems and building structures. Lightning risk analysis is presented, as well as protection basics used in almost all electronic systems, from medical systems to telecommunications systems: shielding, grounding, and bonding.

Chapters 3 and 4 on lightning protection of communication systems discuss the need to heighten the protection system against the direct and indirect effects of lightning, Lightning Electromagnetic Pulses, or LEMP. The impact of indirect lightning effects becomes increasingly important with the advent of digital microelectronic technology in electrical/electronic systems. The evolution of Internet of Things through radio frequency identification devices, barcodes, smartphones, and the convergence of smart technologies in smart homes, smart industries, smart cities, smart environment, and smart ecosystem in smart people with micro-chips implanted makes new demands on protection measures. These form the smart planet and demand more extensive, interconnected, and highly sensitive lightning protection of future systems. Smart technologies utilizing modern communication and information technologies and applications of wireless sensor networks are in everyday use. Thus, the discussions on protection measures applied in advanced technologies will heighten the need for a professional approach to lightning protection. The protection measures applied in communication infrastructures and installations include due consideration of masts, air-termination rods, earth screening and grounding, and surge protection devices.

Further discussions on lightning threats associated with radio frequency interferences and Lightning Electromagnetic Pulses (LEMPs) that can impinge on the surface of a containing sensitive electronic equipment are described. With the advent of digital electronic technology in fly by wire and the evolution of aircraft function automation, LEMPs threats can have severely damaging effects. The increasing use of microelectronic devices and systems makes modern technology more vulnerable to lightning-produced voltage spikes and LEMP.

In Chap. 5 we take an in-depth look into the lightning flash, its science and the use of mathematical modeling to simulate lightning using computer codes. The chapter develops the distributed transmission line model of the lightning flash. Moreover, it presents a very general method for computing the Lightning Radiated Electromagnetic Fields (LEMP) from the lightning currents and electric charges. The techniques presented in Chap. 5 provide a self-consistent computer-based testbed for computing lightning flash simulation and interaction with engineering systems. It presents a powerful tool to get a handle on both the direct and indirect effects of lightning. The techniques related to lightning currents and radiated electromagnetic fields are crucial for the development of future lightning simulation, understanding, and testing. The lightning electrodynamics material presented in Chap. 5, combined with the computer-based electrostatics tool presented in Chap. 2, provides an all-encompassing and highly versatile testbed for pre-lightning and lightning electrodynamics study and design for both structures and electrical and electronic systems.

In Chap. 6 we look into the localization of lightning from two kinds of waves the lightning flash produces. These are, namely, the acoustic wave associated with thunder and the electromagnetic wave associated with LEMP. New techniques are introduced for finding where lightning actually occurred, using either the acoustic wave measured or by using the measured LEMP. The development of these techniques used with artificial intelligence helps us to localize lightning as well as to recognize the lightning generated wave.

In Chap. 7 we go into details of two special, but different types of lightning protection scenario. These are the protection of electric power systems and the protection of aircraft. The chapter focuses on modeling and computer-based analysis to investigate aspects of lightning interaction normally not accessible to measurements. As the aerospace industry expands into both manned and unmanned commercial and military vehicles using materials such as carbon fiber composites, preventing electric field enhanced aircraft initiated lightning strikes becomes a major concern. The aircraft-lightning environment threat is heightened further with the latest state-of-the-art digital electronics in command, control and automation, and communication systems. For an aircraft to be air worthy, aircraft manufacturers need to provide the overall assurance of adequate lightning protection. This process requires certification plans for tests done on components or systems of components. The protection of aircraft against lightning strike can be categorized into the following steps (i) determine lightning attachment zone; (ii) determine systems and components which are likely to be damaged by lightning; (iii) set lightning protection standards for systems and components; and (iv) confirm the rationality of the protection design by the use of laboratory tests.

Chapters 3, 4, and 7 on structural lightning protection discuss protection systems for building infrastructures against direct and indirect lightning effects. Lightning threats can be contained through proper protection measures such as surge protection devices, and proper shielding and grounding practices as defined in various standards on lightning protections. With climate change changes in lightning activity intensity and pattern, new electric storm environments that will be encountered by electric power and electronic systems, and both ground and airborne structures.

The Unique Contribution of this Book

There is increasing threat posed by lightning and its effects on electrical and electronic systems as well as structures. Although, for almost a century, protective equipment has been developed for lightning, it still does so much damage that insurance claims on lightning damage figure very high in developed countries. Electric grid and electronic systems are regularly destabilized or damaged by lightning and electric corrosion associated with several lightning surge hits to a given equipment or system. This book seeks to bring out the best insights of almost a century of lightning research and development of lightning protection systems, gathering the important understanding, ideas, and protection methods from published researchers and engineers in several lightning hot spots. We are in particular debt to advanced work, volumes of work, done in the USA, Sweden, UK, Japan, and Switzerland. Moreover, this book provides several important details and aspects missing in the literature, including:

(a)   A focused presentation of lightning protection of ground and airborne systems and structures, rather than producing lengthy descriptions of lightning threat and protection. This book also highlights the sub-microsecond changes in the return stroke and their threats to both Ultra High Frequency (UHF) communication and electronic systems as well as the steep rate of rise of currents due to these changes which pose a severe threat to high voltage and low voltage electric systems. These need to be incorporated into our understanding of lightning-caused disruption and damage, in high voltage testing, protection, and standards.

(b)   A self-contained electric charge and electromagnetics based model of the lightning channel and return stroke, where computation of currents and voltages not easily accessible to direct measurements are calculated from the computational testbed which may be used for design and protection of both ground systems such as power systems, as well as airborne structures such as aircraft. In contrast the focus of the past decades has been on field measurements to get a handle on lightning return stroke currents by curve fitting rather than by self-contained scientific models.

(c)   Lightning models that are well-attested by physics, together with a detailed study of the physics of both electric gas discharge and electromagnetic fields. Instead of using curve fitting models, the book presents a scientific model that exactly follows what happens in real time, by first determining the lightning currents and voltage shock waves, then determining the lightning electromagnetic pulse radiated by the fast moving high-current pulse. An easy to employ and accurate electromagnetic computation technique is presented.

(d)   A detailed and self-contained analytical tool to determine the electric stress and electric potential induced by lightning in both airborne and ground structures close to a thundercloud. This computational analysis and design application is essential for hardening structures against the electrostatic stress in a thunderstorm environment, designing structures with clearly identified zones that are most prone to generate lightning or be hit by lightning. This tool helps reduce the probability of lightning strikes.

(e)   The details of studying the sub-microsecond changes in lightning generated currents and LEMP are in relation to lightning damage, testing, protection and shielding.

(f)   The development of a well-tested physics/computer based tool to test for a lightning withstand and protective system, based on the authors' work. n is to. The need for this is obvious since many important aspects of lightning cannot be tested for in the High Voltage Laboratory (e.g. wheeling into any HV Laboratory a large Power Substation Transformer, Fighter Aircraft or Airbus Commercial Aircraft.). Nor is it possible in an HV laboratory to generate the important sub-microsecond changes and threats to structure and to performance-critical microelectronic equipment, floating structures hit by lightning, etc. In this book we present several results, not accessible to measurements or by HV Laboratory testing. If there should be a companion volume to this book, we expect to discuss the comprehensive studies on pre-lightning strike threat, design and protection, direct lightning strike effects, and crucial data on currents and voltages that are not accessible to measurements, as well as radiated LEMP and indirect effects.

This book seeks to cover every important area and detail of lightning protection, including the protection of electric power systems, electronic equipment, low voltage systems, structures, and grounding. The above are a few of the important points, in a book which we hope is scattered with critical engineering and research insights and challenges.

The authors wish to thank the DEHN, Germany for generously allowing us to make use of material and figures from their arguably the most detailed book on critical areas of lightning protection: *Lightning Protection Guide*, in which they have so helpfully shared their years of experience and minute details of systems they have developed for lightning protection. The authors are also thankful for the generosity of NASA, USA and NSSL_NOAA, USA, who have an enthusiasm not only for high tech, cutting edge research and exploration, but also to encourage education and knowledge; in this book, we have used some amazing photographs of lightning which they have made available. We are thankful to the Springer publishers for the pleasant opportunity to work with them on this book.

Finally, it has been an honor and a pleasure to write such a book as this that the authors hope will serve a useful guide to readers—including students, engineers, researchers and general readers—on the lightning flash and lightning protection measures.

The heavens declare the glory of God; the skies proclaim the work of his hands......

(The Book of Psalms)

Southampton, UK   Paul Hoole
                                        (D.Phil. Eng. Oxford University)

Michigan, USA   Samuel Hoole
                    (Ph.D. Carnegie Mellon, D.Sc. (Eng.) Lond.,
                                                    IEEE Life Fellow)

Contents

1 Introduction to Lightning and Lightning Protection   1
  1.1   The Lightning Flash: General Characteristics and Damage
         Caused   1
  1.2   The Leader Stroke   11
  1.3   The Return Stroke   15
        1.3.1   General Description   15
        1.3.2   The Empirical Model   16
        1.3.3   Lightning Return Stroke Models   19
  1.4   Lightning Radiated Electromagnetic Pulses (LEMP)   20
        1.4.1   Computation of Radiated Electromagnetic Pulses   20
        1.4.2   Calculating Rate of Rise of Currents
                from Measured Electric Fields   23
  1.5   Electromagnetic Waves   24
  1.6   Lightning Protection: An Introduction   27
        1.6.1   Lightning Effects   27
        1.6.2   Effects of Lightning on Aircraft   32
        1.6.3   Lightning Effects on Electric Power Systems
                Network   34
        1.6.4   Substation Protection Systems   36
        1.6.5   Rolling Sphere Method Applied in Substation
                Protections   37
        1.6.6   Lightning Protection Methods for Buildings
                and Infrastructures   38
  1.7   Lightning, Climate, Upper ionosphere, and Other Planets   40
        1.7.1   Effect of Temperature on Lightning   41
        1.7.2   Effect of Lightning on Troposphere   42
  1.8   Summary   45
  Bibliography   46

2   Thunderstorms and Pre-lightning Electrostatics   51
     2.1   Introduction   51
     2.2   Formation of Thunderclouds   53
     2.3   The Climatology of Lightning   55
            2.3.1   Cloud Electrification   55
            2.3.2   Cloud Electric Charge Formation   55
     2.4   Negative Lightning Discharge Process   57
            2.4.1   The Negative Lightning   57
            2.4.2   The Electric Discharge Process   58
     2.5   Lightning-Aircraft Electrostatic Interactions   59
            2.5.1   Two Types of Attachment Initiation   59
            2.5.2   Aircraft-Triggered Lightning   59
            2.5.3   Aircraft Intercepted Lightning   61
     2.6   Probability of Lightning Strike to Aircraft   61
            2.6.1   Factors Affecting Probability   61
            2.6.2   Probability Dependence on Aircraft Size   61
            2.6.3   Probability Dependence on Flight Profile   62
            2.6.4   Probability Dependence on Geographic Area
                     of Operations   62
     2.7   Thundercloud Induced Electrostatic Charges   63
     2.8   Pre-lightning Flash Electrostatics of Thunderstorms:
            Analysis   64
            2.8.1   The Electrostatic Fields   64
            2.8.2   Aircraft and Electric Dipole Placements   65
            2.8.3   Determining the Electric Charges Induced
                     on an Aircraft and the Electric Fields Generated
                     Around an Aircraft Body   67
            2.8.4   Analysis of the Airbus A380 Aircraft Results   69
            2.8.5   Zoning   72
            2.8.6   A F16 Military Aircraft Flying Between Two
                     Charged Centers   72
     2.9   Electrostatic Fields of Pre-lightning Thundercloud
            Environment   74
     2.10 Electrostatic Computation and Evaluation:
            A Computer-Based Tool   79
     2.11 Personal Lightning Safety   79
     Bibliography   80

**3   Lightning Protection of Domestic, Commercial, and Transport
     Systems**   85
     3.1   General   85
     3.2   Lightning Protection of Houses   87
            3.2.1   An Overview   87
            3.2.2   Choosing Service Entrance Surge Protectors
                     (SPDs)   90

   3.2.3 Surge Current Rating 91
   3.2.4 Ground Potential Rise 92
   3.2.5 Signal Protectors 93
   3.2.6 Inter-System Bonding 93
   3.2.7 Special Purpose Protectors 93
 3.3 Boats 94
 3.4 Photovoltaic (PV) Systems 96
 3.5 Frequency Converter Protection 99
 3.6 Networks and Interactive Services 100
 3.7 Wind Turbines 101
 3.8 Historic Buildings 103
 Bibliography 104

**4 Practice of Lightning Protection: Risk Assessment, External
Protection, Internal Protection, Surge Protection, Air
Termination, Down Conductor, Earthing, and Shielding** 105
 4.1 Introduction 105
 4.2 General Principles of Lightning Protection 106
 4.3 Risk Management 108
   4.3.1 Introduction 108
   4.3.2 Risk Assessment: Basics 108
   4.3.3 Advanced Risk Assessment 110
 4.4 Inspection of Lightning Protection System 113
 4.5 Internal Lightning Protection 114
   4.5.1 Surge Protection Measures 114
   4.5.2 Lightning Protection Zones 116
   4.5.3 SPM Management 118
 4.6 Equipotential Bonding for Metal Installations 120
   4.6.1 Prologue 120
   4.6.2 Equipotential Bonding for Metal Installations
      at the Boundary of LPZ0$_A$ and LPZ1 120
   4.6.3 Equipotential Bonding for Metal Installations
      at Boundary of LPZ 1 and LPZ 2 121
   4.6.4 Protective Equipotential Bonding 121
   4.6.5 Earth-Termination System for Equipotential
      Bonding 122
   4.6.6 Protective Bonding Conductors 122
   4.6.7 Equipotential Bonding Bars 123
   4.6.8 Integrating Pipes in Equipotential Bonding System 123
   4.6.9 Testing and Monitoring Equipotential Bonding
      System 123
   4.6.10 Supplementary Protective Equipotential Bonding 123
   4.6.11 Minimum Cross Section for Equipotential
      Bonding Conductors 124
   4.6.12 Equipotential Bonding for Power Supply Systems 124

	4.6.13	Equipotential Bonding for Power Supply Systems at the Boundary of LPZ0$_A$ and LPZ1	124
	4.6.14	Equipotential Bonding for Power Supply Systems at the Boundary of LPZ0$_A$ and LPZ2	126
	4.6.15	Equipotential Bonding for Power Supply Systems at the Boundary of LPZ1–LPZ2	127
4.7		Equipotential Bonding for Information Technology (IT) Systems	127
	4.7.1	Introduction	127
	4.7.2	Equipotential Bonding for IT Systems at the Boundary of LPZ0$_A$ and LPZ1	128
	4.7.3	Equipotential Bonding for IT Systems at the Boundary of LPZ0$_A$ and LPZ2	128
	4.7.4	Equipotential Bonding for IT Systems at the Boundary of LPZ 1 and LPZ 2 and Higher	129
4.8		Protection of Antenna Systems	129
4.9		Protection of Optical Fiber Installations	130
4.10		Telecommunication Lines	131
4.11		Choosing Internal Lightning Protection System: Type of Surge Protection Devices (SPDs)	132
4.12		External Lightning Protection	134
4.13		Air-Termination Systems	138
	4.13.1	Isolated and Non-isolated Air-Termination Systems	138
	4.13.2	Air-Termination System for Buildings with Different Types of Roof	139
	4.13.3	Air-Termination System for Building with Gable Roofs	139
	4.13.4	Air-Termination System for Buildings with Flat Roofs	140
	4.13.5	Air-Termination System for Buildings with Metal Roofs	140
	4.13.6	Air-Termination System for Buildings with Thatched Roofs	141
	4.13.7	Air-Termination System for Buildings with Inaccessible Roofs	141
	4.13.8	Air-Termination System for Buildings with Green Roofs	142
	4.13.9	Air-Termination System for Steeples and Churches	142
	4.13.10	Air-Termination Rods Subjected to Wind Loads	142
	4.13.11	Safety System and Lightning Protection	143
4.14		Down Conductors	143
	4.14.1	Determination of the Number of Down Conductors	143
	4.14.2	Down Conductors for a Non-isolated Lightning Protection System	144

4.15   Earth-Termination System   145
4.16   Manufacturer's Test of Lightning Protection Components   148
4.17   Shielding of electrical and electronic systems against LEMP   149
        4.17.1   Magnetic Field Calculations for Shielding   149
        4.17.2   Calculation of the Magnetic Field Strength in Case
                  of A Direct Lightning Strike   150
        4.17.3   To Determine the Magnetic Field in Case
                  of Nearby Lightning Strike   151
        4.17.4   Implementation of the Magnetic Shield
                  Attenuation of Building/Room Shield   152
        4.17.5   Cable Shielding   153
References   155

5 **Lightning Physics, Modeling, and Radiated Electromagnetic**
   Fields   157
   5.1   Introduction: The Need for Computer-Based Testbeds
          for Lightning Testing   157
   5.2   Lightning Return Stroke   160
          5.2.1   Electromagnetic Wave Nature of the Lightning
                   Return Stroke   160
          5.2.2   Lightning Return Stroke Models   163
   5.3   Analysis of Experimental Data of Lightning Return Stroke   164
          5.3.1   Background   164
          5.3.2   Lightning Current and Electromagnetic Field
                   Measurements   165
          5.3.3   The Empirical Models: Lumped Circuit Model
                   and the Curve Fitting Model   168
   5.4   The Distributed Circuit, Transmission Line Model (DLCRM)   171
          5.4.1   Background to the DLCRM   171
          5.4.2   The Transmission Line Dispersion Relation   173
          5.4.3   Numerical Solution of the Transmission Line
                   Wave Equation   176
          5.4.4   Return Stroke Velocity and the Transmission Line
                   Model   178
   5.5   Negative Cloud to Ground Earth Flash Return Stroke:
          Simulated by the DLCRM   181
          5.5.1   Background   181
          5.5.2   LRS Currents from DLCRM Simulation   182
          5.5.3   Calculation of the Electric and Magnetic Fields
                   Radiated from the Lightning Currents   183
          5.5.4   Computed Electromagnetic Field Pulses LEMPs   192
          5.5.5   LRS Electric and Magnetic Fields Calculated
                   from Currents Obtained from DLCRM Simulation   192
          5.5.6   Summary   197
   5.6   A Case Study: Lightning Interaction with Aircraft   198

        5.6.1   Aircraft and Lightning Protection 198
        5.6.2   Computation of Lightning Currents and Voltage
                 on An Aircraft 199
   References 204

6 Localization and Identification of Acoustic and Radio Wave
**   Signals Using Signal Wavefronts with Artificial Intelligence:**
**   Applications in Lightning** 209
   6.1   Introduction 210
   6.2   Methodology: Test Signals and Wavefronts 211
        6.2.1   Methodology for Acoustic Signals 211
        6.2.2   Methodology for Radio Wave Signals 212
   6.3   Test Results 218
        6.3.1   Test Results of Acoustic Signal Model 218
        6.3.2   Test Results of Radio Wave Model 219
   6.4   An Array Antenna for Direction and Identity of Lightning
        Radiated Signals 223
   6.5   Application of the Perceptron ANN for UHF Lightning
        Flash Detection 226
   6.6   Conclusion 230
   Bibliography 230

7 Lightning Electrodynamics: Electric Power Systems
**   and Aircraft** 233
   7.1   Introduction 234
        7.1.1   Lightning and Electric Power Systems 234
        7.1.2   Lightning and Aircraft 236
   7.2   Circuit Elements Used in Back Flashover and Shielding
        Failure Performances 237
        7.2.1   Preamble 237
        7.2.2   Tower Surge Impedance 237
        7.2.3   Shield Wire Surge Impedance 238
        7.2.4   Tower Ground Resistance 239
        7.2.5   Conductor Circuit Elements 240
   7.3   Lightning Fash Parameters 240
        7.3.1   Ground Flash Density 240
        7.3.2   Number of Lightning Strokes to the Line 241
   7.4   Simulations of Lightning Flash to a Transmission Line 242
        7.4.1   Back Flashover Analysis for 500 kV Transmission
                 Line 242
        7.4.2   Sub-microsecond Analysis of Conductor Back
                 Flashover Current at Substation Tower 243
        7.4.3   Sub-microsecond Analysis of Shielding Failure 245
        7.4.4   Back Flashover, Shielding Failure Current
                 Parameters, and Mitigation of Flashovers 246
        7.4.5   Summary 247

7.5   Lightning Flashover on Transmission and Distribution Lines   247
        7.5.1   Back Flashover   247
        7.5.2   Shielding Failure   248
        7.5.3   Probability and Intensities of a Flashover   248
        7.5.4   Factors Influencing Lightning Strike
                 to Transmission Lines   249
7.6   Protection Measures to Reduce Impacts of Lightning
        on Transmission Lines   252
7.7   Lightning-Aircraft Electric Circuit Models   254
        7.7.1   The Basic Equations   254
        7.7.2   The DLCRM Parameters of the Lightning channel   259
        7.7.3   The DLCRM Parameters of the Aircraft   260
        7.7.4   The F16 Military Aircraft and Lightning Strike   266
        7.7.5   The Airbus A380 Commercial Aircraft
                 and Lightning Strike   271
7.8   Swept Stroke Mechanism   274
7.9   Metallic Versus Carbon Fiber Composite Aircraft   274
7.10   Significance of Lightning Testing Standards
        and Certifications   276
        7.10.1   Procedural Requirements   276
        7.10.2   Direct Lightning Effects Protection   277
        7.10.3   Indirect Lightning Effects Protection   278
        7.10.4   Scaling Test Method   279
        7.10.5   Measurements on Aircraft Struck by Lightning
                 in Flight   279
        7.10.6   Airport Lightning Protection   280
        7.10.7   Lightning Engineering: The Present and the Future   280
Bibliography   283

**Appendix: STAT2ARC2EMP: A Computer-Based High-Voltage
Testbed for Electrostatic and Transient Current Threats to Ground
and Airborne Structures and Equipment: For Arcs and Lightning
Flashes**   289

Index   295

Chapter 1
Introduction to Lightning and Lightning Protection

Abstract In this chapter we introduce the entire subject of the book from both engineering and physics perspectives. A brief presentation of the general nature of lightning flashes is followed by describing, with simple models, the two main parts of the lightning flash. Namely, the leader stroke and the return stroke. The electromagnetic phenomena related to lightning is also presented. First the electromagnetic waves along the lightning channel are analyzed considering the lightning channel as an electric plasma channel with free electric charge particles moving in it; We study the electric parameters of the lightning channel, including its electric conductivity. Models of the lightning flash are briefly presented. Lightning protection is summarized in this chapter considering aircraft interaction with lightning and aircraft protection zones, and protection of electric power systems. In addition, the protection of electronic systems and devices is also considered.

1.1 The Lightning Flash: General Characteristics and Damage Caused

Lightning engineering is an increasingly important discipline due to the increase in lightning damage to electronic and microelectronic systems that are operated at very low voltage and current levels. The electronic systems include computers, communication systems, medical equipment, security and safety equipment, military systems, and monitoring devices. Relatively small lightning induced voltage surges and slightly increased current flows can damage and disrupt the function of these sensitive systems used in navigation, military technology, biomedical systems and many other transport, business and service systems and smart homes applications in ground and airborne systems and devices. Furthermore, since electrical power and communication, and command and control systems are interconnected and cover a large space, the entire system is simultaneously exposed to lightning-caused electric voltage and current threats. When large machinery to handheld devices are electronically monitored and operated, the entire interconnected system hardware is exposed to instability and damage if the microelectronic systems should be interfered with or burnt by lightning flash voltage impulses. Present and future smart cities are particularly vulnerable to lightning-caused malfunction and damage.

As much as thirty percent of damage in electrical power and electronic systems is caused by over-voltages due to switching and lightning flashes. The remaining seventy percent of damage is due to water, human error, fire, sundries, theft and storms. Damage due to lightning surges far exceeds that due to switching surges. Lightning damage is caused by direct lightning strikes to installations and structures, as well as by indirect effects of lightning where the electromagnetic pulse radiated by lightning (lightning electromagnetic pulse, LEMP) induces surge voltages and currents in distant, electrically unconnected electric power lines and electronic systems. The effects of LEMP are similar to those due to nuclear electromagnetic pulses (NEMP). The ratio of the number of direct lightning strike surges to indirect lightning surges is about 1: 600, but the ratio of damage caused is about 1: 2 since direct strikes are far more severe than the surges induced by indirect effects of LEMP. In other words, say there are 100 direct lightning strike surges in a system of a city each year. Then the indirect lightning effect caused surges will be about 600,000. However, if the number of damages caused by direct lightning strike to the telecommunication installations of the city is 50,000, then the damage caused by indirect effects will number 100,000. The cost of lightning-caused damage to electrical and electronic systems and devices runs into tens of millions each year for a moderately sized, technologically advanced country. Insurance payout due to lightning has reached such high proportions that insurers only pay for damages to hardware, and even that only if it is a first event. After the first damage, they expect the customer to improve lightning protection to prevent further damage.

Consider first a few examples of lightning-caused damage to hazardous areas. Outdoor or underground (e.g. diesel station) Storage tank flammable material is susceptible to catch fire when lightning strikes the tank or the ground nearby. In 1965 a solid petrol tank roof was struck by lightning. Once the petrol tank roof voltage was elevated above the lightning flash voltage of a million volts, the large volt drops between the tank and the wires of the measuring cable, which was at earth potential, resulted in an electric arc flashover between the roof and the cable. The electric arc which is at a very high temperature fired the explosive mixture. The whole tank exploded and was burnt. In Netherlands, in 1975, a lightning flash to a tree close to a kerosene tank resulted in a flashover between the roots of the tree and the underground earthing system of the tank. Once the earthing conductor voltage increased to millions of volts, there was a flashover between the earth conductor and the line running from the thermostat measuring the temperature inside the tank. The flashover ignited the kerosene-air mixture inside, which resulted in an explosion and a fire. In 1984 in Herne, the potential of the measuring cable entering the tank was raised due to lightning flashes, and the potential drop between the measuring line and the ground conductor caused an arcing flashover inside the alcohol chemical plant resulting in a fire. In 1995, an Indonesian oil refinery tank was hit, and the tank caught fire. Due to poor grounding of the earthing system this fire resulted from a lightning-produced arc. Neighboring tanks also caught fire causing a major oil crisis in the country. In 1996, a lightning strike to a petrol tank in USA set fire to multiple

Fig. 1.1 Multiple lightning flashes over a built-up city *Credit* NOAA_NSSL With permission

tanks. The basic reason for these damages to explosive installations is the potential drop of about one million volts that develops between the enclosing Faraday cage like tank and the single cable coming into the tank, where the cable is connected to monitoring and measuring devices inside the tank.

Figure 1.1 shows multiple lightning flashes from the cloud to ground (CG). The downward, that is cloud to ground, direction of the initial electric breakdown (the leader stroke) is indicated by the downward pointing branches of the lightning flash. As cities move towards greater use of Internet of Things (IoTs) and smart cities, the threat of lightning induced impulses poses a greater threat to microelectronic system based transport, safety, security, communication, navigation, and commercial systems. If the thunderstorm is in the vicinity of an airport, any aircraft that are landing or taking off may be struck by lightning, as well as aircraft parked outside the hangars. A lightning strike to a commercial aircraft taking off from the Tokyo airport showed one part of the lightning channel to originate from the radome of the aircraft and move up towards the thundercloud. With the branches of the lightning segment pointing upward, the indication is that the aircraft imitated the lightning flash due to large accumulation of electric charges at the radome resulting in an electric field greater than the breakdown electric field for air, which is about 30 kV/cm (or 3 MV/m) at ground level (for dry air but lower in wet or moist conditions). Moreover, the second portion of the lightning channel extended from the fin of the aircraft down towards the ground and had branches pointing downwards. This indicates that the second part of the lightning flash also originated from the aircraft, specifically from the aircraft tail, and moved towards the ground. When connections with the thundercloud above the aircraft and the ground below are completed, then the high-current return stroke (e.g. 300,000 Amperes, with rise times of the order of one microsecond) which radiates intense light, is initiated. The aircraft structure, as well as the internal power, electronic, control, navigation, and information technology systems and equipment need to be well protected against adverse effects of aircraft-lightning electrodynamics. In 1987, in Rundschau, lightning struck a Boeing 747 aircraft with 225 passengers, when it entered into a thunderstorm zone close to Newark, New Jersey airport. The four lightning strikes to the aircraft within a few minutes damaged the autopilot, radio communication to the airport and the weather radar. The captain and the co-pilot had to exercise immense effort to keep the aircraft flying because the elevator control was also damaged. Air to air communication with a nearby British Airways aircraft was used to safely land the aircraft. Landing gear brakes had to be used since the braking thrust reversals of the four engines were also damaged by the lightning strikes. About the structure of the aircraft, parts of the tail fins were missing, and hundreds of fire damages to the aircraft shell and wings were also found, where lightning had attached itself to the aircraft, or was burnt by the dragging of the lightning stroke over the surface of the structure.

In 1985 a lightning-caused fire accident of the Perishing II rocket in Germany killed three army personnel and injured nine others. The fire was caused by electro-static sparks produced by the thundercloud electric fields in the propelling charge of the motor. Apollo 12 rocket and the Saturn V rocket were struck by lightning in 1964, 36 s after lift-off. The Saturn V was struck by lightning when it was 2000 m above ground, with a connecting strike to the ground platform. In 1987 lightning struck the 78-million-dollar Atlas Centaur rocket 51 s after take-off, sending it out of control. Lightning struck the nose of the rocket. Both the rocket and the 83 million Pentagon satellite payload it was carrying had to be destroyed over the Atlantic Ocean. The lightning strike which penetrated the rocket, by making a 5 cm hole on its nose, disrupted the main computer which gave false commands to the driving engineer resulting in a failed trajectory. In 1987, lightning struck three small research rockets at the NASA base in Wallops, tripping the ignition mechanism. The three rockets had a common earthing system. After lifting off after the ignition was switched on by the lightning induced currents, the three rockets fell into the Atlantic Ocean. The normal practice is not to trigger the take off of a satellite launch vehicles when there is thunderstorm activity in the vicinity, just as aircraft are usually prevented from taking off or flying under thunderclouds.

Lightning also strikes small passenger planes and control towers. In Fig. 1.2a is shown a lightning strike path through a military aircraft flying at striking distance from the thundercloud. More frequently lightning strikes commercial aircraft when it takes off or landing, and under the thundercloud. In 1995 the radar control station at Changi Airport, Singapore was directly hit by lightning, and it took four hours to start up the system with the backup system. In 1993 in France, an Airbus was struck when taking off, the nose was broken, and the radar was affected, and the aircraft had to be landed in emergency. In 1992, the almost impregnable lightning protection system with 32 lightning arresters in the airport control tower was bypassed by a lightning strike knocking out the control tower for two hours. The fire control system was also set on fire. In 1996 lightning strike to the German meteorological measuring system

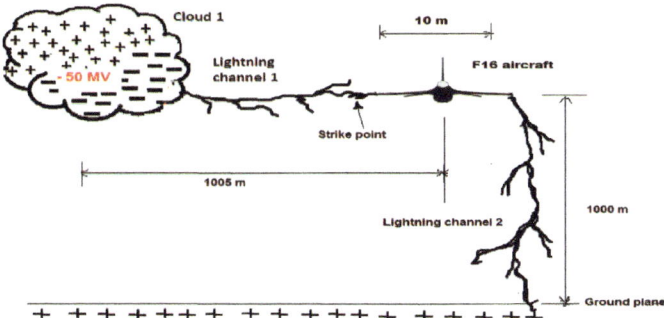

(a)

(b)

Fig. 1.2 Lightning flash to **a** Lightning strike to an aircraft; the lightning channel connects the thundercloud and ground through the aircraft. **b** Lightning strike to an elevated structure, such a electric power towers and buildings. *Credit* NOAA_NSSL, USA. With permission. Endeavor space shuttle pad hit by lightning *Credit* NASA, USA). **c** Lightning flash triggered by a rocket fired towards a thundercloud. Photograph of a space vehicle struck by lightning (*Credit* NASA, USA). **d** Lightning flash triggered by a laser beam fired towards the thundercloud, An experimental system set up in Switzerland

at Dusseldorf made it malfunction, resulting in a temporary shutdown of the airport to flights.

Figure 1.2b shows a lightning strike to an object elevated from the ground. Here the forked lightning channel is seen above the tower, such as a telecommunication tower. Lightning strikes electric power line towers as well as the bare power lines held up by the towers which run for several hundreds of kilometers. Multiple lightning channels from the strike point to a power line indicate that there is not only one flash, but following the first stroke, are subsequent strokes to the same point on the power line through the now ionized channels, imitating a multiple number of destructive high voltage transient pulses that will travel along the line in both directions, that is,

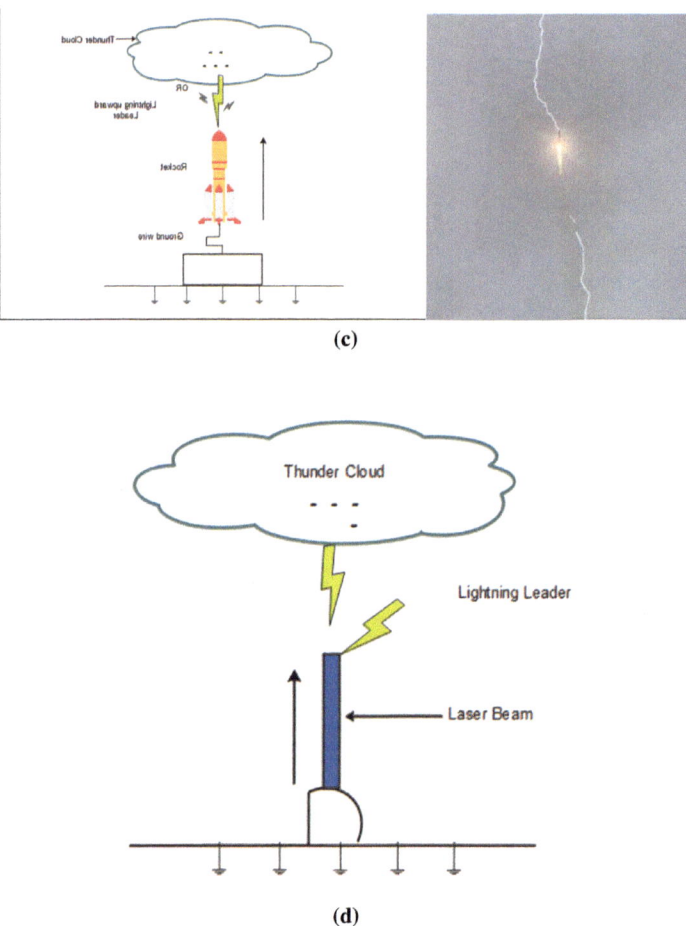

Fig. 1.2 (continued)

towards the power generating station at one end and towards the power substation at the other end, at which the transmission voltage is stepped down to lower voltages for electric power distribution. In Fig. 1.2c is shown an artificially triggered lightning strike. A rocket which has a light, flexible, grounded conducting wire connected to it, is fired towards a thundercloud. The rocket, thereby, takes the ground potential close to the thundercloud, thus increasing the electric field at the tip of the rocket at ground potential to make it become very large until it launches a leader stroke towards the thundercloud. Then the first return strike occurs through the wire and the rocket, melting or destroying both with its heat. In Fig. 1.2d, instead of a rocket attached to a ground conductor wire, a laser beam is shot towards the thundercloud, which takes the ground potential closer to the thundercloud. Here the first return stroke occurs through the conducting laser beam. These artificially triggered lightning strikes are used to make measurements on lightning at a controlled, instrumented, fixed point on earth.

About thirty percent of electric power failures in the USA are due to lightning flashes. The damage caused can exceed five billion dollars. Lightning-caused electric power failure and damage to both the power apparatus and the consumer installations connected to the power grid are much higher in poorly protected electric grids in developing countries. In 1977 a 345 kV power line close to New York was struck by lightning. With the whole city plunged into an electric power cut, it took one day to restore electric power, with a loss of 350 million dollars to the city. Two 345 kV power lines in Minnesota went out of service due to lightning and subsequent overheating of power lines sagging down to touch trees, caused further electric short circuits and failures. For over nineteen hours, the chain reaction at other interconnected state power grids resulted in eight states plunging into loss of electricity. In technically developed countries, such as France, close to five percent of all insurance claims in a year is lightning related. In the telecommunication industry this can be close to ten percent of all insurance payments. In the USA about 50% of the annual 200,000 forest fires are due to lightning. In the summer of 1999 about 2000 forest fires were caused by lightning flashes resulting in 400 million dollars of property damage. With the increase in renewable energy sources used for the generation of electricity, lightning protection of both wind power stations and solar power installations becomes more important. In 1988 a particularly severe thunderstorm in Sweden triggered 1400 alarms, and the police radio and telephone exchange was disrupted. The 130 kV power system failed, and the emergency generators were not started up because the control computers were damaged. The low voltage mains distribution, the control room, and computer terminals were damaged. In 1993 the rotor wings of the large wind generator in Helgoland, Germany, were destroyed, causing damage with up to DM 800,000 spent on repair. Although the annual number of lightning flashes over the ocean is much lower than the annual number of lightning flashes over land, offshore oil platforms need lightning protection because of the tall structures and the special type of material handled.

Figure 1.3a shows a heavily branched cloud-to-ground lightning flash. The branches of the lightning flash point downwards, giving it the appearance of an inverted tree. The downward pointing branches indicate that the leader stroke of the flash traveled from cloud towards ground, and the intense return stroke traveled from the ground towards the cloud. The return stroke not only runs towards the thunder cloud, but also along the branches thus neutralizing the electric charges deposited in the branches of the leader stroke. Such lightning flashes are called cloud-to-ground (CG) lightning flashes. This is the most common type of lightning flash to ground. The upward going leader, from tall earth structures or aircraft radome (or nose) for instance, is called a ground to cloud (GC) flash. In GC flashes, the branches point upward towards the cloud. In Fig. 1.3b is shown a ground to cloud (GC) lightning flash, with the branches pointing upward. Such upward GC flashes are produced by tall buildings and towers. Lightning protection of the structure of buildings and its surrounding environment is important. Moreover, the electrical and information

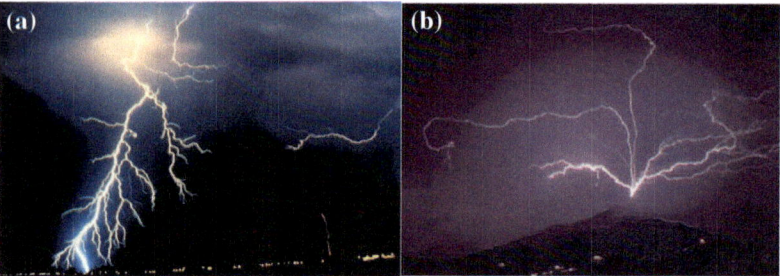

Fig. 1.3 **a** A heavily branched lightning strike to ground. A simultaneous, horizontal cloud to air flash well above ground, like an inverted tree. A cloud-to-ground (CG) lightning flash. *Credit* NOAA_NSSL With permission. **b** An upward lightning stroke, where the branches point upwards, like an upright tree. A ground to cloud (GC) lightning flash. *Credit* DEHN. With permission

technology (IT) equipment inside the buildings needs to be protected from the lightning currents and voltage impulses, as well as from radiated electromagnetic pulses (LEMP) produced by the lightning flash.

Figure 1.4 shows long lightning flashes stretching across the sky. One such cloud flash in Oklahoma terrain stretched to a distance of 350 km. Such flashes could be between electric charge centers within a single cloud, between the thundercloud and air or lightning flashes between two large thunderclouds. It is expected that unusually intense lightning flashes, as well as long flashes that may last for several seconds (instead of the conventional one second flash) will increase with climate change, especially with the warming up of the earth's surface. Much research into lightning strikes and lightning strike parameter prediction for severe lightning flashes with climate change is an urgent need for the protection and preservation of electrical, power, telecommunication and emergency electronic systems (e.g. medical surgery and intense care unit electronic/computer systems), as well as for human safety.

The microphysical and thermodynamics-based nonlinear processes of the atmospheric disturbances and anthropogenic enhancements of heat emission play a crucial

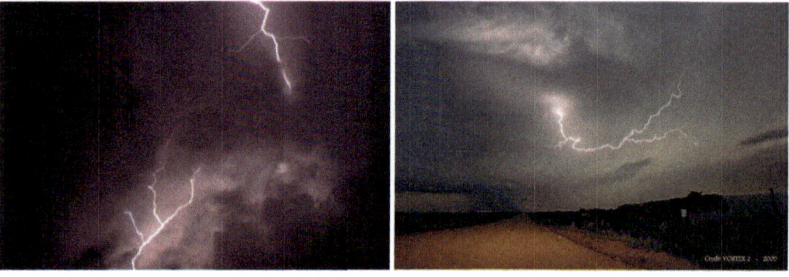

Fig. 1.4 Lightning flashes within a thundercloud (the intracloud or IC flash) or cloud to air flash (cloud to air or cloud to cloud or CC flash). *Credit* NOAA_NSSL, USA. With permission

role in the cloud-to-ground electrification. Evidence has shown that surface temperature rise from heat generated through anthropogenic activities is a key factor in driving lightning activities. Such evidence points to an inevitable risk of cloud-to-ground (CG) flashes that have been observed over major cities around the world. The risk is heightened further as a result of unprecedented weather patterns due the effects of climate change. There is strong interactions between climate change and the electrical processes of the earth's atmosphere (i.e. the troposphere, extending from the earth's surface to a height of 14 km) and beyond (notably, the ionosphere, the layer that stretches from 90 to 400 km above the earth). Moreover, it is conventional to link climate change with all extreme weather events including high frequencies and intensities of lightning flashes. Lightning inception criterion is still a subject of debate and research. Lightning activity is more continental than oceanic, with continental updrafts at 50 m/s producing thunderclouds compared to the 10 m/s updrafts over the ocean. Intense lightning activity is seen to prefer dry climates (e.g. Africa) rather than wet climates (e.g. South America), although both regions may be close to the earth's equator. Thus, lightning flash, especially the CG flash, induced voltage and current transients pose serious threats to ground and airborne vehicles, structures, and systems. The need for mitigation of lightning's direct and indirect effects continues to drive the protection systems to structures, their contents and systems to a higher level.

There are, to date, no devices or methods capable of modifying the natural weather phenomena to the extent that they can prevent lightning discharges. Lightning flashes are hazardous to people, to the structures (buildings, towers, aircrafts, etc.) and their contents and installations. This is the overarching reason why protection measures in aircraft, structures and systems become vital against both the direct and indirect effects of lightning. The need for protection, the economic benefits of installing protection measures, and the selection of adequate protection measures should be determined in terms of risk management.

Lightning interaction with structures is categorized as direct effects and indirect effects. The direct effects of the lightning stroke (or flash) comprise high return currents. The current peak magnitudes are of the order of several tens of kilo amperes. A value of 200 kA and up to 500 kA has been reported. The four specific effects of lightning current due to direct effects that are considered to be of high severity in producing damage are: (1) the peak current, which is the high-current pulse flowing through a conducting surface. It is responsible for the voltage induced on the conducting surface of magnitude $v = iR$, where i is the current pulse, and R is the resistance of the surface; (2) The maximum rate of change of current. This is dependent on the current steepness which gives rise to an electromagnetically induced voltage $v = M \frac{di}{dt}$, where M is the mutual inductance of the loop of conductors; (3) The integral of the current over time, $Q = \int i \, dt$, which is the electric charge transferred and is responsible for the mechanical force and the heating effects; and (4) The integral of the current squared over time $\frac{W}{R} = \int i^2 dt$, where W is the energy dissipated into a 1 Ω resistor (R) which is referred to as the specific energy or the action integral. The resistance R is the temperature-dependent D.C resistance of the conductor and R/W is the specific energy which is responsible for the melting effects.

The indirect effects of lightning threats are due to the radio frequency interferences and lightning electromagnetic pulses (LEMPs). The LEMPs can induce disruptive voltages ($v = Ldi/dt$) and currents ($i = Cdv/dt$) that can adversely impact electrical and electronics systems through resistive and/or electromagnetic couplings. The advent of digital electronic technology in electrical/electronic systems and the evolution of Internet of Things (IoT) through radio frequency identification devices, barcodes, smart phones, and the convergence of smart technologies in smart homes, smart industries, smart cities, smart environment, and smart ecosystem in smart people with micro-chips implanted forming the smart planet by integrating modern communication and information technologies will all heighten the requirements for a professional approach to lightning protection. LEMPs threats can have serious damaging effects. The electrical and electronics systems are susceptible to LEMPs at frequencies between 1 and 500 MHz and produce internal field strengths of 5 to 200 V/m or even greater. Internal field strengths greater than 200 V/m of pulse widths less than 10 μs can absorb lightning-induced voltages and currents ranging from several tens to thousands of voltages, say from 50 V and 20 A to over 3000 V and 5000 A. Susceptibility of electrical/electronic system to LEMPs has been suspect as the cause of "nuisance disconnects," "hardovers," and "upsets" in electronic systems. Generally, such malfunctions in digital electronics systems occur at lower levels of EM field strength than that which could cause component failures, if no proper shielding or protection system is utilized.

Because of the lack of detailed knowledge of the lightning strokes, little theoretical work has been done. Where the physics of the leader stroke or return stroke is taken into consideration, the underlying theories frequently contradict even some of the known, measured behavior of the strokes. The bulk of the work on the leader stroke has been to determine the velocity of the leader—each school of thought assuming different processes to dominate. We examined the leader stroke theories, to gain an understanding of the path over which the return stroke travels. Work on the return stroke was examined, to learn from the ideas found therein and their limitations, examining why no satisfactory solution or agreement regarding the physics of the return stroke has been found to date. A common weakness has been to prescribe unreasonably large, stored energy in the leader in order to obtain observed return stroke velocities of the order of 10^8 m/s, close to a third of the velocity of light in free space. The lightning flash is a challenging problem in physics and engineering, but it is not easy to tackle to full satisfaction any one of the many distinct stages which make up the majestic flash, from thundercloud formation, to lightning initiation, to the leader stroke and return stroke, or the subsequent dart leader and return strokes. That each stage plays an important role in discharging the thundercloud, thus balancing the electrical changes which take place during fair-weather conditions, is clear.

1.2   The Leader Stroke

The fair-weather volt drop of 250 kV between the upper atmosphere and the earth may be represented by a battery. Shown in Fig. 1.5 is also the electric circuit model of the fair-weather electric circuit: voltage source represented by a DC battery connected through the thundercloud resistance R to the fair-weather atmospheric resistor R_f (200 Ω), which in turn is connected to the ground through the fair-weather capacitance Ci (0.7 F). The fair-weather CR time constant is 2 min. This is the time taken for the 250 kV electric potential difference between the earth and the upper atmosphere to charge the fair-weather capacitor Ci. The charging current Ic is about 1250 A. When the thundercloud replaces the fair-weather environment, the large, approximately 50–100 million volt drop (50–100 MV) between the thundercloud and the earth generates electric breakdown (at about 30 kV/cm electric field stress, sometimes as low as 18 kV/cm) and the leader stroke moves between the earth and the thundercloud to create a short circuit. The short circuit results in the intense, destructive return stroke current wave.

The nature of the leader is represented by an RC circuit (Fig. 1.5) triggered by a constant voltage source, for which

$$RI(t) + \frac{1}{C} \int I(t)dt = V(t) \tag{1.1}$$

which on differentiation gives

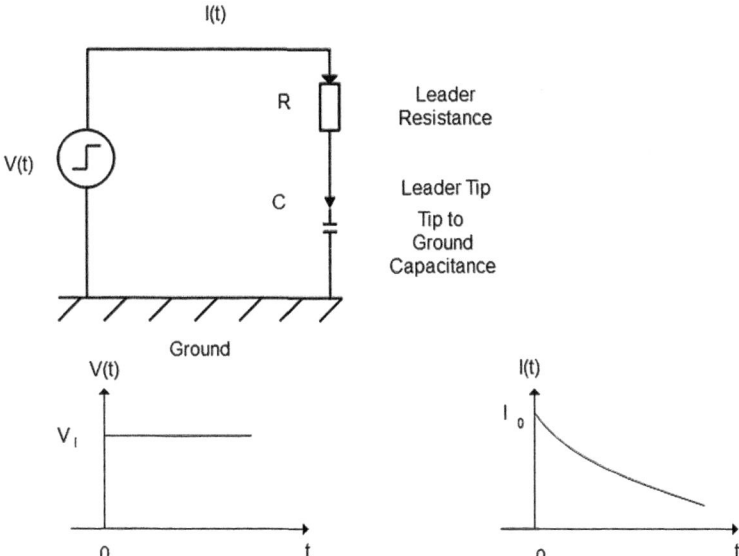

Fig. 1.5 A crude circuit model for the leader stroke

$$I(t) = e^{\left(-\frac{t}{RC}\right)}\left[\int \frac{1}{R}e^{\left(\frac{t}{RC}\right)}\frac{dV}{dt}dt + I_o\right]. \quad \text{or} I(t) = I_o e^{(-t/RC)} \qquad (1.2)$$

After the initial rise of V, we have dV/dt = 0. Thus the current is simply Io exp(-t/RC) and is sketched as in Fig. 1.5b.

The fair-weather electric field between the upper ionosphere and the earth at a height z may be given as

$$E(z) = -\left(-93.8 \, e^{(-4.5278z)} + 44.4 \, e^{(-0.121z)}\right)V/m$$

for height z less than 60 km. At ground level, with $z = 0$, E is about 150 V/m. At $z = 10$ km, E is 4.5 V/m and at 30 km it drops to 0.3 V/m and to 1 μV/m at very high altitudes close to 60 km. Under the thundercloud, with its lower region at a height of 1 km from ground, the electric field at ground level could rise to about 50 MV/km, yielding 50 kV/m. The high electric field produces the lightning leader stroke which results in the high-current lightning return stroke. The energy in a return stroke is in the region of 10^9 to 10^{10} J. Can this lightning energy be used to provide electricity to homes? Consider a 100 W bulb burning for one month. The energy it requires is $100 \times 3600 \times 24 \times 30 = 0.26 \times 10^6$ J. This means just to light up a 100 W electric bulb, it will take about 10^4 years of lightning energy to light that electric bulb for one month. The reason for this is that although electric power in lightning is high, the electric energy is very low, since it lasts for only about half a minute. Moreover, since much of the lightning energy is dissipated into heating in the lightning leader channel, the energy delivered to earth is much lower.

The electric charges in the lower part of the thundercloud determining the electric charge polarity of the lightning flash varies from country to country and season to season. A positive electric charge concentration in the lower part results in a positive lightning flash, with positive electric charges lowered to the ground by the return stroke. When the lower part of the thundercloud has a surplus of negative electric charges, this results in a negative flash with the return stroke lowering negative electric charges from the thundercloud to the ground. In the USA, for instance, monthly positive lightning flashes are about 90, whereas the negative flashes per month are about 50. In summertime, there are more negative flashes (20 per month) than positive flashes (15 per month). In Japan the number of monthly positive flashes (66 per month) is much higher than negative flashes (29 per month). In France on average there are 81 positive flashes each month and 34 negative flashes each month.

There are two difficulties facing the theoretician studying lightning and laboratory spark leader strokes: (1) There is scant knowledge of transport, ionization and recombination coefficients and of thermal and electrical conductivities of the processes involving kinetic energy dissipation by chemical, thermal and radiative means; (2) Theoretical calculations of unknown parameters such as the electron temperature, calculations that are based on mere assumptions. The problem still remains intractable. In the extensive work on leader channels using the time invariant fluid equations for electric plasmas, some of the assumptions made are quite drastic—for example, it is assumed that there are no conduction, convection, or displacement currents flowing in and around the leader; however, current peaks of 1kA and average currents of 100–300A have been reported for leaders. Further it is assumed that the diameter of the leader is over 5 m, hence postulating that the leader motion is due to electrostatic forces between the cloud and the tip. However, such a shock wave theory, in addition to the unacceptable assumptions made above, is difficult to defend for the following three reasons:

(1) In a sphere—plane electrode system the electric field will exponentially drop, as the observer moves from the sphere to the electrode. In such a situation the energy for the progress of the streamer to a short distance from the thundercloud may be provided by the cloud. But the major part of the gap must be governed by the streamer using stored potential energy in the pocket of charges accumulated at the tip, which is constantly refurbished by conduction currents flowing down from the cloud;

(2) The theory does not properly explain the progress of a positive streamer. In order for a positive streamer to exist, the electron pressure must pass through a sharp maximum in the positive streamer wave front such that the electron diffusion process allows the electrons to move against the electric field. Such a high temperature it is claimed is achieved by electron–electron collision, which is rather doubtful since this would require very large electron density and collision frequency. Even if electrons could be heated to high temperatures, the question of how much energy is used for velocity reversal and how much lost to the electric field has to be addressed; and

(3) The shock wave theory of the leader requires a leader radius larger than 5 m. However, the leader radius for long laboratory sparks is 1–5 mm, with a thinly ionized region surrounding it. Generally, the thin corona streamer develops into a highly conducting leader in about 75 μs, during which time both current and the temperature changes are observed as the channel expands from 1 mm to say 2 mm.

The leader tip is preceded by quite complex processes. Some choose to ignore these discharge processes which precede the leader. The view that there is a pilot leader preceding the main leader appears to be correct; the pilot leader must be made up of corona and glow phenomena. Collisions between accelerated free electrons and atoms generate more free electrons, which result in the highly ionized, electric plasma environment around the leader tip. Electron acceleration depends on the electric field at the tip of the leader. As to how a 1 m diameter corona region first collapses into a glow region and then into a thin leader channel is not clear. This transition is observed to occur with the arrival of a rekindling wave. There is not a sufficient amount of data on long DC sparks, these being very difficult to generate, and hence we assume that some of the observations made on leaders with impulsive voltages are also applicable to DC sparks. We take it that the lightning leader diameter is about 5 mm, with high conductivity which is invariant once having fully developed. In the return stroke considerations, we are not interested in the initial expanding stage of the leader.

Various relationships between the leader velocity and leader current have been claimed. One such relationship suggested for a 16 m long spark at about 2.5 MV is v $= \mathbf{I} + 0.95$ cm/μs where v and $\mathbf{I}$ are average velocities and currents, respectively. The leader currents observed for laboratory sparks are of the order of 1 A, giving a velocity of about 2 cm/μs. This velocity for the positive spark is about an order smaller if the current is taken to be 200 A and compared to a negative leader step velocity of 0.5 m/μs or dart leader velocity of about 0.06 m/s (0.2 m/μs from triggered lightning measurements). A fundamental difference which must determine the higher currents and velocity in the lightning case, compared to the long DC laboratory spark, must be the very high potential of the thundercloud, leading to restrikes which are associated with intense ionization and large current pulses.

When downward leaders in cloud-to-ground flashes approach a tall object or structure, it rapidly changes direction and strikes largely the tip of the tall object. In such lightning flashes one can observe a sharp kink, or change of direction as the lightning leader sharply turns towards the tall, grounded structure. The current-striking distance is related to the stepped leader charge Q (Coulombs). The relationship between the peak return strikes current I (kA) to the electric charge transferred is empirically defined by $I = 10.6\, Q^{0.7}$. The relationship between the striking distance d (meters) and the peak current I (kA) is $d = 10\, I^{0.65}$. The upward leader that rises from the tip of the tall structure to encounter the downward leader heading towards it could be 20–100 m long. In rare cases where there are two tall structures that launch upward leaders towards the down coming main leader, there will be two different stepped leaders producing two different return strikes, giving rise to a forked lightning.

Upward leaders in ground to cloud flashes are initiated from very tall earth structures, and structures or trees in the mountaintops or artificially initiated lightning where a rocket with a ground conductor attached to it is fired towards a thundercloud. The 3×10^5 m/s upward leader first initiates a dart leader that travels down to ground initiating a first return stroke going from ground to cloud, which largely looks like a subsequent return stroke of a cloud-to-ground flash. In rare cases, the upward going leader may initiate a first return stroke that travels from cloud to ground. When the upward going leader enters into a thundercloud, there are strong intracloud flashes that lower 30 C–300 C electric charges into the dart leader that carries it towards the ground. Fast introduction of conductors, including an aircraft, rocket, nuclear explosion produced conductive matter, or plumes of water that sprout from an ocean bomb blast, can also initiate an upward leader that travels towards the cloud. When rockets with ground wires trailing behind them are fired towards a thundercloud, the rocket height H meters at which the upward leader is launched depends on the electric field E kV/m at ground before the rocket is launched. The relationship between E and H is roughly, $H = 3900\, E^{-1.33}$. The velocity of the rocket and the type of the ground wire attached to it does not appear to have a major effect on the lightning flash properties. However, the presence of melted conductor inside the lightning plasma channel and the very good grounding provided at the earth end give rise to sharply rising wavefronts in rocket initiated lightning flashes compared to natural lightning flashes.

1.3   The Return Stroke

1.3.1   General Description

The lightning leader strokes are largely generated at the thundercloud electric charge center and move downwards towards the ground, and at the leader channel terminations at ground are generated the return strokes of a cloud-to-ground (CG) flash. Figure 1.3 shows both CG and GC lightning flashes, where the return stroke travels from ground to cloud (CG, Fig. 1.3a) and from the thundercloud to ground (GC flash, Fig. 1.3b). Figure 1.6 shows the intracloud (IC) flash, where there are multiple lightning flashes from the electric charge centers, e.g. negative charge centers) from one part of a thundercloud to the electric charge centers (e.g. positive electric charge centers) of opposite polarity in another part of the thundercloud. Here in Fig. 1.6 multiple, spidery shaped multiple bright flashes that are seen in Fig. 1.6, namely, multiple return strokes, travel from one electric charge center towards another.

The lightning return stroke therefore originates at the point at which the leader stroke contacts the ground (CG flash) and moves upwards along the ionized leader stroke channel. However, when a tower or ground object of 150 m height is under the thundercloud, about 23% of the lightning leader strokes originate at the upper tip of the ground object or tower, and these move upward towards the thundercloud. In this case, the lightning flash is called a ground to cloud (CG) flash. This return stroke

Fig. 1.6 An Intracloud (IC) lightning flash between multiple electric charge centers inside the thundercloud. Simultaneous clod to ground flash *Credit* NASA, USA

originates at the cloud electric charge center (or at the round?) and moves downward towards the ground (or upward towards the cloud?). When the object is 200 m tall, 50% of the leader strokes originate at the earth object or tower. The percentage of ground to cloud flash increases to 80% of flashes when the height is 300 m to 91% when the object is 400 m tall, and to 98% when the height is 500 m.

Once the first return stroke has completed traveling from ground to cloud in a cloud-to-ground (CG) flash ionized channel, there could be a subsequent leader and a subsequent return stroke occurring. In between the first return stroke and the subsequent leader, smaller electric voltage pulses called the M-components may travel from the cloud-to-ground transporting electric charges from the thundercloud to ground.

During the electrically intense return stroke strong electromagnetic pulses called the lightning electromagnetic pulses(LEMP) are radiated out from the return stroke ionized channel. These LEMP may induce large electric voltage pulses on distant electric lines and circuits. Up to 5 MHz the strength of LEMP is inversely proportional to the frequency (f). In the range of 100 MHz–1 GHz the strength of LEMP is inversely proportional to f^2.

1.3.2   The Empirical Model

Both the lightning leader stroke and the return stroke are columns of ionized gas, in which when the free electrons are subject to an electric potential (such as the thundercloud voltage), the electrons collectively move in one direction, resulting in the flow of electric current. Matter exists in four states. These are, namely, solid, liquid, gas, and plasma. Plasma in electrical science, is ionized gas. The thundercloud produced electric breakdown results in electrons freed from oxygen and water molecules, producing leader strokes, and subsequent return strokes. In nuclear fusion technology, electric plasma is produced and the free ionized particles are confined to the plasma region by strong magnetic fields externally produced in parallel to the plasma column. Nuclear fusion plasma immersed in an external magnetic field is called magnetized plasma. But the lightning channel is an unmagnetized plasma channel, where at high pressure the channel particles may in a high-pressure internal explosion be thrust out of the ionized column. That expulsion as a shock wave is heard as thunder. The return stroke currents flow along the electric plasma channel of the leader. The ionized gas channel (i.e. the plasma channel) acts like a conductor for the lightning return stroke current.

We may identify two different approaches to the modeling of the return stroke. The first approach is to specify the current-time and current-height characteristics. Parameters such as the peak current, the time constants, and the velocity of the return stroke may be obtained from ground measurements of currents and/or the LEMP. The double exponential empirical description of the return stroke current is the most widely used, because it is simple and easily generated in a high voltage laboratory.

Fig. 1.7 A simple circuit
model for the return stroke

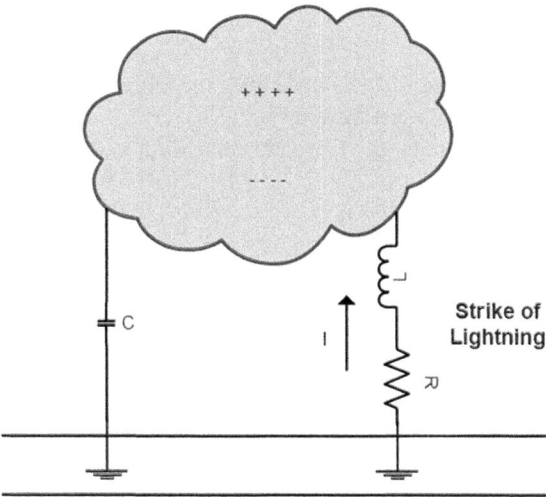

However, it is open to question whether such models are true to the physical processes
and whether they may rightly be called models of the return stroke.

We may best illustrate this lumped LCR circuit model by a very crude, intuitive
model of the return stroke, as shown in Fig. 1.7.

In the case of the return stroke, for the simple model shown in Fig. 1.7 (a), we
assume that all the cloud charge is transferred to the leader as the leader is connected
to the ground. For the lumped circuit shown in Fig. 1.7a, Kirchhoff's law gives

$$L\frac{dI}{dt} + RI + \frac{1}{C}\int I dt = 0. \tag{1.3}$$

Differentiating (1.3) gives

$$L\frac{d^2I}{dt^2} + R\frac{dI}{dt} + \frac{I}{C} = 0 \tag{1.4}$$

which has the solution form

$$I(t) = Ae^{\left(\frac{-R-K}{2L}\right)t} + Be^{\left(\frac{-R+K}{2L}\right)t}, \tag{1.5}$$

where

$$K = \sqrt{\frac{R^2 - 4L}{C}}. \tag{1.6}$$

Using, the initial condition I(t) = 0 when t = 0 we obtain

$$I(t) = V_o\left[e^{\left(\frac{-R-K}{2L}\right)t} - e^{\left(\frac{-R+K}{2L}\right)t}\right] \qquad (1.7)$$

The form of (1.7) resembles the empirical Bruce-Golde lightning return stroke model and is sketched in Fig. 1.7b.

Differentiating (1.7) and setting di/dt = 0, we have the rise time given by

$$t_T = \frac{L}{K} \log B\left(\frac{R+K}{R-K}\right) \sec s \qquad (1.8)$$

which is a strong function of L/R, for $R^2 >> 4L/C$ and K ~ R. We note that the important parameter t_T depends on the careful estimation of L/R. Some LCR circuit models for the return stroke set L/R = 0.0 so that such distributed LCR models are unreliable for rise time estimation. In the case of erroneous setting of parameters, for instance, ignoring the L parameter at ground level, singularity points appear in the calculations. In computation it is important to keep the time step $\Delta t << L/R$, and the accuracy is easily checked by ensuring that there are computed points on the wavefront.

Return stroke velocity. The velocity with which the lightning return stroke current and voltage surges travel from ground to cloud, in a cloud-to-ground lightning flash, is about 10^8 m/s. The value is less than the velocity of light (3×10^8 m/s). The velocity of a surge traveling along a lossless electric conductor is equal to the velocity of light. The return stroke current I = dq/dt, where q is the electric charge collapsing along the lightning return stroke channel as it is rapidly lowered to ground by the return stroke traveling upwards in a cloud to ground flash. The velocity of the return stroke wave is v = dl/dt. where l is the length traveled by the return stroke. With q being the electric charge deposited along a channel of length dl, current I = qv. Unit mismatch. I becomes C m/s Hence for velocity v = 10^8, and q = 10^{-3} C/m check unit, we get I = 100 kA, the return stroke current. We have assumed that the velocity v is constant from ground to cloud. But in lightning flashes, the velocity v of the rerun stroke decreases with height.

There are two factors that affect the velocity of the return stroke. (I) First, the longitudinal resistance of the lightning channel, which is not zero. The lightning plasma channel, or the ionized channel, is not lossless, but a lossy conductor with a finite value of R of about 1 Ω/km. The maximum electric field along the lightning channel, or the longitudinal electric field, falls from about 100 kV/m in the first few microseconds. But as the lightning channel diameter expands the electric field falls to a value of about 10 kV/m. The resistance of the channel may increase to about 30 Ω/km in 20 to 30 μs after the return stroke first reaches a point along the channel. (II) Secondly, there is a further reason why the lightning return stroke velocity is less than the velocity of light. When considering the lightning return stroke channel, at the center it is very narrow, a highly conductive core the diameter of which is a few mm to a cm. This inner core is surrounded by a larger region of dispersed, conductive ionized particles. This sheath therefore has a conductance. The outer sheath conducting region further adds to the surge impedance, which from free space values of 300 Ω could increase to 500 Ω. This outer corona sheath could further impede the propagation of the return stroke and reduce its velocity.

1.3.3 Lightning Return Stroke Models

Although a variety of lightning models have been discussed in the literature, the two most important models from an engineering perspective are the following two. (1) The curve fitting model obtained from electromagnetic field measurements and then trying to guess what the electric current that is generated should look like, by a curve fitting technique. The curve fitting method is used extensively by lightning researchers in USA and Sweden. The method is to construct the lightning current pulse by a curve fitting method to get the shapes of lightning electromagnetic pulses (LEMP) measured at ground level. This is an indirect, curve fitting method, that is not strictly based on the physics of the lightning flash. (2) The other, the most important model in our view, is the distributed transmission line model. The lightning flash was initially modeled as a lumped circuit as described in Sect. 1.3.2. Taking the lumped circuit model a step further, the return stroke is modeled by the distributed LCR transmission line model. For the model to have self-consistency, it is necessary to ascertain that it is scientifically valid to represent the return stroke by a transmission line, and if there is a satisfactory case for such a model, with the elements of the line to be determined from basic principles. The models, unless carefully calculated L, C, and R parameters are used, may yield erroneous solutions for a CR network instead of an LCR network. The CR network is a diffusion model and not a fast electromagnetic wave model. In Chap. 5 is presented a self consistent LCR model of the lightning return, which is solved using the finite difference time domain method (FDTD).

There have been suggestions that it is important to include resistances which vary with frequency or time t, as the channel expands. Braginskii's model of a hot plasma channel is used to obtain radius r (proportional to $t^{1/2}$). Moreover, a curve fit is used to obtain conductivity σ following Plooster. Discussion on the lightning leader models using shock wave-like phenomena and their shortcomings also apply to using the Braginskii's and Plooster's works. The degree of freedom in estimating r–t and σ–t characteristics are larger. $R = 1/6\pi r^2$, where R, σ, r are the per unit resistance, conductivity and the radius of the channel, respectively. The distributed LCR electric circuit model to represent the lightning return stroke wave as an electromagnetic wave yields, as shown in Chap. 5, results that adequately match all known measurements, including the convex wave front of the return stroke, without going into additional frequency or time domain calculations of channel radius and conductivity.

Now in the case of the lightning channel where high currents repeatedly flow through the channel, both during the leader and the return stroke phases over a considerable length of time, the channel would have reached a saturation level in terms of its conductivity. During the leader phase itself the discharge carries average currents of the order of 300 A for a negative leader, and above 1000 A for a positive leader. It is reasonable to assume that the high-current leader resembles a high-pressure arc in its final stage, with the current mainly determined by the external circuit. The idea is further supported by the fact that the estimated temperature of the leader is 20,000° K, and for the return stroke the peak temperature is about 25,000° K (the average is about 20,000° K). It is therefore reasonable to expect a negligible amount of change in the lightning channel conductivity during the return stroke phase.

The general picture of return stroke modeling work is that there has been a trend to put in lots of details which may not be significant in reality but are characterized by a serious lack of agreement with fundamentals. The questionable models include postulating a conducting shell around the return stroke channel, and equating the energy dissipated to establish a laboratory arc to the energy used in establishing the lightning return stroke. Fluid equations for electrons in a laboratory leader have been extended to the return stroke. A standing wave for the return stroke has been assumed without any discussion of how the wave is formed in the first place. Data from laboratory sparks has been used to model the lightning return stroke without considering the different processes taking place, and in putting in details that may lead to drawing wrong conclusions from computer procedures which are questionable (e.g. in the time step used which may allow the fast moving wave to spill over to the next spatial increment before a single time step is over.) Large capacitances have been prescribed to obtain low velocities for high frequency signals along transmission lines. Others have prescribed electromagnetic models of the return stroke, as described above, using a curve fitting technique to project back from measured return stroke radiated electric and magnetic fields.

1.4 Lightning Radiated Electromagnetic Pulses (LEMP)

1.4.1 Computation of Radiated Electromagnetic Pulses

The most severe lightning flashes have return strokes of the order of 200 kA, with an electric charge of the order of 100 C and $T_1/T_2 = 10/350$ μs where T_1 and T_2 are the front time and half fall-off time. The action integral, that is the specific energy, is about 10 MJ. The return stroke front time to half fall-off time ratio values is given by $T_1/T_2 = 10/350$ μs. Subsequent return strokes of severe lightning strikes have a maximum value of 50 kA, rate of rise of 200 kA/μs and T_1/T_2 of 0.25/350 μs. The very short rise times of 0.25 μs give rise to high values for dI/dt. This results in large radiated electromagnetic fields or LEMP. These radiated LEMP may couple to loops to induce destructive voltage and currents in electric power systems to microelectronic circuits. Thus, one needs to try to avoid loops in electric circuits. Continuing currents lower electric charges up to 200 C over a duration of 0.5 s. The total flash charge is of the order of 300 C. The voltage induced in a voltage loop of area A by a magnetic field B radiated by the lightning return stroke is given by V = A dB/dt. A current I flowing along the lightning channel produces a magnetic flux density $B = \mu_o\ I/(2\pi r)$, at a distance-r from the flash channel, with $\mu_o = 4\pi \times 10^{-7}$. Let us do a simple calculation on induced voltage. At a distance of 10 m from the flash, $B(t) = 2 \times 10^{-8}\ I(t)$. For $dI/dt = 10^{11}$ A/s, the voltage induced in a 10 cm $\times$ 10 cm square loop of area A $= 10^{-2}$ m^2, is $V = (\mu_o\ A/2\pi r)\ dI/dt = 2 \times 10^{-8}\ dI/dt = 2 \times 10^3 = 2$ kV, which is relatively large. In addition to the electromagnetic pulses radiated by the lightning leader and return strokes, there are also smaller radiations of electromagnetic pulses called the M and K pulses. The K pulses are produced by electric processes that take place between two return strokes, that is during the inter-stroke period or the dark period after the ground flash is over. The M pulses occur mostly during the luminous, 200 A or so continuing current that immediately flows after the, say 100 kA, return stroke. The M and K changes are non-threatening, and similar to each other. The M process radiate more frequent submicrosecond pulses than the K process. These processes are of some interest in radio noise studies.

Calculation of the electric and magnetic fields radiated from the lightning return stroke are important for two main reasons: (1) any new return stroke (and leader) models must be tested to verify that the fields determined from such models are consistent with the general radiated electromagnetic pulses (LEMP) measured on ground. (2) To be able to determine the LEMP at any height above the ground and distance from the lightning flash. Although the current-time characteristics at different heights of the lightning channel postulated from field measurements do not give a valid return stroke model, a self- consistent return stroke model ought to give LEMP which generally resembles the observed, ground LEMP. Such a return stroke model may be used to determine the LEMP at any point in space surrounding the return stroke model, including points at which an aircraft which is landing or taking off may be located in space, points that are inaccessible for measurements.

The methods of LEMP calculations may be classified into four different classes.

(i) The old dipole moment method. This method has been widely used to determine fields far from the lightning channel.

(ii) The method based on integral equations for electric and magnetic fields. The integral formulation is most suitable when the field at a limited number of points is required. A widely used approach, it formulates the integral expression of magnetic and electric fields radiated from the entire length of the channel. What is unfortunate about this method is that it is difficult to visualize the contribution to LEMP fields made by the distance-R dependent electrostatic term (dependent on the electric charge, and decays as $1/R^3$ where R is the distance from the channel), the induction or intermediate term (dependent on current, and decays as $1/R^2$) and the radiation term (dependent on the rate of change of current, and decays as $1/R$). Furthermore, there appears to be a fundamental error in the expressions as they stand. The constants of integrations do not result from the return stroke. This constant may be arbitrarily taken to be 3100 A. The leader charge has not been accounted for in the calculations reported.

The leader has an electric charge distribution of the order of 0.3 mC/m, for a total charge of say 1 C distributed along the leader trunk of 3 km length. The average charge lowered by a subsequent stroke (without branches) is about 1.4 C. Ignoring this charge in LEMP calculations will give higher field values close to the flash; in particular, for positive flashes where the charges involved are higher, the appearance of bipolar fields (+ve field changing to −ve, or vice versa) near the flash will be suppressed. In physical terms, ignoring the charge deposited along the leader amounts to not considering the return stroke discharging the charged leader channel. In order to rectify this error, an effort is made to place point charges at the tips of the elements of a discretized leader channel, whence adding a new term to the former expression. The magnitudes of these charges are determined by considering the difference in currents between two adjacent segments to reside in a sphere and integrating it over time. Such an assumption is difficult to defend, and it should be remembered that the charge is distributed over the whole length of the channel and does not reside in globules.

(iii)   The integral technique reported in this book, the best and most scientifically accurate method reported to date, includes accurate integral method LEMP calculations linked to the currents and distributed electric charges along the lightning channel calculated using the distributed LCR model of the return stroke electromagnetic wave. The channel is discretized, during the LCR model for return stroke current calculations, made up of many segments of specified length (e.g. 100 m segments for a 1000 m lightning channel) to solve the LEMP by numerical integration. This book, moreover, includes the technique for the calculation of electrostatic fields before a lightning strike, with or without an aircraft present in the vicinity. This allows the lightning currents, electric charges, voltages and LEMP to be calculated with external bodies or structures, such as aircraft and buildings or towers present in the vicinity, and become attached to the lightning leader-return stroke phases. Thus, it provides a self-consistent, low memory, fast and accurate computer-based testbed for lightning simulation and lightning interaction and testing.

(iv)   A fourth method for determining fields is numerically to solve the differential formulation of the problem using the finite element method. It is best used when fields over the whole space around the lightning channel are required, and external objects such as an aircraft are present in the vicinity. But it is an expensive method if LEMP fields at only a few points are required. The generality of the differential formulation makes it difficult to identify the contribution of the different field terms. The finite element method is a promising tool for three-dimensional and transient field calculations on aircraft.

1.4.2 Calculating Rate of Rise of Currents from Measured Electric Fields

With a notable increase in electric field measurements in the late 1970s and 1980s, there have been attempts at the calculation of the rate of rise of return stroke current using measured electric fields E, from the expressions for E and B given in the well-known forms:

$$E_Z = \frac{\mu_0}{2\pi} D \int \frac{di}{dt} dz. \tag{1.9}$$

The electric field at ground level is approximately expressed as

$$E_Z = \frac{\mu_o}{2\pi D}\left[Iv - \int i \frac{di}{dt} dz \approx \frac{\mu_o}{2\pi D} Iv \right] \tag{1.10}$$

at ground, where v is the return stroke velocity (assumed constant along the channel) and I the current (also assumed constant). With the knowledge of dE_Z/dt at the wave front, dI/dt has been determined—giving very high values for dI/dt, even as high as 280 kA/μs. The average measured value of dI/dt for negative subsequent strokes is 40 kA/μs, for first strokes it is 12 kA/μs. Since wave front of E_Z measured at far distances (e.g. 100 km) will be significantly modified by the earth, trees, buildings, etc. fields measured close to the lightning strike point, at say 1 km were considered. However, only radiation field component was assumed to contribute to the wave front, and thus normalized it to 100 km by using 1/R decay. The contribution of the near (electrostatic) and intermediate fields to the overall electric fields measured was ignored. Hence the erroneous. large dI/dt values obtained using near field values, and then they were normalized to distant field values for the computation of dI/dt from measured dE/dt.

Three further points to note are (a) (1.10) only applies for a return stroke wave front with a constant dI/dt. In actual fact E_Z is proportional to $dI/dt.v$ in agreement with (1.10). However, measured LEMP electric field E wave fronts have a sharp rise for only about 90 ns, giving rise to the calculated value of 280 kA/s which is an overestimate, since it has been assumed that the whole E_Z wave front has a 45 v/m/μs rate of rise. (b) The second note is a word of caution in interpreting measured electromagnetic fields. If the value 45 V/m/μs was obtained from measurements made near to the flash, the 19 V/m/μs contribution of leader has to be deducted. Furthermore, the near and intermediate components of the fields contribute about 20 per cent or more of the fields. If the fields were measured at distances of about 30 km, the 9 V/m/μs contribution of cloud pulses has to be deducted. (c) Without proper velocity measurements, the validity of the use of (1.10) is limited by the fact that it is velocity dependent. The return stroke velocity ranges from 50 m/μs to 150 m/μs. Very few correlated measurements exist. We have also noted that there is a large discrepancy in the rate of rise of ground electric fields reported. And inaccuracies in field measurements need careful consideration.

1.5   Electromagnetic Waves

The four Maxwell equations that form the basic mathematical foundation to electromagnetic fields yield the electromagnetic wave equation in conducting medium,

$$\nabla^2 E \frac{1}{\varepsilon \nabla \rho} = \mu \varepsilon \frac{d^2 E}{dt^2} + \mu \frac{dJ}{dt}, \tag{1.11}$$

where E is the electric field, ρ was the volume electric charge density, J is the electric current density, ε the permittivity of the material, and μ the permeability of the material.

Using (1.11) and rearranging we obtain

$$\nabla^2 E - \mu \sigma \frac{dE}{dt} - \mu \varepsilon \frac{d^2 E}{dt^2} = \frac{1}{\varepsilon \nabla \rho}. \tag{1.12}$$

We assume that $E(r,t) = E\exp(j(\omega t - k.r))$ and also remember that the electromagnetic wave is a transverse wave $k.E = o$ where k is unit vector in the direction of travel; thus (1.12) reduces to

$$k^2 E = k_o^2 \left(1 + \frac{\sigma}{i \omega \varepsilon_o} \right) E. \tag{1.13}$$

where k is wave number and $k_0 = (\omega^2/c^2)^{1/2}$ and $c = 1/\mu_0 \varepsilon_0$ the velocity of light. E on left and right in (1.13) will cancel, yielding,

$$k^2 = k_o^2 \left(1 + \frac{\sigma}{i \omega \varepsilon_o} \right), \tag{1.14}$$

where k is complex. But when $\sigma/\omega \varepsilon_0 \gg 1$, k^2 is predominantly imaginary. In this case the attenuation distance, the distance over which the amplitude decreases by 1/e and is roughly given by the usual formula for the skin effect,

$$d = \left(\frac{2}{\mu_o \sigma \omega} \right)^{\frac{1}{2}} \tag{1.15}$$

for $\sigma = 4000 \ \Omega^{-1} \ m^{-1}$, f = 1 MHz say, d = 8 mm and d = 25 cm for f = 1 kHz with k = 1/d + j 1/d. In a poor conductor, $\sigma/\omega\varepsilon_0 \ll 1$, k = ω/v where wave velocity v = $1/(\mu\varepsilon_0)^{1/2}$.

In the discussion above the magnetic force exerted on an electron is ignored in comparison to the force exerted by the electric field; the ratio magnetic force/electric force = v/c $\ll$ 1, when the electron velocity v is much less than the velocity of light, c. The term "fully ionized gas" is somewhat loosely used for the lightning channel where there is about 10 per cent ionization. For an electron density of about 10^{24} /m^3, with the density of gas molecules for air at standard temperature, pressure is 2.5×10^{25} m^{-3}, it is appropriate to point out the implications of taking the lightning channel to be a fully ionized gas.

For a weakly ionized gas, the number densities of electrons and positive ions are considerably less than the number of the neutral particles. The electronic motion is modified by collisions with the neutral particles. For this Lorentz gas, the Langevin equation ignoring the effect of the magnetic field of the electromagnetic wave,

$$M_e \frac{dv}{dt} + M_e \gamma_{em} v = -eE \tag{1.16}$$

$$v = -\frac{e}{Me} \frac{1}{(\gamma_{em} - i\omega)} E. \tag{1.17}$$

Now the electron current density

$$J = \sigma E - N_e ev. \tag{1.18}$$

And hence from (1.18) and (1.17) the conductivity is given by

$$\sigma = \frac{N_e^2}{M_e} \frac{1}{(\gamma_{em} - i\omega)}. \tag{1.19}$$

The conductivity has a real and an imaginary part. The DC value of conductivity is $N_e e^2/M_e \gamma_{em}$.

If we substitute (1.19) in (1.14) and extract the real and imaginary parts,

$$k^2 = k_o^2 \left[\left(\frac{1 - \omega_p^2}{\gamma^2 + \omega^2} \right) - \frac{1\omega_p^2 \gamma}{\omega(\gamma^2 + \omega^2)} \right]. \tag{1.20}$$

Dropping the subscript for the collision frequency and remembering that the plasma frequency is given by $\omega_p^2 = N_e e^2 /M_e \varepsilon_0$ when $\gamma = 0$, we have the familiar dispersion relation

$$k^2 = k_0^2 \left(\frac{1 - \omega_p^2}{\omega^2} \right), \tag{1.21}$$

where no dissipation takes place in the plasma. The situation given by (1.19) and (1.21) is important if the lightning channel is a very poor conductor, in which case the channel is like a dielectric rod, or, if the electron temperature is very high. The latter situation would arise if there is a very high electric field (about 1 MV/m) to be associated with return stroke, in which case there will be electron run away with the electrons moving quite independently of other particles so that $\gamma = 0$. The possibility of this cold plasma type of motion occurring in lightning needs further discussion.

The first case of dielectric line plasma is more to be associated with low current discharges, as in corona or glow discharges. In this case low frequencies cannot propagate along the column, the actual dispersion curve depending on the ratio γ / ω, according to (1.20). In lightning leader and return strokes, the low frequency signals dominate, whereas the high frequency signals are not very significant. The dielectric analogy where electrons are like bound charges, each being associated with a positive ion to form an electric dipole, is probably true for lightning initiation and inter-stroke processes; it is these processes which will radiate in the frequencies above 1 MHz. For the return stroke the static conductivity may be employed without appreciable error, well into infrared frequencies if necessary.

In a fully ionized isothermal gas, the collision frequency assumes another meaning from that in the Lorentz gas. First, when electrons and ions are the only constituents, γ_{ef} ($>\gamma_{em}$) is the predominant frequency, so that conductivity σ is smaller than when neutrals are present. Secondly, the presence of large numbers of electric charges requires that the frequent distant encounters must be included. It is this conductivity in which we shall be interested. The boundary between weakly and strongly ionized gases depends on collision cross section area and gas temperature.

Considering a temperature of 20,000° K, with a gas density of about 10^{25} /m^3, the electron-ion collision frequency is about 10^{15} s^{-1}. At such a high electron-ion collision frequency we may take $N_e/N_n > 10^{-2}$ to be for a strongly ionized gas. A useful test for local thermodynamic equilibrium is

$$N_e(cm^{-3}) > 1.75 \times 10^{14}(T_e(ev))^{\frac{1}{2}}(E_i^{z-l}) \qquad (1.22)$$

from which it is reasonable to take for a discharge column in air, mainly consisting of N$_2$ (V$_i$ = 15.5 V) and O$_2$ (V$_i$ = 12 V), and a fair measure of oxygen atoms (V$_i$ = 13.5 V), and NO (V$_i$ = 12 V), at N$_e$ = 10^{18} cm^{-3} and T$_e$ = 20.000° K (1.7 eV) to be at local thermodynamic equilibrium. In this situation the population of excited states is mainly determined by collision with free electrons.

The electron density of 0.5 to 1 × 10^{18} cm^{-3} is taken to be a good estimate, since this was calculated from Stark profiles of Ballmer series of hydrogen-hydrogen due to decomposition of water vapor—which is primarily dependent upon charged particle number densities and only slightly dependent upon particle energies. The values obtained using Saha's equation at temperatures determined by line intensities for nitrogen plasma show good agreement.

1.6 Lightning Protection: An Introduction

Over the past decade there has been an increasing interest in lightning and lightning protection for several reasons, including the proliferation of microelectronic
equipment and IT systems in mission critical systems as well as in everyday use
in banks, industries to homes. Both high voltage power systems and low voltage
networks need lightning protection. Personal protection of people is also highly
critical in thunderstorm environment and from electrocution by lightning currents
flowing through the ground, electric conductors and water systems. Lightning strikes
to power lines produce large, fast transient voltage and current surges which trickle
down to IT systems, military command and control systems as well as to several other
microelectronic equipment and control systems. Moreover, aircraft may be struck by
lightning when it is parked on ground, landing and taking off or in military operations
where the aircraft has to keep close to ground and when the atmosphere is electrified
by a thundercloud. Unusual phenomena have been recently observed which includes
a lightning flash which stretched to over 350 km over Oklahoma, USA, and in 2016
about 300 reindeer in Norway were killed by a single lightning strike to ground.
Severe thunderstorms may soon become more common if the temperature signature
of the earth surface with climate change continues as at present. Whereas a single
lightning phenomenon was expected to last only for one second, it has been recently
observed that a single lightning event may last as long as seven seconds, packing in an
immense amount of energy and repeated strikes at one location or to one object. The
energy and intensity of lightning may continue to increase causing damage and electronic rust, as well as increasing threat to human life. We explore here the protection
of electronic equipment, structures and in house systems from lightning. We will
also consider lightning related Electrostatic Discharge (ESD) threat to aerospace
vehicles and microelectronic systems. This is especially so with the increased use of
non-metallic, composite material for the aircraft body. Moreover, we will summarize
the important lightning techniques used in the protection of electric power systems
and houses.

1.6.1 Lightning Effects

Lightning can initiate forest fires, endanger life, destroy electric equipment and cause
power line faults. It plays a vital role in atmospheric chemistry. It was found that
during the summer months lightning activities increase NO_X by 90% and ozone by
more than 30% in the free troposphere.

The damage caused by lightning mostly occurs at some point where a cloud-to-
ground leader stroke terminates on a tree, a structure such as buildings, or a conductor
situated on an elevated position like high rise wiring systems. Electric control systems
malfunction and fail, and expensive microelectronic equipment to super high voltage

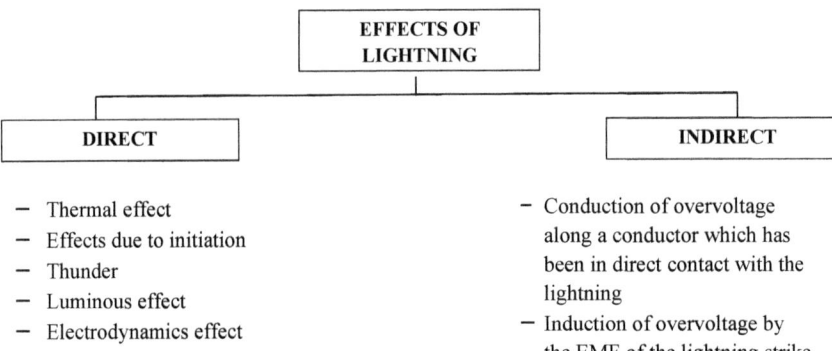

Fig. 1.8 The direct and indirect effects of lightning

(SHV) apparatus are destroyed by lightning strikes. But the lightning strike also results in other kinds of after-effects that impact living things (Fig. 1.8):

- Visual effect (lightning flash)
- Acoustic effect
- Thermal effect
- Electrodynamic effect
- Biological effect

Generally, these lightning effects are classified into two types of accidents or disasters which are a direct lightning strike accident which relates to all the effects mentioned above when lightning directly strikes upon an object or a being. The second classification is due to indirect lightning effects where the after effect of the lightning strikes causes problems to any nearby object or being which is due to (i) electric induction, (ii) conduction, and (iii) magnetic coupling of lightning currents and voltages in the ground to other grounded electric conductors and systems. Consider first the direct effects.

Visual effect: this effect is due to the extremely high brightness emitted by the lightning strikes which can injure any person or damage visual measuring devices such as a light sensor. With a close enough distance from the striking zone, this will cause a short-term blindness in eyesight. In certain instances, a fatal permanent eye blindness may occur. Other than the eye, electronics also may be affected where a light sensor can be burnt due to overdose of light received from the lightning strike. Hence, any devices with light sensors such as a camera may suffer malfunction.

Acoustic effect: this is normally caused by thunder emanating from the expanding lightning channel during the return stroke, where the sudden shock wave from the overheated lightning channel causes a rapid release of pressure (2–3 times atmospheric pressure) which breaks the sound barrier and creates a sonic boom. Exposure to the sonic boom from thunder may lead to a short-term loss of hearing or in the worst case, damaging the ear drums. Other organs such as the heart and lung too may be affected by the pressure wave. Fracture of glass and disruption of microphone and transducers are also caused by thunder.

Thermal effect: since lightning strike emits extremely high heat (higher than the heat emitted by the sun), it results in melting holes of varying sizes on high resistant materials that are attached to the lightning channel. On other materials with low conductivity or high insulation (e.g. trees), the large amount electrical energy released by the lightning strike is instantly converted to heat energy. This quick transformation of heat can burn any materials, sometimes causing flammable liquid or gases to be ignited and turning water into steam at high pressure in a short amount of time, which can result in an explosion of the container with flammable material.

Electrodynamic effect: lightning currents through the magnetic fields produced may become strongly coupled to other current carrying conductors. In this situation, the effects of the interactions between conductors and other equipment occur due to large magnetic fields produced by a lightning strike's current. This in turn causes a considerable amount of mechanical forces which may be attractive or repulsive. The mechanical forces are stronger when the distance between the conductors is smaller and the current flowing through them is higher.

Consider now the indirect effect of lightning strikes. The indirect effects have become more important due to the increasing use of highly sensitive microelectronics. These microelectronic devices are vulnerable to the transient overvoltage caused by lightning. The overvoltage can either be atmospheric in origin (such as lightning) or of industrial origin (from man-made equipment such as electric motors). The atmospheric (e.g. lightning) overvoltage is more harmful than industrial overvoltage. The indirect effects of lightning are categorized as follows:

(1) Conduction: this is from an overvoltage that flows along a conductor or apparatus which has been directly hit by a lightning strike. This sudden rush of electricity may have a very destructive effect on other systems, apparatuses, and humans at a distance since most of the lightning energy will spread through the entire power system network or grid causing threats to all the other devices or systems the strike point is connected to.

(2) Induction: this is caused by the radiation of the electromagnetic field (LEMP) produced by the lightning strike. The return stroke currents flowing along the lightning channel resemble a transmitting wire antenna carrying currents. The LEMP radiated may be picked up by other conductors that act like a receiver antenna. Therefore, under the effect of the sudden fluctuation in current, the wiring cables, air ducts which act as aerials, and antennae may receive through induction a large portion of the lightning destructive overvoltage and energy. This is also the reason that putting the network and power system underground does not guarantee full protection from a lightning strike, since LEMP can penetrate the ground or be generated by lightning currents flowing inside the ground.

(3) Lightning effects transferred through the ground: this happens when a lightning strike hits the ground, upon which an overvoltage and current can rise up from

the ground in order to find a more preferable path (in other words, more conductive path) to flow. This may cause a sudden overvoltage in a nearby grounding conductor. It can lead to a backflow of currents from earthling conductors and shields into the tanks, electric and electronic devices, and apparatuses, posing a threat. Moreover, the large voltage drop caused in the ground, with the voltage rapidly dropping from the point of lightning contact with ground, causes circulation of currents through the legs of creatures such as cows, as well as across differently grounded electrical installations, which is fatal.

With the disastrous effects, direct or indirect, of lightning strikes on living beings and material things (such as electronics, power grids, trees, and man-made structures), the importance of lightning protection is obvious and it must address the control, avoidance or diversion of high currents, high voltages, high temperatures, and high-pressure waves that are associated with lightning flashes.. Loss of the impacted equipment will be costly for the utility or user. If the lightning protection system is weak or poorly maintained and not upgraded, loss of lives, massive loss of revenues, crimes, and social disruption are frequent short-term effects. In addition we have long-term damage to electrical and electronic equipment and systems which is unavoidable.

According to Ohm's Law.

$$V = I.R \tag{1.23}$$

V = Voltage Drop, in V.
I = Peak current of lightning strike, in A.
R = Earth Resistance, in Ω.

Assuming that the resistance of the object struck (e.g. ground) is constant, the voltage at the point of strike follows the time domain pattern of the lightning current. Other than current and voltage, the charge Q of the lightning current is also an important element in the characteristics of a lightning strike. The charge of the lightning current plays a role in energy conversion where the energy W (electrical energy) will be converted into another form of energy (mostly heat energy) at impact point. The formula for Q and W can be seen below.

$$Q = \int i dt \tag{1.24}$$

$$W = Q.V \tag{1.25}$$

Q = Charge of lightning current, in C.
i = current of lightning strike, in A.
W = energy conserved in a lightning strike, in J.
V = voltage drop of the lightning strike near the impact zone, in V.

Therefore, the electric charges and the energy conversion at the lightning strike point result in extremely high heat which will cause the impact point to melt or burn.

However, the efficiency of energy conversion from electrical to heat largely depends on the resistance at the impact point. This energy conversion is given by

$$\frac{W}{R} = \int i^2 dt \qquad (1.26)$$

$$W = R. \int i^2 dt = R. \frac{W}{R} \qquad (1.27)$$

R = resistance of conductor (temperature dependant), in Ω.
$\frac{W}{R}$ = specific energy, or action integral.

The calculation implies that all the heat generated by the conversion of energy is dissipated in the ohmic resistance of the impact point. Furthermore, it is also expected that there is no perceptible heat exchange with the surrounding due to the extremely short duration of the conversion process. Table 1.1 shows the temperature rises of different materials that are used in lightning protection as well as their cross-sections as a function of specified energy.

Apart from specific energy, the electrodynamic forces F generated by the current I in the conductor that was struck by the lightning on a long, parallel conductor of length l, at a distance d, can be estimated by using the formula:

$$F(t) = \frac{\mu_0}{2\pi}.i^2(t).\frac{l}{d} \qquad (1.28)$$

F(t) = Electrodynamic force, in N.
i = current within a conductor, in A.
μ_0 = Magnetic field constant in air, (in $4\pi.10^{-7}$ H/m).

Table 1.1 Temperature rise ΔT in K of different conductor materials

Cross section (mm²)			4	10	16	25	50	100
Material	Aluminum W/R (MJ/Ω)	2.5	–	564	146	52	12	3
		5.6	–	–	454	132	28	7
		10	–	–	–	283	52	12
	Iron W/R (MJ/Ω)	2.5	–	–	1120	211	37	9
		5.6	–	–	–	913	96	20
		10	–	–	–	–	211	37
	Copper W/R (MJ/Ω)	2.5	–	169	56	22	5	1
		5.6	–	542	143	51	12	3
		10	–	–	309	98	22	5
	Stainless Steel W/R (MJ/Ω)	2.5	–	–	–	940	190	45
		5.6	–	–	–	–	460	100
		10	–	–	–	–	940	190

Fig. 1.9 Representation of
Potential Gradient Area

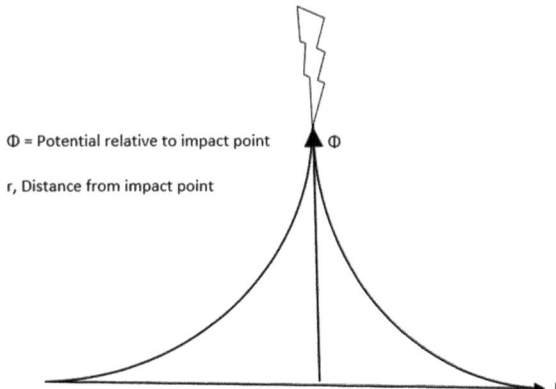

Φ = Potential relative to impact point Φ

r, Distance from impact point

l = conductor length, in m.
d = distance between two parallel conductors, in m.

The force between two conductors can be attractive if the two currents flow in the same direction and repulsive if the two currents flow in the opposite directions. The force is proportional to the magnitude of current and inversely proportional to the distance between the conductors. Thus, the specific energy dissipated by the lightning strike will cause stresses upon the impacted conductor to cause deformation.

When a lightning strike hits the ground, large amounts of current flow through the ground, neutralizing them. However, this action also causes the conductive surface of the impact area to form a potential gradient area, as shown in Fig. 1.9. If a living being (either human or animal) is inside the potential gradient area, an upwards voltage is formed and can cause an electric shock towards that being which could potentially kill it or cause other negative effects which can be referred to the effects of lightning strikes section. But, this risk can be reduced if the ground's conductivity is higher which then, flattens the potential gradient area making the victim less likely to get electrocuted.

Most lightning properties are beyond normal human comprehension. The cloud-to-ground discharge has an enormous magnitude of voltage that is tens of millions of volts or more. The maximum discharge of currents in each strike may vary from several thousand Amperes to 200,000 Amperes or even more. The current increase to this high magnitude lasts for a short time of about few millionths of a second (microsecond), and the primary part which usually has the highest possible value of current on each strike usually lasts even less than a thousandth of a second.

1.6.2   Effects of Lightning on Aircraft

Both commercial and military aircraft in flight are subject to many atmospheric disturbances of which lightning is no exception. As commercial aircraft are scheduled to fly fixed routes, it is often difficult to avoid thunderstorm formation along their paths. It is statistically reported that on average, every commercial aircraft is struck by lightning once every year. The flight path is an influential factor that increases the probability of lightning strike rate on an aircraft. That is, a lightning strike to an aircraft is a function of both the aircraft flight path and altitude, and the thunderstorm formation altitudes. Aircraft at a low altitude either in ascending or descending phases have an increased probability of being struck by lightning.

As the aerospace industry expands into both manned and unmanned commercial and military vehicles, preventing electric field enhanced aircraft initiated lightning strikes and protections against serious damage and accidents become a major concern to the aerospace industry. When an aircraft flies into the environment of an electrified cloud, it enters into an enhanced electric field region surrounding the cloud, which in most cases has a large negative charge center in its lower region. The electric fields will induce an electric dipole charge over the body of the aircraft, with positive electric charges on the top surface of the aircraft and negative electric charges on the underbelly of the aircraft, resulting in an electric dipole charge structure. These can be sufficiently enhanced to result in electric discharges, for instance, resulting in positive leaders emanating from the radome of the aircraft. With this, at another extremity of the aircraft, that in the tail part, a negative leader may develop from electrostatic discharges occurring at another electric field enhanced part of the aircraft body. The negative leader will move towards the ground or another nearby thundercloud or electric charge center in the air. It is important to determine the electric field enhanced areas of the aircraft in order to design preemptive measures to reduce lightning strike risks, even to design and to maintain the aircraft to reduce electric field enhancements in these high-risk areas. A knowledge of the electric charges induced on the aircraft body and the electric field distribution is also essential to decide on the safe placement of sensitive microelectronic systems associated with aircraft measurement and navigational systems.

For an aircraft to be air worthy, the aircraft manufacturers need to provide the overall assurance for adequate lightning protections. This process requires certification plans for tests done on components or systems of components such as the airframes, power and electrical wirings and components, fuel systems and components, avionics and communication systems such as the radar, and other control and automation components. The certification plans for the tests are done either within the aircraft manufacturers' laboratories or the component suppliers' laboratories. In a nutshell, the protection of aircraft against lightning strike can be summarized in the following steps (i) determine lightning attachment zone; (ii) determine systems and components which are likely damaged by lightning; (iii) set lightning protection standards for systems and components; and (iv) confirm the rationality of the protection design by the use of tests. The following parts of the aircraft experience electric fields that have potential to cause electrostatic discharge or electronic circuit flashovers: the radome, the wing tip, the wing surface, and the stabilizer tip.

In future aircraft the metal body surfaces will have their body materials increasingly replaced by composite materials, and fiber glass. Determining the lightning strike effect zones using prestrike electric field stress, becomes more critical to these non-electric shield materials. The most susceptible return stroke zones, the first strike zones, are the extremities of the aircraft. All the areas of the aircraft surfaces where a first return stroke are likely during lightning channel attachment with a low expectation of flash hang on, that is the lightning flash remains than being attached to the aircraft as the aircraft moves. The current at these zones of attachments may exceed 200 kA. Zone 2 are the aircraft surfaces where a subsequent return stroke is likely to be swept with a low expectation of flash hang on. The current in Zone 2 can exceed 100 kA. Zone 3 includes those surfaces not in Zones 1 and 2, where any attachment of the lightning channel is unlikely, and those portions of the aircraft that lie beneath or between the other zones and/or conduct substantial amount of electrical current between direct or swept stroke attachment points.

1.6.3   Lightning Effects on Electric Power Systems Network

Electric power transmission and distribution grids are routed for miles in open fields. Thus, they are prone to lightning strike. A lightning strike on structures such as a high voltage overhead transmission line can induce voltage and current surges whose amplitudes far exceeding the peak values of the nominal operating levels. The amplitudes are in the order of 1000 kV and 100 kA or more in the transmission line. The values of the peak rate of rise can reach measured and modeled values of 100 kA/µs. An overhead earth wire provides protection against direct lightning strikes in diverting the current and or voltage pulses to ground through the tower footing resistance. The tower footing resistance should be as low as 10 Ω or even less for more negative reflection from the tower base to reduce the chance of a voltage flashover at the top of the tower.

However, in the event of shielding failures, back flashover, and or an induced voltage on a transmission line when lightning strikes a nearby object, high current and voltage pulses will reach the terminal equipment such as a transformer at substations. (That is, the cloud charge induces a charge on the line which is attracted to a point closest to the cloud, and when the cloud charge flashes to a nearby object, the charge on the line is released from the Coulomb forces holding it and runs in both directions). In such cases, surge protection devices (SPD) are required to divert the major part of the energy of the surge to ground via surge diverters or absorbers, or by modifying the waveform to make it less severe via surge modifiers. Fuses which depend on exploding wires, or melting wires, are too slow to act for the high speed lightning surges that travel along conductors. Surge diverters (or lightning arrestors) generally consist of one or more spark gaps in series, together with one or more nonlinear resistors in series. Silicon Carbide (SiC) was the material most often used in these nonlinear resistor surge diverters. However, Zinc Oxide (ZnO) is being used in most modern day surge diverters on account of its superior volt-ampere characteristic. An ideal lightning arrester should: (i) conduct electric current at a certain voltage above the rated voltage; (ii) hold the voltage with little change for the duration of overvoltage; and (iii) substantially cease conduction at very nearly the same voltage at which conduction started. Figure 1.10 gives an illustration of lightning protection system with placements of shield wire, SPDs, circuit breakers, grounding systems, and the air terminals.

An absorber is in series with the line, such as a conducting PbO. As the surge runs through it, the property of PbO changes with heat, it becomes a nonconductor and kills the surge. A divertor is connected from the line to the ground like a spark plug. The surge makes the gap break down and diverts it to the ground. It is reusable. But an absorber is not reusable, because its composition changes.

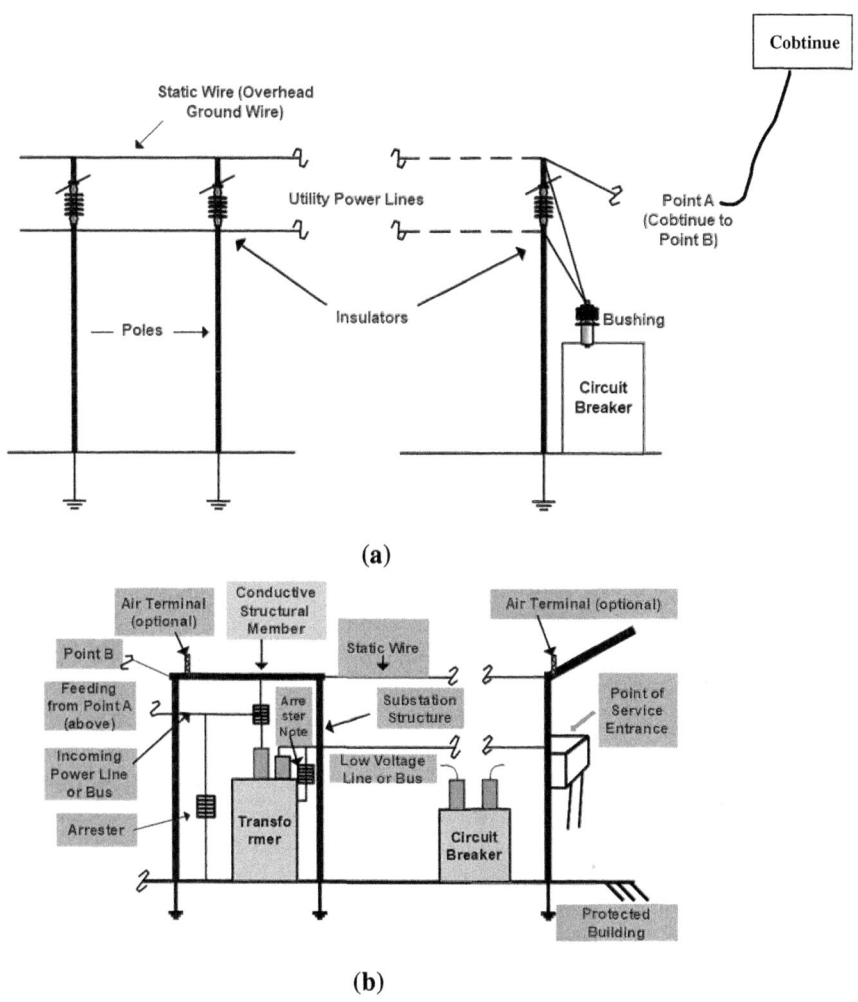

Fig. 1.10 Illustrations of power network protection **a** Transmission line and **b** substation protection systems *Credit* Adapted from DEHN

1.6.4 Substation Protection Systems

Substations accommodate some of the most expensive equipment such as the power transformers, current transformers, voltage transformers, and relays. Protection systems using arresters, absorbers, and breakers only protect the equipment from traveling waves induced by lightning. Protection against a direct lightning strike requires masts and shielding wires, as shown in Fig. 1.10a. The lightning protection of a substation utilizes three methods: using masts, using shielding or static wires, and/or using both masts and shielding wires. However, breakage of shielding wires (due to lightning current or poor maintenance) can cause catastrophic faults in substations when it snaps off. Further, another disadvantage of using shielding wires is high cost in comparison with the using of masts. Moreover, a mast attracts lightning flashes more easily than the shielding wire when the tip is made small. Thus, the application of a mast in substations is preferred to shielding wires for lightning protection for substations. A general arrangement for a power substation protection is shown in Fig. 1.10b. The requirements for the two different lightning protection mechanisms are discussed below (i) A shield wire lightning protection system will be generally used in smaller substations of lower voltage class, where the number of bays is fewer, the area of the substation is small and the height of the main structures is of normal height. The major disadvantage of shield wire type lightning protection is that it causes a short circuit in the substation or may even damage the costly equipment in case of its failure (i.e. snapping off); (ii) A lightning mast: this type of protection is generally used in large, extra high voltage substations where the number of bays is more. It has the following advantages, (a) It reduces the height of the main structures, as peaks for shield wire are not required, and (b) It removes the possibility of any back flashover with the nearby equipment or structure during discharge of lightning strokes.

Further, electrical substations require an earth mat for good grounding. Grounding high frequency signals or currents of the lightning currents require special care in order to keep the high frequency grounding resistance small enough. Vertical grounding rods that conduct lightning currents carry currents at frequencies up to 100 MHz, compared to the low 50 Hz or 60 Hz electric power frequency. At higher frequencies the currents flow only over the outer surface (determined by the frequency dependent skin depth) of the vertical rod, thus increasing the grounding impedance because of the reduced surface area over which the currents flow at high frequencies.   The earthing system provides a low resistance return path for earth faults within the plant, which protects both personnel and equipment. The earthing system provides a reference potential for electronic circuits and helps reduce electrical noise for electronic, instrumentation, and communication systems. The earthing system also provides a low resistance path (relative to remote earth) for voltage transients such as lightning and surges/over-voltages. Another requirement for substation earthing is to provide for equipotential bonding which helps prevent electrostatic build up and discharge, which can cause sparks with enough energy to ignite flammable atmospheres. Special consideration must be given to the protection of the increasing amount of electronic systems, such as the distribution static compensator (D-SATCOM), that form the critical systems of the electric power grid.

1.6.5 Rolling Sphere Method Applied in Substation Protections

The application of the rolling sphere method involves rolling an imaginary sphere of radius S over the surface of a substation. The sphere rolls up and over (and is supported by) lightning masts, shield wires, substation fences, and other grounded metallic objects that can provide lightning shielding. A piece of equipment is said to be protected from a direct stroke if it remains below the curved surface of the sphere by virtue of the sphere being elevated by shield wires or other devices. An equipment that touches the sphere or penetrates its surface is not protected. The basic concept is illustrated in Fig. 1.11 based on IEEE Standard 998–2012, "IEEE Guide for Direct Lightning Stroke Shielding of Substations."

   The calculation for the rolling sphere method is based on the electro-geometrical model. It is summarized in the following equations based on a 69 kV substation. The striking distance, S with respect to lightning strike peak current, I_s is calculated by using Eq. (1.28).

$$S = 10 \cdot I_S^{0.65}, \tag{1.28}$$

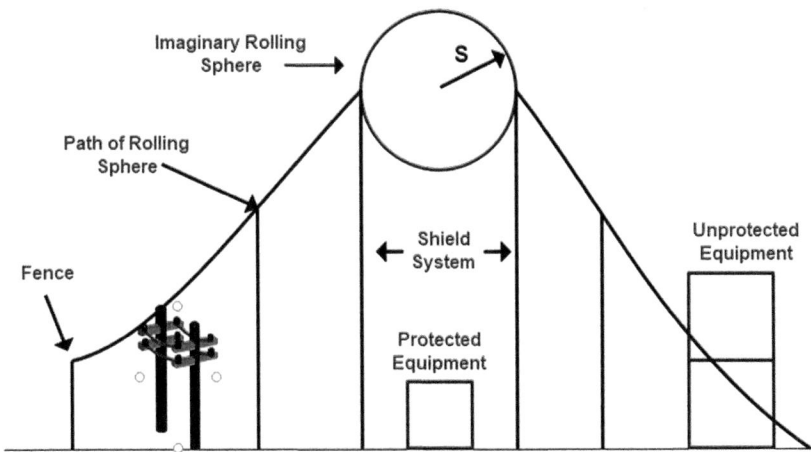

Fig. 1.11 Illustration of rolling sphere protection method. *Credit* Adapted from DEHN

where I_S is a function of the surge impedance (Z_s) and the basic impulse level (BIL). For a 69 kV system for which an assumed $Z_s = 300\ \Omega$ and BIL: $= 350$ kV. Then

$$I_S = 2.2 \cdot \frac{BIL}{Z_s} \tag{1.29}$$

from which we get a value of I_S of 2.567 kA. Using the calculated current stroke value, the radius of the sphere can be computed which is 18.45 m. This is the length of the last leader as it strikes the mast. It is also the length of the upward leader from the mast tip as it meets the downward leader from the cloud at the stroke point.

1.6.6 Lightning Protection Methods for Buildings and Infrastructures

The lightning protection system of buildings and infrastructures is categorized into three as illustrated in Fig. 1.12. It covers (i) protection for buildings and installations against direct strike by lightning; (ii) protection systems against overvoltage on incoming conductors and conductor systems; and (iii) protection systems against the electromagnetic pulse induced by lightning striking a nearby object as an indirect effect.

The protection against direct lightning strike requires air-termination rods with good bonding to ground through down conductors via metallic structures on the external building structures as specified by grounding standards. Figure 1.11 gives and illustration of the rolling sphere requirements for buildings with areas indicated as shown that need air terminals to shield the building from direct lightning strike. Further, a good bonding to ground is necessary to provide an equipotential ground plane for all components within the building for single integrated earth termination systems for structures, combining lightning protection, power and telecommunication systems. Figure 1.13 gives an illustration of the equipotential bonding of an installation and the application of SPD in protection of building components (Fig. 1.14).

Figure 1.14 shows the complete protection system zones within a building. Air-termination methods in protection of building structures requires the electro-geometrical method of rolling sphere to determine the safe zones from lightning strikes. The air-termination method requires the use of rods spanned by wires or cable, and mesh conductors. Further, the air terminal requires the electro-geometrical method of a rolling sphere to determine the safe zones from lightning strikes.

The protection systems cover both external protection devices using air terminals, and down conductors to ground. This protects the building from both direct and induced voltages and currents from lightning striking the nearby objects. The interior protection requires the necessary grounding of the building circuits and utility piping. The SPD devices are also used indoors for the protection of lightning induced LEMPs on appliances.

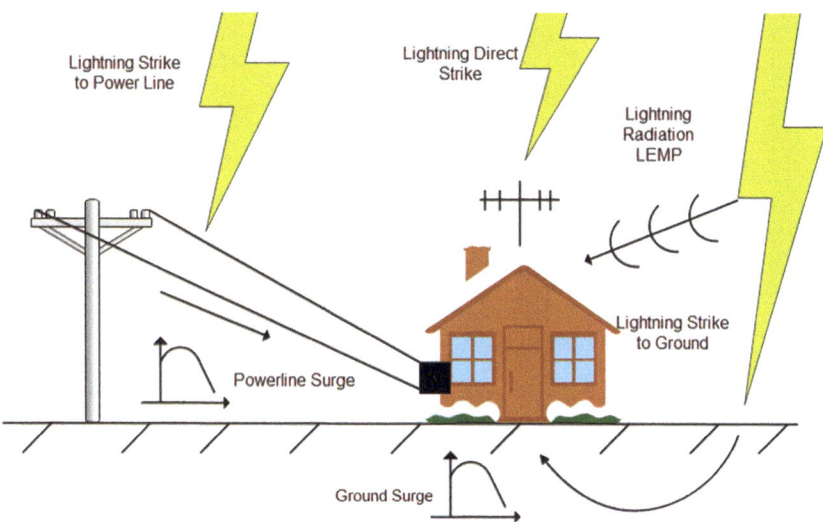

Fig. 1.12 Illustration of lightning strike through direct hit, through incoming conductor, and induced through objects nearby

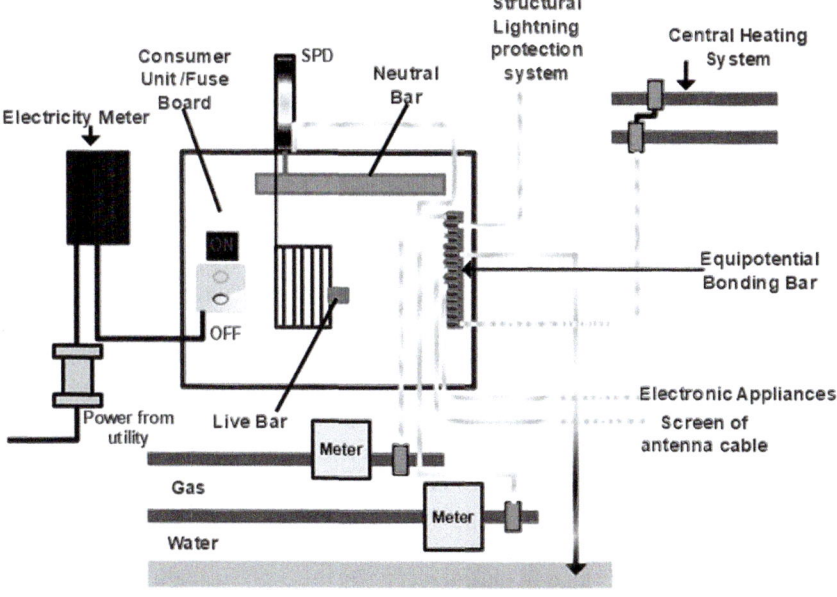

Fig. 1.13 Equipotential bonding of building components (*Credit* Adapted from DEHN)

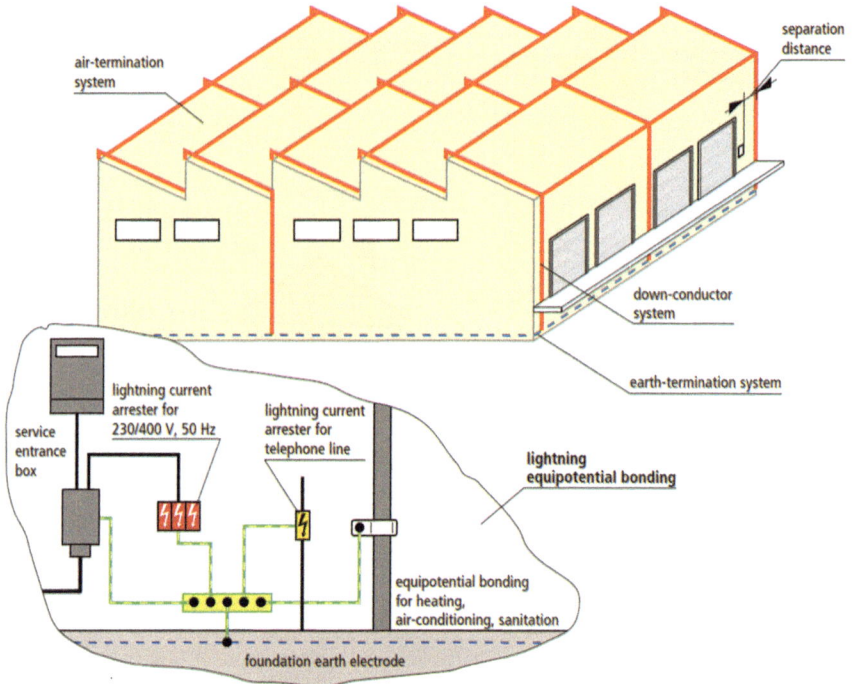

Fig. 1.14 Lightning protection system. (*Credit* From DEHN, with permission)

1.7 Lightning, Climate, Upper ionosphere, and Other Planets

Lightning is a good indicator of the intensity of atmospheric convection. Hence lightning can be related to the earth's atmosphere where there is the greatest instability. This atmospheric instability occurs either due to the heating of the boundary layer by solar radiation or by mixing of air of different densities. There is an organized pattern to form the unstable regions which is originally driven by the heating pattern of the earth's surface by the sun. Therefore, if the climate system is changed, the regions of convection will be changed ultimately changing the lightning patterns around the globe. Meteorological Measuring Systems are used to globally monitor lightning activity, which is also a direct measurement of climate change. Monitoring is carried through all three stages of the thunderstorm life cycle, namely, the developing stage, the maturing stage and the dissipating stage. Over the developing stage the towering cumulus cloud builds up and there may be occasional lightning activity. It is during the maturing stage that we get heavy rain, strong wind, hail, frequent lightning and tornadoes. During the heavy rainfall of the dissipating stage lightning activity may drop, but it continues even after rainfall ceases.

Due to solar heating there will be rising temperature around the globe with maximum occurring in tropical regions. The region of rising air along the thermal equator is known as the Inter Tropical Convergence Zone (ITCZ). The thermal equator is not a constant as is the geographical equator. The thermal equator moves to the northern and southern hemisphere with the seasonal changes. As the land to ocean ratios are different in the two hemispheres and as the heat capacities are also different, the width of the ITCZ will also change.

The tropical monsoons consist of moist oceanic air, which results in heavy rainfall with low rates of lightning. Intense lightning can be observed in a dry environment. That is why more intense lightning activities can be observed in the African continent. Lightning activity in the North Pole has dramatically increased in recent times, as well as the annual number of lightning flashes over many countries, including the USA. Increase in the number of electrically severe thunderclouds results in increasing number of severe tornadoes produced by such super thunderclouds. The greater the air mass density difference, the greater the atmospheric instability which will result in a greater intensity in these storms.

1.7.1 Effect of Temperature on Lightning

By using different time scales (diurnal, daily variations, intra-seasonal, semiannual variations, annual variations, etc.) it was observed that there is a positive relationship between temperature and lightning with lightning activity increasing for every degree the surface warms up. The daily, regional, averaged surface temperature over Africa when compared with the regional lightning activity, showed that lightning activity increases with surface temperature. For tropical lightning surface temperature is a key factor determining daily lightning activity, while no relationship was observed in the long term over the last 50 years.

Water vapor absorbs infrared radiations emitted from the earth's surface and it is the primary natural greenhouse gas influencing the climate. The earth's climate is more sensitive to the changes of water vapor in the upper troposphere which naturally has a low level of water vapor. Recently it was found out that thunderstorms deposit a large amount of water vapor in the upper troposphere. Lightning activity over Africa was seen to vary with the specific humidity of the upper troposphere. It may be stated that lightning can be used to monitor the intensity of the deep convection. Water vapor, cloud cover, ice water content, and ice particle size have different impacts on earth's radiation balance and many studies have proved lightning activity has a direct correlation with the above.

1.7.2   Effect of Lightning on Troposphere

Lightning activity will produce nitrogen oxides and ozone, which is another green-house gas. Recent studies show concentrated NO_X in thunderstorm anvils and concentrated ozone in downwind. Taking exact measurements of the concentration of these gases and building a relationship with lightning parameters is a difficult task. But lightning is the main source of NO_X and these gases play a vital role in earth's climate.

It is evident that global temperatures are increasing and the cause is the increase of greenhouse gases. The modeling of future lightning activities showed that the greatest warming due to greenhouse gases will occur on the equatorial upper troposphere and not on the surface, and moisten the upper troposphere due to the deep convection. But a paradox occurs as the upper atmosphere warms up, and atmospheric lapse rate becomes more stable tending to inhibit future convection which will reduce the amount of thunderstorms. But many climate model stimulations conclude that lightning activities will increase in a warmer climate with 10% increase in lightning activity globally for every 1 °C warming. It is shown that in a doubled CO_2 climate, the updraft strengthens, and drying in a warmer climate reduces the frequency of the thunderstorms but those that do occur are very intense.

Global temperature in recent times has increased by about 2–5 °C depending on the region of the planet earth. The annual increase in temperature on average is 0.25 °C per decade. Nitrogen oxides NOx (NO and NO_2) break into Nitrogen and Oxygen molecules through electron impact and very high temperatures in the lightning channel. About 10 eV energy electrons inside lightning channels are able to break N_2. Electrons at energy higher than 5 eV are able to break O_2. In a region with 10^8 ions in a 0.5 mm radius streamer, electron collisions can produce about 5×10^{11} NO molecules. Moreover, high temperatures of the order of 25,000 °K inside the lightning channels also result in the formation of NO molecules, through the reactions of $O + N_2 = NO + N$ and $N + O_2 + NO + O$. However, NO can be destroyed by thermal collisions too: $NO + N = N_2 + O$, $NO + O = N + O_2$ and $NO + NO = N_2O + O$. In general, a total of about 10^{25} NO_x molecules may be found in a 10 km long lightning channel. This production of NO_x occurs not so much during the return stroke phenomena, but during the production of streamers, corona and M-components.

Lightning activity over ship routes are expected to double due to particles emitted from ships. In cities lightning flashes to earth are expected to increase in the range of 45–80% during summer. This is partially due to an increase in urban cloud condensation nuclei concentration. A fourfold increase in summertime lightning activity is expected. Over growing cities, due to convergence of surface winds due to the heat island effect and high air pollution, may increase lightning flash density from 1 per km^2 to 4 per km^2. In some cities, lightning activity over cities has increased by 60–70% over the built-up areas of a city. A decrease in positive lightning flashes by 7–8% has been observed over such cities. Due to increase in pollution particles over the city, cloud-to-ground flashes were seen to increase by about 50%. Lightning is a major cause of damage to trees and forests through direct hits or by setting forests on fire. In the Western USA 40% of forest fires over an area of 0.4 ha are due to lightning, with 76% in Alaska, with an area larger than 200 ha. Over the years between 1973 and 2013, with climate change, it was found that there was a global mean of fire season increased in duration by 18.7%, with an increase of 10.8% in burnable areas. With the increase of wind turbines in 90 countries from a capacity of 514 GW to 5476 GW in 2050, damage to the rotor, burns, punctures, tip damage and edge debonding of blades is expected significantly to increase with the increased number of wind turbine installations and increase in atmospheric temperature and frequency of severe lightning strikes. It is known that the rotor blades of the wind turbines initiate lightning at 3 s intervals over periods of hours. The rotating blades increase lightning strike vulnerability.

The most destructive type of lightning is the cloud-to-ground discharges. Since they might discharge large currents to the earth through buildings, living beings, etc., only about 25% of global lightning events are cloud-to-ground lightning. There are several types of cloud-to-ground discharges. They are negative downward, negative upward, positive downward, positive upward lightning flashes and bipolar lightning flashes. From these about 90% of lightning flashes are downward negative lightning flashes which transport negative charges in the downward direction of the cloud. Figure 1.15 shows lightning and electric discharges in the upper atmosphere above the thundercloud. Lightning cloud discharges paint beautiful pictures in the sky. They do not make contact with the ground. They can be either an inter-cloud discharge, intracloud discharge or an air discharge. Out of these three categories intracloud discharges are more frequent. Cloud discharges are less destructive to human life and animals since they do not make contact with the earth. But cloud discharges can be destructive to aircrafts, space crafts and complex electronic devices. Lightning mapping arrays are used to study lightning. These are three-dimensional arrays which are used to map lightning. Red sprites, blue jets and elves are three types of transient luminous events which are associated with cloud discharges to the higher atmosphere. These are rarely observable to the naked eye. Red sprites are not very bright. They are red in color and the period of red sprites are is just a few seconds. Blue jets have been witnessed by pilots and the period of blue jets is less than a fraction of a second. Elves occur in the ionosphere, during the occurrence of a cloud-to-ground discharge. The period of elves is less than a thousandth of a second. Elves are of an expanding disk shape, as shown in Fig. 1.15a. In Fig. 1.15b is shown the global lightning activity, showing very little activity over the sea (say, 5 lightning flashes/km^2/year) and as many as 150 to 250 lightning flashes/km^2/year in some countries close to the equator. The highest number of flashes are found in Africa and South America. In some regions such as North India, lightning activity peaks around the monsoon rain season (in May, for instance), whereas in other countries lightning activity is uniform throughout the year.

Sprites. Consider now the upper ionosphere of earth. The troposphere extends from 10 km altitude to 15 km. Lightning flashes occur from the top of a thundercloud to the upper atmosphere. There are further lightning like activities in the upper ionosphere. First, there are the Sprites which are jelly fish shaped, as shown in Fig. 1.15. They

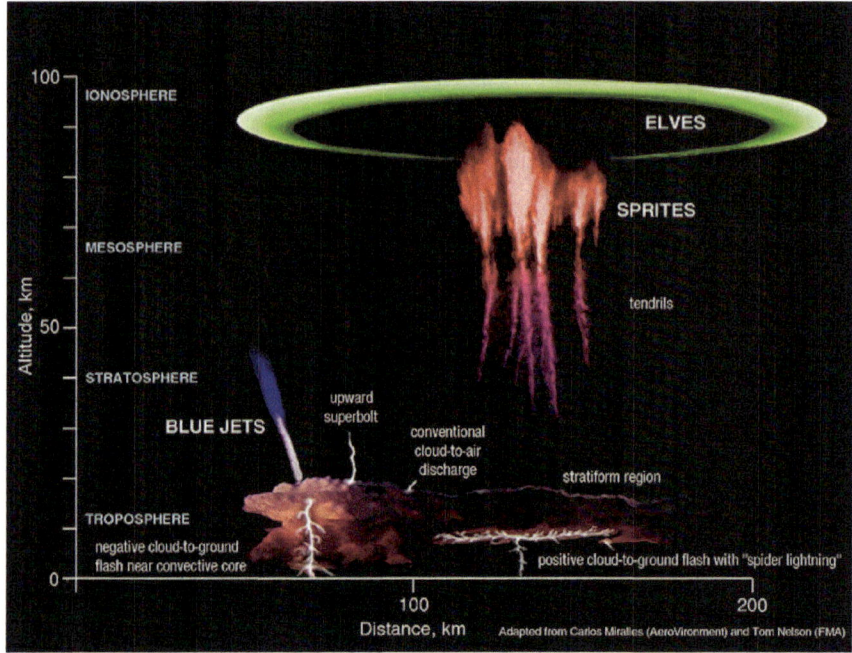

(a)

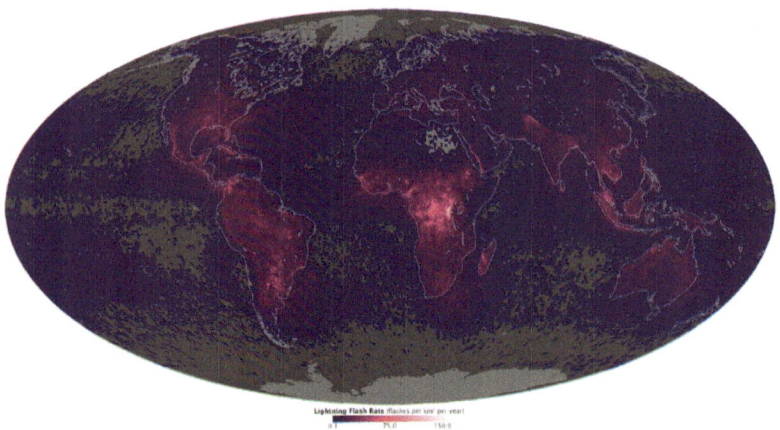

(b)

Fig. 1.15 a Types of Lightning (*Credit* NOAA_NSSL, USA. With permission. https://www.nssl.noaa.gov/education/svrwx101/lightning/types/) **b** Global Lightning activity, with as many as 150 lightning flashes per square kilometer per year in the light pink areas (*Credit* NASA, USA)

have an oval shaped body with numerous tendril-like electric discharge channels. The Sprites occur at altitudes of 70 to 90 km. The main electric discharge body is reddish in color, and the tendrils tend to be bluish in color. The total length of Sprites is 5 to 30 km with the individual tendrils being 10 m in diameter. The entire volume is about 10,000 km^2. It lasts for about 10 to 100 ms. The Sprites are thought to be produced by lightning activity in the earth's lower atmosphere. The upper atmosphere is a good conductor. The upper region of a thundercloud produces a strong electric field in the upper atmosphere, which move the free electrons in the upper atmosphere. The thundercloud electric charge builds up slowly, so that the movement of the electrons is slow. However, when the lightning flash suddenly depletes the thundercloud of electric charges, the electric field in the upper atmosphere suddenly collapses. This results in a fast, sudden collapse of electrons producing rapid ionization, that appears as sprites. Inside the upper region of a thundercloud, at an altitude of 10 km, the total electric charge could be about 100 C.

Blue Jets. In addition to Sprites, there are also blue jets and gigantic jets that appear in the upper atmosphere. These travel up from the upper region of the thundercloud, originating from 15 to 18 km and moving up towards the upper atmosphere to heights of about 40 km. The blue jets are conical shaped, and travel at about 10^5 m/s, with a cone angle of 15° and last for 200 to 300 ms. When lightning flash rate is high, gigantic blue jets are produced which travel from the top of the thundercloud at 15 km to all the way up to the ionosphere at 70 km. They travel at a velocity of 10^5 m/s. Blue jets are largely positive discharges.

Elves. When electromagnetic waves (electric field E and magnetic flux density B) impinge on the ionosphere, they produce the Lorentz force F = q (E + v x B) on free electric charges q inside the ionosphere moving at velocity v. These electric charges moving along the upper layer of the ionosphere produce a current that moves out in ring shape. These currents radiate visible electromagnetic fields when the moving electrons collide with N_2 and O_2 with the excited atoms radiating light. These ring shape red light radiations are called elves and last for about 1 ms. There are other non-visible energetic radiations also associated with lightning, including X ray and antimatter from within thunderclouds. These radiations occur due to very large electric fields inside the thunderclouds which accelerate electrons to run away speeds. These electrons, which are at energy levels of 100 meV, radiate X-rays and γ-rays. These strong electric fields are generated at points where positive and negative streamer tips meet.

1.8 Summary

Direct and indirect lightning effects on a power systems network, and building and system structures have become serious threats. The effects of climate change coupled with the anthropogenic enhancements of heat emission play a crucial role in the cloud-to-ground electrification which is experienced in major cities today. The threats can be contained through proper protection measures such as the SPDs, and proper shielding and grounding practices as defined in various standards on lightning protections highlighted. Future changes in lightning activities and severity with climate change may force a reevaluation of current lightning protection standards and techniques. However, with the probabilistic nature of lightning phenomena, the protection measures many not provide 100% protection in shielding the direct and indirect effects of lightning flashes. Thus, new measures in protections and standards will have to drive protection to a higher level for the mitigation of lightning threats.

The need to harden aircraft, power systems, structures, electronics, computers, electronics devices and other information communication systems against severe electric storm has been highlighted in this chapter. Among the many things highlighted is that electric storms remain a complex phenomenon that cannot be controlled and or be prevented. Undoubtedly, it remains a serious threat to aircraft, power systems, electronics, and communications systems. As an aircraft in flight near ground can become a path of the electric storm discharge circuit either through a triggering mechanism process or through interception, the direct and indirect effects pose severe threats to flight safety. These threats are heightened further with aircraft industries continually modifying and or adopting new designs into the aircraft such as the CFC in airframe design and the latest in the state-of-the-art digital control, command, and communication systems. As climate change may result in lightning flashes with more electric voltage and current pulses, and electric power transmission is pushed to higher voltages and taller towers, the number and severity of lightning strikes to electric power systems and surges that trickle down to low voltage commercial, industrial and smart city installations will increase. Microelectronic circuits and systems that operate at very low battery voltage ARE??? more vulnerable to small voltage and current spikes due to the lightning radiated electromagnetic pulses (LEMP).

Bibliography

Airbus A380, Aircraft Characteristics Airport and Maintenance Planning, Airbus Technical Data and Support and Services (2014)

Aircraft Lightning Environment and Related Test Waveforms, SAE ARP 5412 Committee Report, April (2012)

A Antonescue S Stefan 2011 The urban effect on the cloud-to-ground lightning activity in the Bucharest area Rromania, Romanian Reports Phys. 63 2 535 542

Barry, J.D.: Ball Lightning and bead Lightning, Plenum (1980)

Bazelyan, E.M., Raizer, Y.P.: Lightning Physics and Lightning Protection, IOP Publishing (2000)

Car, S.K., Liou, Y.A.: Analysis of cloud-to-ground lightning and its relation with surface pollutants over Taipei, Taiwan, Annales Geophysicae, Vol.32, pp. 1085–1092 (2014)

Catlin, X.L.: GAPS Guidelines, A Publication of Global Asset Protection Services LLC, A Publication of Global Asset Protection Services LLC, [Online], Available: http://xlcatlin.com/~/media/gaps/522_0.pdf (2015)

Chalmers, J.A.: Atmospheric Electricity, Pergamon, 2nd Edn (1984)

Cooray, V.: An Introduction to Lightning. Springer (2015)

Cooray V. (Ed.): The Lightning Flash, IET (2003)

Cooray, V. (Ed.): Lightning Protection (2010)

Cooray (Ed.), Lightning Electromagnetics (2012)

DEHN + SÖHNE, "Lightning Protection Guide," 3rd updated edition December (2014)

Dung, L.V., Petcharaks, K.: Lightning protection systems design for substations by using masts and MATLAB. World Acad. Sci. Eng. Technol. Int. J. Math. Comput. Phys. Electr. Comput. Eng. **4**(5), (2010)

Dong, Z.Y., Luwen, C.: Climate lightning activity and its correlations with meteorological parameters in South China, XV International Conference on Atmospheric Electricity, pp. 1–14, June (2014)

EHV AC Substations, Layout, Equipment, Bus arrangements. http://srldc.in/var/NRC/PSTI%20course/vol2_PS%20Operation/2EHV%20susbattions_psjg_apr11.doc

Farias, W.R.G., Pinto, W.O., Naccarato, K.P., Pinto, I.R.C.A.: Anomalous lightning activity over the Metropolitan Region of São Paulo due to urban effects," Journal of Atmospheric Research Vol. 91, pp. 485–490, Lightning Physics and Lightning Protection, IOP Press (2009)

Fisher, F.A., Plumer, J.A., Perala, R.A.: Lightning Protection of Aircraft", Lightning Technologies Inc., second edition (1999)

Fisher, J.J., Hoole, P.R.P., Pirapaharan, K., Thirukumaran, S., Hoole, S.R.H.: Cloud to ground and ground to cloud flashes in lightning protection: and future severe lig htning and climate change. International Conference on Lightning Protection (ICLP), pp. 440–445, IEEE Xplore Digital Library (2014)

Fisher, J., Hoole, P.R.P., Pirapaharan, K., Hoole, S.R.H.: Applying a 3D dipole model for lightning electrodynamics of low-flying aircraft. IETE J. Res. **61**(2), 91–98 (2015)

Fisher, J., Hikma, S., Hoole, P.R.P., Sharip, M.A.R.M., Pirapaharan, K., Al-Khalid Hj Othman, Julai, N., Harikrishnan, R., Hoole, S.R.H.: Observations on the electrostatic discharge threats to aircraft body and to aerospace electronics. J. Telecommun. Electr. Comput. Eng. (JTEC) (2017)

Fisher, J., Hoole, P.R.P., Pirapaharan, K., Hoole, S.R.H.: Applying a 3D dipole model for lightning electrodynamics of low-flying aircraft. IETE J. Res., 1–8 (2015)

Fisher, J., Hoole, P.R.P., Pirapaharan, K., Hoole, S.R.H.: Parameters of cloud to cloud and intra-cloud lightning strikes to CFC and metallic aircraft structures. Proceedings International Symposium on Fundamentals of Engineering, IEEE Xplore Digital Library (2017)

Fisher, J., Hoole, P.R.P., Pirapaharan, K., Thirukumaran, S., Hoole, S.R.H.: Cloud to ground and ground to cloud flashes in lightning protection: and future severe lightning and climate change, Lightning Protection (ICLP), International Conference pp. 440–445. IEEEXplore Library (2014)

Fisher, J., Hoole, P.R.P., Pirapaharan, K., Thirukumaran, S., Hoole, S.R.H.: Three dimensional electric dipole model for lightning-aircraft electrodynamics and its application to low flying aircraft, Lightning Protection (ICLP), International Conference, pp. 435–439, IEEEXplore Library (2014)

Gamerota, W.R., Elismé, J.O., Uman, M.A., Rakov, V.A.: Current waveforms for lightning simulations. IEEE Trans. Electromag. Comp. **54**(4), 880–888 (2012)

Golde, R.H.: Lightning Protection, Arnold (1971)

Golde, R.H. (Ed.):Lightning, Vol 1: Physics of Lightning, Academic Press (1977)

Golde, R.H.: Lightning, Vol. 2: Engineering Applications, Academic Press (1977)

Hasse, P.: Overvoltage Protection for Low Voltage Systems, IET (2000)

Heidler, J.F., Flisowski, Z., Zischank, W.: Ch. Bouquegneau, Mazzetti, C.: Parameters of lightning current given in IEC 62305-background, experience and outlook. 29th International Conference on Lightning Protection, Uppsala, Sweden, 23–26 June (2008)

Hess, R.: The electromagnetic environment. The Avionics Handbook, CRC Press LLG (2001)

Hoole, P.R.P., Coowar, F.: Statistics on tropical lightning and its interaction with power systems. Electr. Power Syst. Res., 63–77 (1989)

Hoole, P.R.P., Pirapaharan, K., Kavi, M., Fisher, J., Aziz, N.F., Hoole, S.R.H.: Intelligent localisation of signals using the signal wavefronts: a review. Lightning Protection (ICLP), International Conference, pp. 474–479 IEEE Xplore Library (2014)

Hoole, P.R.P.: Simulation of Lightning attachment to open ground, tall towers and aircraft IEEE Trans. Power Delivery 8 2 732 738 (1993)

Hoole, P.R.P.: Modeling the lightning earth flash return stroke for studying its effects on engineering systems IEEE Trans. Magn. 29 1839 1844 (1993)

Hoole, P.R.P., Hoole, S.R.H.: Guided waves along an un-magnetized lightning plasma channel IEEE Trans. Magn. 24 6 3165 3167 (1988)

Hoole, P.R.P., Hoole, S.R.H.: Simulation of lightning attachment to open ground, tall towers and IEEE. Trans. Power Delivery 8(22), 732–740, Institute of Electrical and Electronics Engineers (1993)

Hoole, P.R.P., Hoole, S.R.H.: Charge simulation method for the calculation of electromagnetic fields radiated from lightning. JJ Conner CA Brebbia Eds Boundary Element Technology Computational Mechanics Publications Southampton 153 169 (1986)

Hoole, P.R.P., Hoole, S.R.h.: finite element computation of magnetic fields from lightning return strokes. In: Cendes, Z.J. (Ed.), Computational Electromagnetics, North Holland, pp. 229–237 (1986)

Hoole, P.R.P., Pearmain, A.J.: A review of the finite-difference method for multidielectric electric field calculations. J. Electr. Power Syst. Res. 24(11), 19–30. Elsevier (1992)

Hoole, P.R.P., Thirukumaran, S., Hoole, S.R.H., Harikrishnan, R., Jievan, K.: Ground to cloud lightning flash currents and electric fields: interaction with aircraft and production of ionospheric sprites. Proceedings The 28th International Review of Progress in Applied Computational Electromagnetics, 6 p., Michigan, USA (2012)

Hoole, P.R.P., Pirapaharan, K., Hoole, S.R.H.: Waveguide and circuit EM models of lightning return stroke currents. J. Japan Soc. Appl. Electromagn. Mech. Japan 19, S167–S170 (2011)

Hoole, P.R.P., Hoole, S.R.H.: A distributed transmission line model of cloud-to-ground lightning return stroke: Model verification, return stroke velocity, unmeasured currents and radiated fields Int. J. Phys. Sci. UK 6 3851 3866 (2011)

Hoole, P.R.P., Thirikumaran, S., Ramiah, H., Kanesan, J., Hoole, S.R.H.: Ground to cloud lightning flash and electric fields: Inter., action with aircraft and production of ionospheric sprites. J. Comput. Eng., Article ID 869452. https://doi.org/10.1155/2014/869452 (2014)

Hoole, P.R.P., Thirukumaran S., Hoole, S.R.H.: A software testbed for electrodynamics of direct cloud to ground and ground to cloud lightning flashes to aircraft. Int. J. Appl. Electromagn. Mech. 47(4), 911–925 Japan (2014)

Hoole, P.R.P., Pirapaharan, K., Hoole, S.R.H.:An electromagnetic field based signal processor for mobile communication position-velocity estimation and digital beam-forming: an overview J. Japan Soc. Appl. Electromagn. Mech. Japan 19 S33 S36 (2011)

Hoole, P.R.P., Pirapaharan, K., Hoole, S.R.H.: Waveguide and circuit EM models of lightning return stroke currents J. Japan Soc. Appl. Electromag. Mech. Japan 19 S167 S170 (2011)

Hoole, P.R.P., Fisher, J., Pirapaharan, K., Al K. H. Othman, Julai, N., Aravind, C.V., Senthilkumar, K.S., Hoole, S.R.H.: Determining safe electrical zones for placing aircraft navigation. Measur. Microelectron. Syst. Static Thunderstorm Environ. Int. J. Control Theory Appl. 10(16) (2017)

Hoole P.R.P., Hoole, S.R.H.: Computing transient electromagnetic fields radiated from lightning. J. Appl. Phys. 61(8), 3473–3475 (1987)

Hoole, P.R.P., Balasuriya, B.A.A.P.: Lightning radiated electromagnetic fields and high voltage test specifications. IEEE Trans. Magn. 29(2), 1845–1848 (1993)

Hoole, P.R.P., Pirapaharan, K., Hoole, S.R.H.: Waveguide and circuit EM models of lightning return stroke currents J. Japan Soc. Appl. Electromag. Mech. Japan 19 S167 S170 (2011)

Hoole, P.R.P., Hoole, S.R.H.: Computer aided identification and location of discharge sources. J. App. Phys. 61 (1987)

Hoole, P.R.P., Hoole, S.R.H.: Computing Transient Electromagnetic fields from lightning. J. Appl. Phys. 61, 3473 (1988)

Hoole, P.R.P., Hoole, S.R.H: Stability and accuracy of the finite difference time domain (FDTD) method to determine transmission line traveling wave voltages and currents. J. Eng. Technol. Res. (2011)

Hoole, P.R.P.: Smart Antennas and Electromagnetic Signal processing for Advanced Wireless Technology: with Artificial Intelligence and Codes, River Publisher (2020)

Hoole, P.R.P., Pirapaharan, K., Hoole, S.R.H.: Electromagnetics Engineering Handbook WIT Press UK (2013)

Hoole, P.R.P.: Smart Antennas and Signal Processing for Communication, Medical and Radar Systems, WIT Press, UK, 2001. (See IEE review of this book close to the end of this document)

Hoole, P.R.P.: Electromagnetic Imaging in Science and Medicine WIT Press UK (2000)

Hoole, S.R.H., Hoole, P.R.P.: A Modern Short Course in Engineering Electromagnetics, Oxford University Press, USA (1996)

IEC, Protection against Lightning Vol. 1: General Principles

IEC Protection against Lightning Vol. 2: Risk Management

IEC Protection against Lightning Vol. 3, Physical damage to Structures and Life Hazard

IEC Protection against Lightning Vol. 4: Electrical and Electronic Systems within Structures

IEC Protection against Lightning Vol. 5: Services

IEEE Guide to Direct Lightning Stroke Shielding of Substations, IEEE Std. 998–1996 (1996)

Industry Document to Support Aircraft Lightning Protection Certifications, U.S. Department of Transport Federal Aviation Administration Advisory Circular, AC No: 20–155A (2013)

IEEE Guide, How to Protection Your House and Its Content From Lightning, IEEE Guide for Surge Protection of Equipment Connected to AC Power Communication Circuits. http://lightningsafety.com/nlsi_lhm/IEEE_Guide.pdf

IEC, Protection Against Lightning – Part 1 General Principles, IEC International Standard, IEC 62305–1, Edition 2 (2010–2012)

Kendall, C., Black, E., Larsen, W.E., Rasch, N.O.: Aircraft generated electromagnetic interference on future electronic systems. Federal Aviation Administration Report No: DOT/FAA/CT-83/49, December (1983)

Kramer, J.: Lightning: Nature in Action, Lerner (1992)

Lightning Protection Standards, BS EN/IEC 62305 Standard for Lightning Protection (2006). http://www-public.tnb.com/eel/docs/furse/BS_EN_IEC_62305_standard_series.pdf

Morgan, D., Hardwick, C.J., Haigh, S.J., Meakins, A.J.: The interaction of lightning with aircraft and the challenges of lightning testing Aerospace Lab J. 5 ALO5-11 1 10 (2012)

Naccarato, K.P., Campos, D.R., Meireles, V.H.P.: Lightning urban effect over major large cities in Brazil. XV International Conference on Atmospheric Electricity, Norman, Oklahoma, U.S.A, 15–20 June 2014

NSSL: Severe Weather 101: Lightning, https://www.nssl.noaa.gov/education/svrwx101/lightning/

NPFA, Standard for the Installation of Lightning Protection System, NPFA-780 (2020)

Petrov, N.I., Haddad, A., Petrova, H., Griffiths, H., Walters, R.T.: Study of effects of lightning strikes on aircraft. Recent Advances in Aircraft Technology, Dr. Ramesh Agrawal (Ed.), ISBN978–953–51–0150–5, InTech, pp. 523–544 (2012)

Pinto, O., Pinto, I.R.C.A.: About sensitivity of cloud-to-ground lightning activity to surface air temperature changes at different time scales in the city of Sao Paulo, Brazil. 20th International Lightning Detection Conference and 2nd International Lightning Meteorology Conference, Tucson U. S. A, April (2008)

Poelman, D.R.: On the science of lightning: an overview. Royal Meteorological Institute of Belgium, Publication Scientifique et Technique, No. 56 (2010)

Price, C.: Thunderstorm, lightning and climate change, 29th Internation Conference on Lightning Protection Conference, Uppsala, Sweden, June (2008).

Rakov, V.A.: Lightning phenomenology and parameters important for lightning protection. IX International Symposium on Lightning Protection, 26th–30th November, Brazil (2007)

Rakov, V.A.: Lightning discharge and fundamentals of lightning protection. J. Lightning Res. **4**(Suppl. 1: M2), pp. 3–11 (2012)

MH Rawoot MFAR Satarkar 2014 Some critical study of lightning impulses on electrical transmission line system Int. J. Curr. Eng. Technol. Spec. Issue 3 217 221

Rakov, V.A., Uman, M.A.: Lightning: Physics and Effects, CUP (2003)

Rakov, V.A.: Fundamentals of Lightning, CUP (2016)

Severson, V., Murray, P., Heeter, J.: Establishing cause and effect relationships between lightning flash data and airplane lightning strike damage, 23rd International Lightning Detection Conference and 5th International Lightning Meteorology Conference, Tucson U. S. A (2014)

Shah, G.S., Bhasme, N.R.: Design of earthing system for HV/EHV AC substation: a case study of 400 kV substation at Aurangabad, India. Int. J. Adv. Eng. Technol. 6(6), 2597–2605 (2014)

Teng, X., Liu, G., Yu, Z., Zhuang, X., Zhao, Y.: Research on initial lightning attachment zone for aircraft. 3rd International Conference on Electric and Electronics, Hong Kong, December (2013)

Thirukumaran, S., Hoole, P.R.P., Ramiah, H., Kanesan, J., Pirapaharan, K., Hoole, S.R.H.: A new electric dipole model for lightning-aircraft electrodynamics. COMPEL: Int. J. Comput. Math. Electr. Electron. Eng. (COMPEL) 33(1/2), 540–555 (2014)

Uman, M.A.: All About Lightning, Dover (1986)

Uman, M.A.: The Lightning Discharge, Academic Press (1987)

Uman, M.A.: The Art and Science of Lightning Protection, CUP (2008)

Uman, M.A., Rakov, R.A.: The interaction of lightning with aerospace vehicles. Progress in Aerospace Sciences, Vol. 39, Elsevier Sciences Ltd, pp. 61–81 (2003)

Yadee, Y., Premrudeepreechacharn, S.: nalysis of tower footing resistance effected back flashover across insulator in a transmission system. International Conference on Power Systems Transients, Lyon, France on June 4–7 (2007)

Yahaya, N.Z., Daud, M.A.: Study of lightning safety distance using rolling sphere method. J. Scient. Res. Energy Power Eng. 5, 266–273 (2013)

Zhang, Y., Ma, M., Weitao, L., Tao, S.: Review on climate characteristics of lightning activity. Acta Meteor. Sinica 24(2), 137–149 (2010)

Chapter 2
Thunderstorms and Pre-lightning Electrostatics

Abstract Lightning flashes are preceded by the development of electrified thunder-clouds and the electrostatic fields that are generated by the electric charges distributed inside the thundercloud. In this chapter we discuss the thundercloud and the theories that seek to explain the origin of the electrically charged thunderclouds. We review the importance of the pre-lightning flash electrostatic fields for lightning protection design, as well as the importance to structures of reducing electric field stress that leads to electric discharges, when electrically stressed by the thundercloud's electrostatic fields. The chapter presents the computation of electrostatic fields in the thundercloud environment, with specific reference to aircraft, which is treated as a floating electrode that moves inside the electrostatic fields of the thundercloud. The numerical results are only illustrative of the application of the techniques, and not final values for the structures considered. The flying, electrically floating aircraft presents a far more complex structure to the thundercloud electrostatic and dynamic lightning interaction than stationary ground structures, such as buildings and electric power substations. The technique developed may handle changes in the pitch and roll angles of the aircraft, or of any other object, in a thunderstorm environment. All the techniques developed for the aircraft may be readily applied to simpler ground structures as well.

2.1 Introduction

The physical nature of lightning is best described as a sudden flow of electrical charge within a cloud or from cloud to air (or to another thundercloud) or from a cloud to ground. That is, for a lightning discharge to happen, there should be an electric breakdown caused by high electric fields, resulting in a flow of electrical charges (i) within a cloud which is referred to as the intra-cloud (IC) flash, or (ii) between clouds which is known as the cloud to cloud (CC) flash, or (iii) from cloud-to-ground (CG) flash, and or (iv) from ground to cloud (or upward) flash, which is referred to as the GC flash. The direct and indirect effects of a lightning flash can adversely sever the operations of ground systems (e.g. an electric power gird) and airborne systems (e.g.

the navigation and control systems of aircraft). Direct effects through lightning flash attachment on structures (e.g. an aircraft or a house) can result in physical damages ranging from puncture/splintering in non-metallic structure, holes and burns, arcing, vaporization, melting, or joule effects such as fire. Similarly, indirect or induced effects through resistive and electromagnetic couplings are capable of generating high voltages and currents within command, control, communication, and power circuits severing operations of mission critical systems if no protection and shielding mechanisms are applied.

It is useful to first examine the physics of the formation of thunderclouds within the troposphere, in particular the cumulonimbus clouds which are the major producer of electric storms. There are several reasons for this. The first and foremost reason is the proximity of these thunderclouds to the earth and the fact that thunderclouds are the most common source of lightning flashes. It t is important to acquire good knowledge of the electrical characteristics of the lightning-producing clouds, in order to model and determine conveniently the magnitude of the capacitance and the electric charges induced. In Fig. 2.1 is shown the distribution of electric charges inside a thundercloud. The electric charge structure is important when designing and protecting structures and electric systems against lightning strikes, both with reference to their geometry, placement, and the materials used. Moreover, knowledge of the thundercloud electric charge structures is important when imitating lightning strikes inside a high voltage laboratory for testing purposes.

The magnitudes of the electric charges are influenced by the size, structure, and topography of the clouds. That is, the electrical structure of the clouds can be conveniently modeled based on the Gaussian imaginary surface in order to determine the

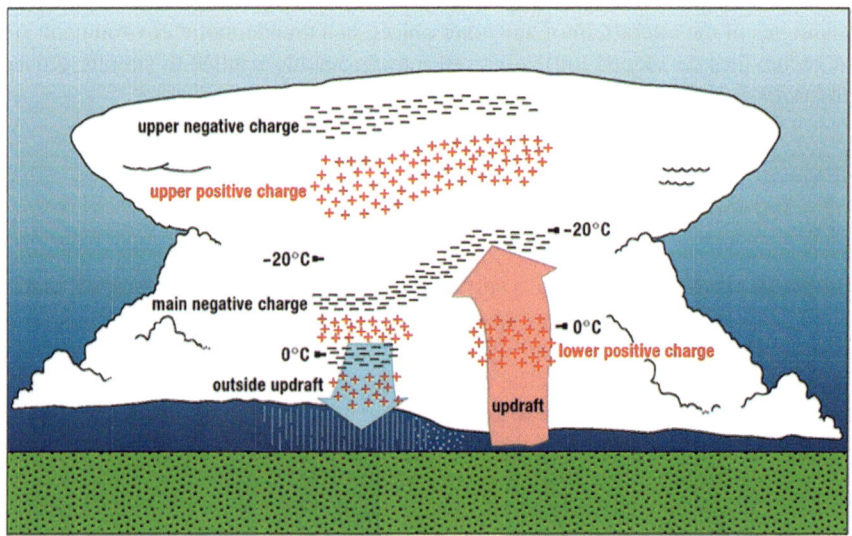

Fig. 2.1 Electric charge distribution in a thundercloud (Credit: NOAA, with permission)

capacitance and the induced electric charge of the thunderstorm clouds. This knowledge will become useful in electric circuit modeling, especially the (i) near-ground electrostatic perturbations, pre-breakdown of aircraft-thundercloud interaction and capacitance, and (ii) near-ground electrodynamics of aircraft interaction with lightning using capacitance determined for both metallic and non-metallic aircraft-body. Lightning is an atmospheric discharge of electricity. Lightning can occur in a thundercloud, volcanic eruptions, snow storm, and dust storms. In a typical thunderstorm of a negative CG flash, when the electric field in an electric charge-concentrated thundercloud exceeds the breakdown value, it will result in an impulsive transfer of electrostatic charge from the cloud base to the ground. The discharge takes place between two opposite charged regions, the base of a thundercloud which is of negative charge region, and the positive charges induced on the ground surface under the thundercloud. The discharge to ground can induce high current and electric potential through direct effects on a grounded structure, as well as on ungrounded structures such as an aircraft in flight. It can also be a source of electromagnetic pulses which cause indirect effects of lightning.

This chapter looks at pre-breakdown and electrostatic fields produced on structures (e.g. power line, aircraft or building housing sensitive electronics) under a thundercloud. In the case of a flying aircraft, the aircraft is a floating structure that is ungrounded and suspended in air between two electrically charged regions, the cloud and the ground. Since lightning strikes are not easily measureable, a predictive modeling is applied to model the pre-lightning discharges. A computational technique is applied based on three-dimensional (3D) dipole modeling of the pre-lightning strike and aircraft electrostatic environment. The 3D dipole model gives an accurate and succinct presentation of a lightning strike on a floating or ungrounded structures such as an aircraft in flight. The same computational technique presented in this chapter may be applied to electrostatic stress, for instance, on an outdoor, ground power substation under a thundercloud.

2.2 Formation of Thunderclouds

The classification of clouds is done using Latin cloud names based on physical appearances or shapes of the clouds. These classifications initially proposed centuries ago, are still used today with some modifications. The modern classification scheme is based on cloud shape and altitude. Generally the three main altitude classes are defined as low, mid-level, and high-level. From the three mentioned classes come the additional classifications based on the cloud types and combinations.

The cumulus clouds have several significant influences on life on earth: (i) they produce about 75% of the rain water that is much needed for survival on earth, (ii) they produce most of lightning flashes that contributes to the ozone layer build up through the ionization process, (iii) they are the sources of the fiercest winds on earth, the hurricane, the tornados, hailstorms, thunderstorms, and the squall lines, and (iv) the cumulonimbus clouds contribute to the overall earth's energy budget through the absorption via water vapor, and reflections of the sun's radiation through other particulates that exist within the clouds.

Cumulus clouds are also the major source of lightning flashes. Lightning is both destructive and needed for a healthy nature. The lightning flash (or electric discharge) contributes a relatively small portion to nitrogen fixation process where the gaseous nitrogen is converted into forms usable by living organisms. Moreover, the negative charges lowered by the lightning return stroke neutralizes the fair-weather buildup of positive charges on the surface of the earth.

The formation of the cumulus cloud sizes range from a relatively small portion of scattered clouds with no precipitations into the huge towering cumulonimbus (the rainstorm clouds). From the context of lightning cloud formation, the cumulus humilis is the first stage of thundercloud formation and then a significant transition into a deeper cumulus called the cumulus mediocris. The next stage of the development process reaching into a towering stage is the cumulus congestus, often termed the towering cumulus. The special feature of the cumulus congestus is a tall tower-like formation with a flat top called the anvil.

Cumulus clouds are necessary for lightning to form. Lightning can occur in a cumulonimbus with precipitation to ground, or with no precipitation to ground often referred to as dry thunderclouds. The lightning occurring in a dry thunderstorm is often the cause of bushfires. Conversely, there have been reported cases of lightning occurring with the absence of cumulonimbus clouds through sandstorm, snowstorm, and volcanic plumes. Figure 2.2 shows a lightning flash associated with volcanic plumes (Fig. 2.2a) and lightning causing a forest fire (Fig. 2.2b). The Lightning flash shown in Fig. 2.2a did not originate from a thundercloud. The lightning flash shown in Fig. 2.2b originated from a thundercloud.

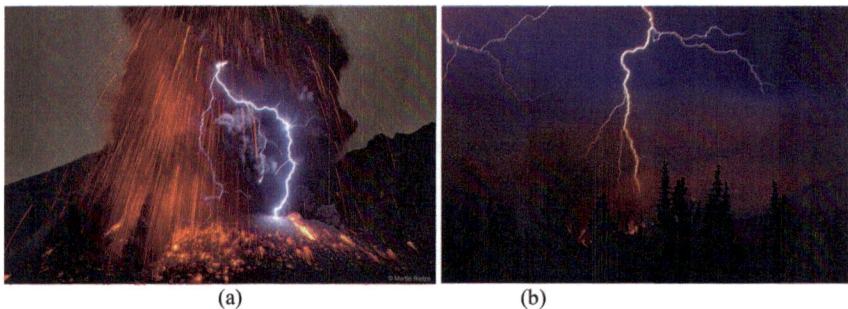

(a)   (b)

Fig. 2.2 **a** Lightning occurring within volcanic plumes. **b** Lightning strike causing a forest fire. (Credit: NASA, USA)

2.3   The Climatology of Lightning

2.3.1   Cloud Electrification

The physical mechanism behind the formation of cloud electrifications still remains debatable. The phenomenology resulting in electric discharges that cause lightning flashes is still a mystery. A number of theories have been proposed to explain the formation of cloud electrification process. However, it is difficult to deduce a definitive explanation due to the range of the distance scales between the micro-scale of the physical processes concerning the cloud hydrometeors (water particles) and the size of the thundercloud for the formation of electric charge distribution and clusters. A look at climatology of lightning will shed some light on the cloud electrification process. It is reported that clouds have to be 3–4 km thick before any electrification process takes place and the depth of the cloud is an influencing factor related to the electric charge and current intensities of lightning. Lightning is associated with convective activity. Cumulonimbus clouds are the largest form of convective cloud and typically produce lightning. Cumulonimbus clouds with lightning activity are generally referred to as thunderclouds. A brief discussion on the cloud electric charge formation and separation is given in the next section.

2.3.2   Cloud Electric Charge Formation

The process of electric charge buildup in thundercloud may be associated with moist air undergoing convection and precipitation resulting in the thundercloud. The convection process can lead to electric charge generation and separation in convective clouds. The presence of strong updrafts and the resulting development of precipitation are instrumental in the formation of an electric field of sufficient intensity for lightning discharge to take place.

There are two general theories that point to cloud electrification. These are the (i) inductive charging, and (ii) non-inductive charging processes. Each process is discussed in brief to shed some light on the mystery of cloud electrification process.

Inductive electric charging: This process will only induce cloud electrification in the presence of some pre-existing electric fields. The pre-existing electric fields that exist, apart from the fields generated by external sources in space such as solar storms and other cosmic radiation sources, will be the fair-weather field and other radiated fields on earth. The existence of a fair-weather field ensures that water particles suspended in the atmosphere in thunderclouds will become polarized.

The inductive charging mechanism is based on the ion capture theory of thunderstorm charge separation. It depends on ions being attached to frozen or liquid hydrometeors in the presence of an electric field, which makes the particles polarized. The lower side of the drop attracts the negative ions and repels the positive ions. In vertical, downward directed field (conventionally defined to be negative electric field), such polarization will cause an excess of positive charges to accumulate in the lower part of the particle, while negative charge will be located in the upper part. While the particle drops it will meet negatively and positively charged particles. Since the lower part is positively charged, negatively charged particles are attracted by the falling droplets, while positively charged particles are pushed away. As a result, the particle grows and becomes more negatively charged. This leads to a cloud with positively charged particles at the upper part and negatively charged particles at the bottom.

Non-inductive electric charging: This refers to those charging processes which are indifferent to the presence of an external electric field, and whose efficiency is not impacted by its strength. The two main mechanisms are: (i) the convection mechanism and (ii) the graupel-ice mechanism.

Non-inductive convection mechanism: This process is where the sources of positive and negative charges are considered to be external, that is, via fair-weather space charges, natural radioactivity near the land surface and cosmic rays near the top of the cloud. The positive electric charges near the ground are carried via warm air updrafts to the top of the growing cumulus. As a result, negative charges—produced by the cosmic rays at the top of the cloud—are attracted and attached to the cloud's boundary. Subsequent cooling and convective circulation result in downdrafts that are assumed to carry the negative charges down the side of the cloud towards the cloud's base. The positive space charges are ingested into the cloud. A negative screening layer forms around the cloud particles on the outside boundary, which moves down the sides towards the cloud base. Additional positive charges are further ingested at the base, and further negative charges flow to the upper cloud boundary to replace the loss of the screening layer charges that flowed to the cloud base along the sides. The lower accumulation of negative charges increases the electric field strength to a magnitude large enough to generate positive corona from ground objects. The corona becomes an additional source of positive charge that feeds into the cloud.

Non-convection graupel-ice mechanism: There is a general consensus that the non-inductive electric charge separation is the dominant mechanism by which thundercloud electric charge centers are formed. This mechanism does not need an external electric field to create a charge on a particle. This electric charge separation mechanism takes place during interactions of ice crystals and graupel particles in the presence of cloud drops. When the rising ice crystals collide with graupel (soft hail), the ice crystals become positively charged and the graupel becomes negatively charged. The updraft carries the positively charged ice crystals upward towards the top of the storm cloud. The larger and denser graupel is either suspended in the middle of the thunderstorm cloud or falls towards the lower part of the thundercloud. The upper part of the thunderstorm cloud becomes positively charged while the middle to lower part of the thunderstorm cloud becomes negatively charged.

Further, it is experimentally found that at certain liquid water content, cloud conditions, and temperature called the reversal temperature T_R, the graupel and ice crystal charge signs reverse. As a result, the smaller ice crystals become positively charged and carried to the upper regions while the larger graupel particles become negatively charged and descend relative to the smaller particles, after collision. Thus, the charge transfers during encounters of ice crystals and graupel will lead to the normal polarity usually found in the observations of terrestrial clouds.

These theories of inductive charging and non-inductive charging seem to be the two generally acceptable theories of cloud electrification despite no proven laboratory experimentations to date to justify these theories. Thus, the two theories remain debatable.

2.4 Negative Lightning Discharge Process

2.4.1 The Negative Lightning

As highlighted in Sect. 2.3, cloud electrification is simply a result of the buildup of electrostatic charges, of different polarities, within the cloud. A fundamental property of electric charge is the force that it exerts on other charges. An electric charge exercises a repelling force on another charge of the same sign as itself and attracts charge of the opposite sign. A region of forces called an electric field surrounds an electric charge. In a cloud-to-ground or negative flash, a positive charge is usually at the top of the cloud with negative charge in the base. There is a cluster of positive charges that resides at the base resulting in the tripole charge buildup. The electric field at the base of the cloud is of the order of 5×10^4 V/m[1]. However, for air to become ionized and gas electric discharge to initiate a leader in the pre-breakdown stage, the electric field intensity has to be above the critical electric field of about 3×10^6 V/m (3000 kV/m) for dry air at sea level and half this value at heights up to 6 km. However, at high altitudes, with reduced air density, the breakdown electric field may occur at 500 kV/m or 600 kV/m.

The negative lightning discharge process is discussed in relation to the cloud-to-ground lightning flash, since it comprises the majority of lightning flashes. Cloud to ground lightning flash makes up about 25% of the global lightning flashes and is referred as a high transient electrical discharge involving a thundercloud and ground. It includes many preliminary discharges or processes such as corona, stepped leader, streamers, inter-stroke process, dart (or dart stepped) leader, first and subsequent return strokes, and continuous current. Typically, a negative lightning flash to the ground may have more than one return stroke and other processes may occur prior to the first stroke, between consecutive strokes and after the last subsequent return stroke.

There are four major stages of a lightning flash. These are the pre-breakdown, the leader, the attachment process as it reaches an object on the ground, and the return stroke. The initial (first-stroke) leader is preceded by an in-cloud process called the preliminary breakdown. The details of mechanism of the preliminary breakdown process are not well known. However, it is believed that the preliminary breakdown process is attributed to the tripole vertical structure of the cloud charges that results in breakdown between the negative charge on the lower part of the cloud and a small pocket of positive charges that resides on that lower portion of the cloud. The formation of the clustered or a secondary small pole of positive charge, which occurs at the base of the thundercloud (see Fig. 2.1), is due to the warmer temperatures at the cloud base and the screening layer charge at the bottom of the cloud is ingested.

2.4.2 The Electric Discharge Process

The preliminary breakdown process generates a leader electric discharge channel which moves towards the ground. It starts as a slow-moving column of ionized air called the pilot streamer. After the pilot streamer has moved down by 30 m–50 m, a more intense discharge called the stepped leader takes place. Note that the 50 m long channel extensions occur rapidly in less than 1 microsecond duration. It takes about 60 ms for the stepped leader to travel a few kilometers from the cloud to near the ground. This corresponds to an average speed of 1×10^5 to 2×10^5 m/sec. Negative electric charge is carried from the main negative charge center and distributed along the length of the leader channel. Currents flowing in the leader channel range from hundreds of Amps to about 1000 Amps. The individual step, or extension of the leader channel, occurs in less than 1 microsecond.

The leader creates a conducting path between the negative cloud electric charge source region and ground. It distributes negative charge from the cloud source region along its path towards the ground. The quantity of positive electric charges residing on the earth's surface becomes even greater. These charges begin to migrate upward through buildings, trees, and tall structures. This upward rising positive charges - known as a **streamer**—approach the stepped leader in the air above the surface of the ground. The point where the leader and the streamer meet is the attachment point which paves way for the first return stroke. The first return stroke current measured at ground typically rises to an initial peak of about 30 kA in some microseconds and decays to half-peak value in some tens of microseconds. The return stroke effectively lowers to ground several Coulombs of electric charge originally deposited on the stepped leader channel including all the branches. It is possible for another leader to travel by the same channel that has been ionized by the stepped leader and the streamer. This leader is referred to as the dart leader which results in a subsequent return stroke. The time interval between the pre-breakdown and the subsequent return strokes could be about 62.5 ms.

The discussion so far has been based on the negative discharge leader from the cloud to ground which makes up about 25% of the global lightning. However, there can be positive leaders as well as a bipolar (both negative and positive) discharges from cloud to ground. Positive discharges make up about 10% of the cloud-to-ground flashes and account for the highest recorded lightning current of about 300 kA. Further, positive flashes usually comprise a single return stroke, compared to the negative lightning flash which produces two or more return strokes. Bipolar discharges of positive and negative polarities often occur in lightning flashes. The bipolar discharges may be divided into three types. The first is associated with polarity reversal in slowly varying (in milliseconds) current components such as those in continuous current components. The second type is characterized by the different polarities of the initial stage currents and the following return strokes. The third type involves return strokes of opposite polarities. Current of different polarities can follow the same ionized channel to ground. But these are initiated by clouds of different charge polarities. Current amplitude as high as 31 kA has been measured for the bipolar discharge.

2.5 Lightning-Aircraft Electrostatic Interactions

2.5.1 Two Types of Attachment Initiation

The aircraft-lightning attachments initiation are of two types. The first is the aircraft-triggered lightning flash and the second is the aircraft-intercepted lightning flash. The two attachment initiation processes are discussed separately in Sects. 2.5.2 and 2.5.3.

2.5.2 Aircraft-Triggered Lightning

An aircraft builds up electrostatic charges just by virtue of flying through the atmosphere as a result of friction or contact with electrified aerosols, dust, water vapor, and other atmospheric particulates. The aircraft lightning interactions begin when the aircraft approaches an electrified space or region of thundercloud formation. The entry of an aircraft into an ambient electric field can be regarded as a sudden introduction of a conductor into an electric field which intensifies the local electric fields around the conductor. This enhances the local electric field buildup. The electric field enhancement will reach a maximum along the aircraft extremities that are oriented towards the ambient fields. Typically, in an ambient field of 100 kV/m, the electric field at the radome could be enhanced to 1 MV/m 1000 kV/m); similarly at tail tips. The electric field at the wing tips could rise to 400 kV/m, and to 200 kV/m at the tips of the turbo engines.

The electric charging of the aircraft produces a potential gradient between it and its surroundings. When the potential gradient builds up to a sufficient level corona discharge results. The corona discharges occur at the extremities of the aircraft and initiate a bi-directional leader that connects the cloud electric charge center electrically to ground, through the aircraft. This is shown in Fig. 2.3a. Hence, there are two distinct phases to lightning-aircraft interaction. The first is the development of streamers and leader sets that develop at the field enhanced parts of the aircraft. The second phase is the high currents produced by the first and subsequent return strokes, once the leaders connect the aircraft to the cloud at one end, and to the ground at

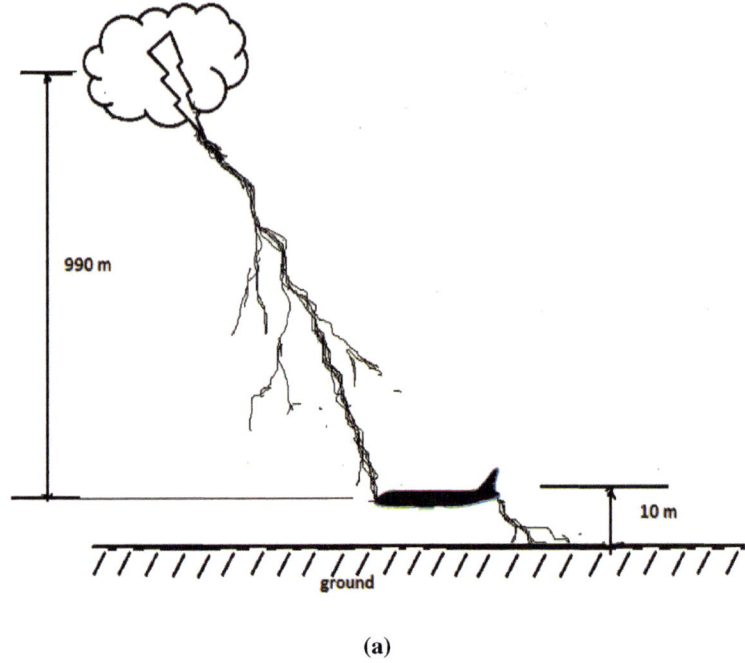

(a)

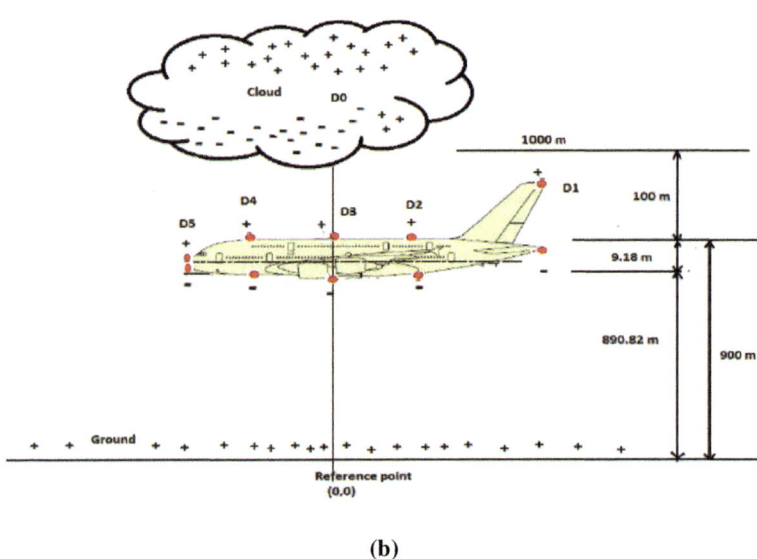

(b)

Fig. 2.3 a Lightning strike to an aircraft. The thundercloud is connected to the ground through leaders at two attachment points on the aircraft (e.g. at radome and tail). **b** Pre-lightning scenario. A 2D polarized dipole along an aircraft (diagram not to scale)

the other end. The second phase, therefore, induces the high energy transient current pulse, subsequent restrikes, and the long duration of the slow currents.

2.5.3 Aircraft Intercepted Lightning

An aircraft can also be exposed to a naturally occurring lightning strike. A naturally occurring lightning strike begins with a leader generated from the cloud base (for a cloud-to-ground lightning strike) and propagates downwards to ground. It may be intercepted by an approaching aircraft, in that the aircraft flies straight into an existing lightning leader. The electric field of the approaching leader intensifies about the aircraft extremities and emanates a leader connecting the naturally approaching lightning leader. Simultaneously, an additional leader emanates from other aircraft extremities connecting the ground (for a cloud-to-ground flash). The point on the aircraft that connects to the leader becomes the attachment point while that on the aircraft extremities which connects the aircraft to ground becomes the exit point (see Fig. 2.3a). The occurrence of aircraft intercepting lightning is very rare compared to that of aircraft triggered lightning incidence.

2.6 Probability of Lightning Strike to Aircraft

2.6.1 Factors Affecting Probability

The probability of an aircraft being struck by lightning depends on three influential factors. These include aircraft size, aircraft flight profile, and geographic area of operations.

2.6.2 Probability Dependence on Aircraft Size

There is a high lightning strike rate probability for large aircraft. This is due to the fact that a large aircraft entering an intense region of thunderstorm electric field would significantly modify the intensities of the field allowing lightning leader formation at a lower ambient electric field than it would be if there were no aircraft.

2.6.3   Probability Dependence on Flight Profile

The flight profile is another influential factor that increases the probability of lightning strike rate to an aircraft. That is, a lightning strike to an aircraft is a function of both the aircraft flight profile and the thunderstorm formation altitudes. Aircraft at a low altitude either in ascending or descending phases have an increased probability of being struck by lightning. It is further stated that aircraft flying short haul between closely situated cities flying mostly at low altitudes stand a higher chance of being struck by lightning. About 63% of lightning strikes to aircraft observed (based on 2700 lightning strikes to aircraft for the period 2002–2009) occurred during aircraft descent while 35% occurred in the ascending stage. Only about 2% of lightning strikes to an aircraft occurred at the cruising stage. The high percentage of strikes occurring during descent could be attributed to the fact that aircraft in flight over long distances on reaching their destinations when encountering a thunderstorm, would have limited fuel supply to reroute to other airports in proximity. Thus, there is no alternative to landing under the thundercloud. Further, there are certain policy restrictions limiting flights, for example, international flights, being allowed to land at certain airports only; e.g. at large international airports). In such circumstances an aircraft would inevitably have to land at the point of destination. Further, aircraft on the ground is usually delayed, and not allowed to take off, if a thunderstorm is hovering over the airport. Such are the probable reasons for the high percentage of lightning strikes to aircraft during descent compared to those while ascending.

2.6.4   Probability Dependence on Geographic Area of Operations

Thunderstorm formation is unevenly distributed with high frequency of lightning strikes occurring along the equatorial regions. Lightning activity is more continental than oceanic, with continental updrafts at 50 m/s producing more thunderclouds compared to the 10 m/s updrafts over the ocean. Moreover, intense lightning activity is seen to prefer dry climates (e.g. Africa) rather than wet climates (e.g. South America), although both regions may be close to the earth's equator. Surface temperature is seen to be a key to driving lightning activity. The average of lightning return stroke current peak is about 30 kA for land, while oceanic lightning strikes have current peaks exceeding 30 kA, since the attachment point on the sea surface has a much smaller resistance than attachment point resistance to land.

2.7 Thundercloud Induced Electrostatic Charges

For aircraft triggered lightning, the return stroke current is induced at the point
where the upward leader from the aircraft extremity meets the downward cloud
leader. Since an aircraft is in flight with no form of grounding, the aircraft leader is
bi-directional, connecting the cloud charge electrically to ground via the downward
leader. The stroke point where the first return stroke current originates can be either
at the ground or on the surface of the aircraft. An aircraft-triggered lightning flash
may be considered using the dipole theory and the corona discharge mechanism.
For a cloud-to-ground negative flash, the charged cloud center and the charge on the
ground can be represented by a dipole with the cloud monopole being negative and
ground monopole positive. The electric field lines prior to breakdown emanate from
the positive charge sinking at the negative charge forming uniform fields that can be
estimated from:

$$E(-) \ = \ E(+) \ = \ \frac{Q}{4\pi \varepsilon_0 \left(H^2 + r^2 \right)}, \tag{2.1}$$

where E is the electric field, Q is the cloud electric charge, which can be determined
from the cloud capacitance modeled on the Gaussian surface of the cloud voltage;
ε_0 is the permittivity of free space, H is the height of the center of the charged cloud
from the ground, and r is the radial distance of, say, a ground observation point from
the charged cloud center. The electric field produced by the negative electric charge
of the thunder cloud, E(–), is equal to the electric field produced by the positive
electric charge image, E(+).

The electric field does not remain uniform when lightning discharge occurs in
the presence of an aircraft as it enters the region between the two electric charges of
the charged dipole. An aircraft entering a charged region would become electrically
polarized. The electric charge build up on the aircraft surface would correspond to
the polarities of both the ground charge and the cloud center. For a negatively charged
cloud center, the charge on the top surface of the aircraft would be positive while
that on the belly of the aircraft surface would be negative. The earth electric charge
will be positive. This forms dipoles on the aircraft surfaces. Figure 2.3b shows the
dipoles shown in small shaded circles (in red) on the top surface and on the belly
of the aircraft. The electric fields will be large on the extremities of the aircraft with
values estimated to be 100 times the ambient electric field. The high electric fields
at the extremities can far exceed the ambient fields, causing a corona breakdown
in the surrounding air producing a bi-directional leader that connects from one end
of the aircraft to the cloud (or the stepped leader descending from the cloud) and
the other end to the ground. When contact is made with the cloud stepped leader,
a high-current discharge is generated that gives rise to the luminous brightness that
is seen during lightning strikes known as the first return stroke current, a rapidly
traveling (at 10^8 m/s) current (and intense light) pulse.

The high lightning current discharge travels from the lightning strike point to the charged cloud interlinking the two oppositely charged regions, thus neutralizing electric charges. The strike point can be located at an aircraft extremity such as the radome, wingtip, tail cones, or engines, and the stroke connects the aircraft to cloud and ground; or the strike point can be on the earth connecting the cloud charge through the aircraft to the earth strike point. The point at which a leader originates on the aircraft surface is referred to as the entry point or the attachment point. The point along the aircraft extremities where the leader propagates towards the ground is called the exit point.

After the return stroke, the lightning flash may end, if the thundercloud electric charge has mostly been lowered to the ground by the return stroke and the continuing current that flows immediately following it. That means that there is not enough electric charge left inside the thundercloud electric charge center to initiate another leader stroke electric discharge. But most negative flashes lead to three or four subsequent leader-return strokes—some even more than 15 subsequent strokes. During this time since the aircraft is moving, the subsequent strokes occur at different points on the aircraft body (producing multiple punctures) with the attached lightning channel being dragged along the aircraft surface. This is, for obvious reasons, called the swept stroke. Thus, if enough electric charge is available in the thundercloud to produce another lightning flash, a continuous leader called a dart leader moves down the return stroke channel from the previous stroke, depositing negative charge along its length. Dart leaders generally deposit less electric charge than stepped leaders. Thus subsequent lightning flashes generally lower less electric charge to the aircraft and to ground and have smaller return stroke currents than the first return strokes.

2.8 Pre-lightning Flash Electrostatics of Thunderstorms: Analysis

2.8.1 The Electrostatic Fields

The lightning flash involves rapidly changing, dynamic electromagnetic fields. However, before the generation of the leader and return strokes, the electromagnetic phenomena are static, that is, largely not changing in time. In this case, there are no magnetic fields, since the electric charges are largely stationary. The static electric field is of great interest to the engineering designer in two respects. First, it is the electrostatic field stress on ground and airborne objects that triggers initial electric discharges on the objects. The electric discharges are initiated at points where the electric (that is, electrostatic) field is large. Hence in all good design practice, the designer seeks to reduce the electric fields generated on objects by the nearby thundercloud which carries large amounts of static electric charges. These thundercloud electric charges generate the large electrostatic fields. Secondly, where the electrostatic fields induce large electric charges on objects and electronic circuits, local electrostatic discharges (ESD) and insulation breakdown may be initiated, causing the equipment and system to malfunction or breakdown. ESD is a major cause of concern to the microelectronic industry, with the use of microelectronic equipment and circuits (e.g. in digital signal processors, digital controllers, and networking), their use being widespread in airborne and ground systems, including critical mission, military, and medical operating systems. Scientists too are interested in the electrostatic phenomena associated with the thundercloud, since the electrostatic fields initiate the initial electric breakdown processes including the streamers, corona and the leader stroke. We explore the electrostatics of the thundercloud by considering the complex situation of an aircraft in the vicinity of a thundercloud.

2.8.2 Aircraft and Electric Dipole Placements

A study of the lightning induced electric field on an aircraft between thundercloud and ground either parked or in cruising, ascending, or descending flight mode is simulated using the dipole method and the cloud charge structure mechanism. The cloud charge is calculated based on the Gaussian spherical surface of the cloud electric charge center, while the charges on an aircraft are computed using the dipole method. The charge on the ground attains the same magnitude as the cloud charge but with opposite polarity, usually a positive polarity (for negative cloud-to-ground lightning flash). For an aircraft traveling below a charged cloud and ground, the totality of charges on its body is electrically neutral. That is, with equal distribution of mono poles of opposite polarities with separation distances equivalent to the aircraft geometry and component/body separation distances. For example, the separation distance of a dipole on the fuselage is simply the diameter of the fuselage separating the mono pole on the underbelly from the top surface of the fuselage whereas a dipole on the wing tip is separated by the thickness of the wing tips or the height of the winglets or sharklets on the wingtips. Thus, modeling an aircraft surface charges using the electric dipoles gives a succinct representation of the charge build up based on the aircraft geometry for the purpose of calculating the electric fields, the aircraft potential, and the capacitance. The electric dipole has charge separation distance d and charges $+q$ and $-q$. The method makes use of elementary theory of electrostatic induction on the distribution of charges within an object that occurs as a reaction to the presence of a nearby charge cluster (Hoole and Hoole, 1996). The analogy is applied to an aircraft as it goes through a charged electric storm causing migration of polarized charges on the surface with positive charges on the top for a negative cloud flash.

In order to model accurately the dipole on an aircraft, accurate dimensions of the aircraft airframe have to be used. In this case study, the A380 Airbus was modeled to assess the pre-breakdown electrostatic charges. Using the known geometry and dimensions of the aircraft, dipoles are placed on the surface as illustrated in Fig. 2.4a and b.

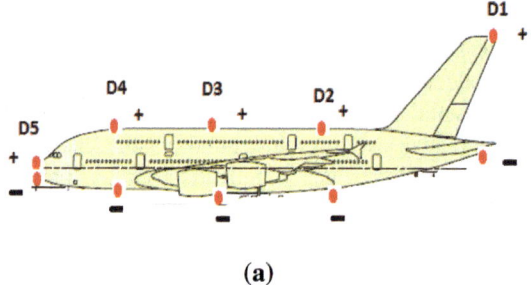

(a)

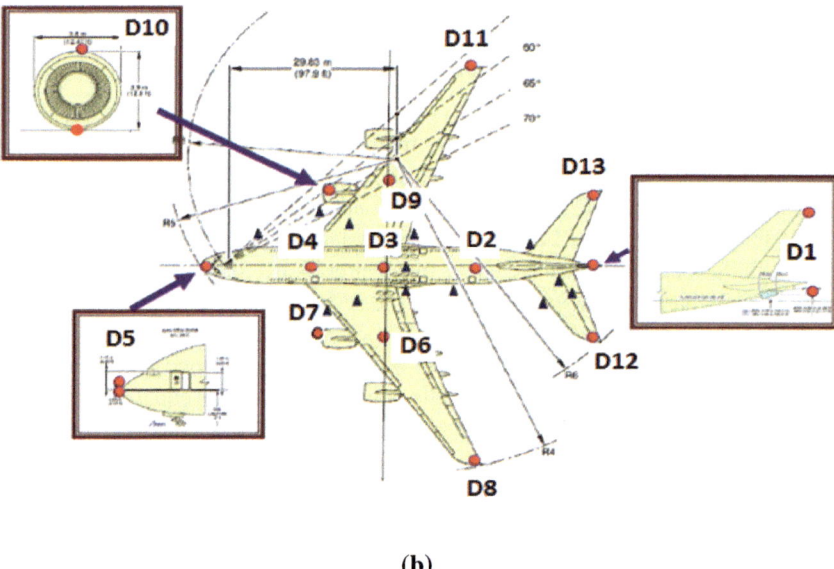

(b)

Fig. 2.4 An illustration of airbus A380 aircraft 3-dimensional 3D) dipole arrangements **a** dipoles along fuselage, rudder, and radome, and **b** dipoles along the wings and engines

There has been extensive research carried out by NASA and other research organizations to categorize and understand the electrical environment surrounding the thundercloud by either flying aircraft close to the electrified clouds and sometimes right into them. Amongst the aircraft used is the NASA Global Hawk aircraft which is mounted with instruments to measure and record electric fields, magnetic fields, electric currents, and voltages induced on the body of the aircraft as well as the internal electrical wiring of the aircraft that connects the communication, control, command, and power system of the aircraft. However, such research activity is expensive and is limited by the aircaraft sizes and payload capacity, and the size of the thunderclouds the aircraft flies into. The 3D dipole method has the advantage that it can give an analysis of the pre-breakdown stage (the electrostatic buildup) as well as the break-down stage (the electrodynamics stage) of any aircraft in a lightning environment with different electric charge structures. The dipole method proposed allows for all kinds of aircraft to be tested and studied in a variety of positions and inclinations with respect to the electric charge centers inside the cloud, before the aircraft is struck by lightning which is the electrostatic stage of the aircraft-thunderstorm interaction. The dipole electric charges are placed on the aircraft surface, and once these electric charges are computed using the technique outlined in Sect. 2.7, the electric fields around the aircraft surface are easily determined, helping also electrically to zone better the aircraft body for structure reinforcement, optimizing the geometry against electric stress, protection measures and where to (and where not to) place sensitive electrical and electronic equipment inside the aircraft. The 3D dipole computational test-bed offers huge advantages in being able to test the whole aircraft with every detail of its body included under a realistically modeled thundercloud with electric charge centers that may be situated in complex arrangements. Both large commercial aircraft as well as smaller military aircraft, and large electric power substations as well as a single, isolated house, may be studied. Hence, more accurate zoning may be done with reference to the probability of lightning strike to different surface areas of say the aircraft or substation or house, which in turn will determine the areas to be most protected against lightning and on how to layout electrical and commu-nication circuits with sensitive electronic navigational, communication and control equipment, as well as the electric power systems apparatus from earth electric power grids to aircraft electric power grids.

2.8.3 Determining the Electric Charges Induced on an Aircraft and the Electric Fields Generated Around an Aircraft Body

The 3D dipole model is used to calculate the aircraft voltage, the electric charge on the surface of the aircraft, and the electric field produced by these charges using Eqs. (2.2)–(2.7). The aircraft surface voltage is given by.

$$V_A = k \cdot q_{AD} \cdot \left(\frac{1}{r_+} - \frac{1}{r_-} \right), \tag{2.2}$$

where k is a constant, q_{AD} is the aircraft dipole charge, V_A, is the aircraft voltage. r_+ and r_- are the distances from the positive and negative mono poles and their images to a selected point on the aircraft surface.

Note that in (2.2), V_A and q_{AD} are unknown terms. The only known terms in the equation are the distances from the dipole to a selected coordinate or point on the surface of the aircraft and the separation distances of the mono poles which is placed according to the aircraft geometry, and the altitude of the aircraft, and the dipole mirror images (replacing the earth). Thus, since the aircraft is at an equipotential surface, the aircraft voltage V_A is the same at all points which makes the coupled equations easier to compute. The cloud charge is computed from the cloud capacitance using a given cloud charge diameter, for instance, 200 m. The cloud potential is taken to be -50 MV for a negative flash. The cloud geometry is assumed to be that of a spherical Gaussian surface.

The aircraft surface electric charge calculation makes use of the coefficients of potential of the electric dipole charges and their mirror images on the ground with reference to a selected observation point on the surface of the aircraft. Since the aircraft geometry is in 3D, three-dimensional distances (x, y, z) are used as defined in (2.3). That is, for a particular point, say p_1, on the aircraft surface, the distance from the center of a dipole to the point p_1 on the surface of the aircraft is given by (2.3). The angle between the midpoint of the dipole and the point p_1 is given by (2.4), with the angle measured with reference to an infinite, perfectly conducting plane, which is the ground plane. A similar equation is used in the calculation for the image dipole charges below an infinite ground plane. The electric dipole charges and their image charges are assigned different variables names in the equations.

$$dis(x_{p1}, y_{p1}, z_{p1}, k) =$$

$$\sqrt{(x_{p1} - x_k)^2 + (y_{p1} - y_k)^2 + (z_{p1} - z_k)^2} \tag{2.3}$$

$$\theta_{p1}(x_{p1}, y_{p1}, z_{p1}) = \left[\cos^{-1}\left(\frac{(y_{p1} - y_k)}{dis(x_{p1}, y_{p1}, z_{p1}, k)} \right) \right], \tag{2.4}$$

where dis stands for the distance between centre of a dipole and an observation point on the surface of the aircraft, and θ is the angle subtended by the monopole and the observation point $p1$ with respect to the ground plane. The general term for the coefficients of potential for the dipole charge is.

$$q_{ADCoeff} = \left(\frac{1}{dis(x_{p1}, y_{p1}, z_{p1}, k)} \right) - \left(\frac{1}{dis(x_{p1}, y_{p1}, z_{p1}, k)} \right) +$$

$$+ \left(\frac{1}{dis(x_{p1}, y_{p1}, z_{p1}, l)} \right) - \left(\frac{1}{dis(x_{p1}, y_{p1}, z_{p1}, l)} \right) \tag{2.5}$$

where $q_{ADcoeff}$ is the coefficient of the charge due to the dipole k on the surface of the aircraft and its image l within the earth. Note that q_{AD} is the dipole charge. Moreover, $q_{ADcoeff}$ is the coefficient of the dipole charge and its image, which are functions of the distance between the dipole charge and its image. The electric charge induced on the surface of the aircraft by the cloud electric charge and the other electric charges on the aircraft surface is given by

$$Q_n = \left(\frac{1}{q_{AD_{Coeff\,n}}}\right)\left((4 \cdot \pi \cdot \epsilon_0 \cdot V_A) - \left(\begin{array}{c} q_{AD_{Coeff1}}Q_1 + q_{AD_{Coeff2}}Q_2 + q_{AD_{Coeff3}}Q_3 + \cdots \\ \\ \cdots + q_{AD}coeff_{\,n-1}Q_{n-1} \end{array}\right)\right).$$

$$(2.6)$$

The equation for the aircraft voltage V_A at any point p on the aircraft surface due to n number of dipole charges is

$$q_{AD_{Coeff1}}Q_1 + q_{AD_{Coeff2}}Q_2 + q_{AD_{Coeff3}}Q_3 +$$
$$\cdots + q_{AD_{Coeff\,n}}Q_n = 4 \cdot \pi \cdot \varepsilon_0 \cdot V_A \qquad (2.7)$$

Rearranging (2.7), the charge Q_n (which becomes the subject of the equation) is given by (2.6). The variable Q_n is then substituted in the equation set for voltage for the next point p_2 in order to eliminate the Q_n. The next charge variable Q_{n-1} is defined and substituted in the equation set for voltage due to the next point p_2. The procedure is repeated for $(n + 1)$ points where n is the total number of dipoles that make up the aircraft dipoles. This is simply the process of solving a set of linear equations by the substitution method. The final equation is a single matrix equation comprising the charge coefficients and the aircraft voltage V_A. The aircraft voltage V_A is the only unknown term in the single matrix equation. This aircraft voltage is computed from the charge coefficients and finally the charges and the electric fields are determined.

A software test-bed was developed to calculate the charges and the electric fields at different points along the aircraft bodies for aircraft at various altitudes and at various distances from the charged cloud center. The results tabulated in Tables 2.1, 2.2, 2.3, 2.4, 2.5 are for the airbus A380 aircraft at various altitudes and distances from the charged cloud.

2.8.4 Analysis of the Airbus A380 Aircraft Results

Tables 2.1 and 2.5 show the results for the computed voltages, electric charges, and electric field strength for the Airbus A380 surface at various altitudes and distances from the charged cloud. The results indicate the areas with the highest electric fields exceeding the breakdown electric fields of 400 kV/m which have a high probability of triggering lightning flash. The breakdown electric fields primarily appear at the aircraft extremities at an average ambient aircraft electric field of about 75 kV/m. The electric field build up at these extremities that exceed the breakdown fields can cause the ionization of the surrounding air thus initiating an upward stepped leader capable of triggering a lightning strike. It is observed that areas of high electric fields include the radome, the wing tips and the middle parts of the wings, and the stabilizers.

Table 2.1 shows the results for an A380 airbus at an altitude of 800 m, that is, at 200 m directly below the charged cloud of -50MV for a negative flash to the

Table 2.1 Electric charges and electric fields for an airbus A380 at an altitude of 800 m directly below a charged cloud at 1000 m altitude

Computed Aircraft Voltage V_A: -2.172×10^7 V			
Dipole location and dipole number	Dipole charge (C/m)	Electric field (V/m)	Comment
Rudder tip	4.416×10^{-3}	2.412×10^5	
Mid-fuselage	3.013×10^{-6}	1.363×10^5	
Radome	1.999×10^{-4}	7.986×10^7	[a]Above Breakdown E-field
Mid-left wing	3.623×10^{-5}	3.514×10^5	
Left wing engine	6.406×10^{-5}	1.276×10^5	
Tip left wing	1.81×10^{-3}	1.252×10^7	[a]Above Breakdown E-field
Mid-right wing	4.875×10^{-5}	4.495×10^5	[a]Within breakdown
Right wing engine	7.489×10^{-5}	1.288×10^5	
Tip right wing	1.63×10^{-3}	1.127×10^7	[a]Above Breakdown E-field
Left stabilizer tip	2.877×10^{-3}	2.873×10^8	[a]Above Breakdown E-field (become possible entry point)
Right stabilizer tip	2.801×10^{-3}	2.798×10^8	[a]Above Breakdown E-field

[a]At high altitude with reduced air density, the breakdown electric field may be as high as 400 kV/m

ground. The aircraft electrostatic potential computed is −21.72 MV resulting in high electric charges and electric fields on the aircraft surface. The dipole electric fields calculated along the A380 aircraft show very high electric fields at the tip of the horizontal stabilizers, the tips of the wings, and at the radome. These electric fields exceeded the specified breakdown electric fields of 400 kV/m. The two extremities with the highest electric fields are most likely to initiate bi-directional leaders towards the charged cloud center to trigger a lightning flash connecting through the other extremities to ground. The left stabilizer is most likely to become the lightning entry point and the right stabilizer to be the exit point. However, with the aircraft moving with respect to the cloud, there is the possibility of a swept stroke path that can develop along the aircraft fuselage. The swept path can either be through the radome or the tips of either wing to ground These two extremities of the wings, like the radome, carry large electric charges and have high electric fields.

It is noted from Table 2.1 that the values of the electric charges and the electric field distribution along the aircraft surface are not symmetrically identical. This is attributed to the non-uniformity of the aircraft geometry and the distribution of points selected along the surface that are used in (2.2)–(2.6) to calculate the coefficients of the charges. This accounts for the slight variations in values of the electric fields and the charges. The electric fields at a point on an aircraft surface will be the vector sum of the collection of the electric fields due to the other dipole charges, the cloud charge, and the image charges. Thus, the slight variations observed in the values of the electric charges and the electric fields calculated.

Table 2.2 shows a similar trend for an aircraft at 500 m altitude with the charged cloud at 1000 m altitude, that is, 500 m directly above the aircraft. The extremities of high electric fields are the two horizontal stabilizers and the radome. The swept path for the lightning flash will be along the stabilizers and the radome. However, with the electrically charged cloud and the aircraft moving, the capacitance, and the charges may vary thus producing changes in the electric fields at the other extremities. The most likely swept path would be along the stabilizers through the fuselage and radome to ground. One of the stabilizer tips becomes CHECK the entry point while the radome becomes the lightning exit point as defined in normative zoning standards.

Further, Table 3.3 shows the results for an aircraft at an altitude of 800 m but at 1000 m distance away from the charged cloud. The charged cloud is at an altitude of 1000 m. The results show the build up of charges initiating very large electric fields at the horizontal stabilizers and the wing tips exceeding the breakdown electric fields of 400 kV/m. The possible entry point is most likely the left horizontal stabilizer. The swept lightning channel will be along the fuselage through the wing tips to ground.

Table 2.2 Electric charges and electric fields for an airbus A380 at an altitude of 500 m directly below a charged cloud at 1000 m altitude

Computed Voltage V_A: -6.642×10^6 V

Dipole location and number	Dipole charge (C/m)	Electric field (V/m)	[a]Comment
Rudder	3.007×10^{-4}	2.501×10^4	
Mid-fuselage	1.423×10^{-7}	2.045×10^4	
Radome	1.31×10^{-5}	1.309×10^6	Above Breakdown E-field
Mid-left wing	2.233×10^{-6}	2.838×10^4	
Left wing engine	3.72×10^{-6}	1.99×10^4	
Tip left wing	1.124×10^{-4}	7.781×10^5	Within breakdown
Mid-right wing	3.316×10^{-6}	3.591×10^4	
Right wing engine	4.335×10^{-6}	1.993×10^4	
Tip right wing	1.003×10^{-4}	6.941×10^5	Within breakdown
Left horizontal stabilizer tip	1.941×10^{-4}	1.939×10^7	Above Breakdown E-field (becomes entry point)
Right horizontal stabilizer tip	1.892×10^{-4}	1.89×10^7	Above Breakdown E-field

[a]At high altitude with reduced air density, the breakdown electric field may be as high as 400 kV/m

2.8.5   Zoning

The results given in Tables 2.4 and 2.5 show the electric fields at the aircraft extremities for the A380 airbus at an altitude of 800 m, but at a distance of 5 km and 50 km from the charged cloud, respectively. The electric field reaches 900 kV/m for the A380 aircraft at a distance of 5 km from the charged cloud center which exceeds the breakdown electric field. This field is capable of initiating a step leader when the aircraft moves close enough to the charged cloud. However, at a distance of 50 km, the electric fields are drastically reduced. This observation shows a distance of 50 km from a charged cloud is a safe distance to fly the aircraft.

The results in Tables 2.1, 2.1, 2.2, 2.3, 2.4, 2.5 were compared with the current practice in zoning of aircraft surfaces and geometrical shapes. The zones identify areas of high probability of lightning attachment depending on the electrostatic field enhancement due to electric charge build up in these areas. Moreover, there are other zones to which the probability of lightning attachment being swept from an original attachment point is high. It is also observed that the swept stroke path would occur along the extremities where large electric field build up occurs. That is, the lightning flash will be swept from one extremity with large electric field along the aircraft body to the other extremity with a large electric field, and discharge to the ground. Further, it is observed that the values obtained in the results may be used to identify and more accurately classify the zones during the aircraft design stages, including for lightning protection design. The electric field enhanced regions include the rudder, the stabilizers, the radome, and the wing tips. Further, the results show a safer distance for large aircraft such as the airbus A380 is about 50 km from the thundercloud, to avoid the generation of electric discharges. Fly by wire and non-metallic body aircraft designers are interested in the enhanced electric field areas of the aircraft body in order to divide the aircraft body into zones where threatening electric field enhancements and lightning strikes are highly probable, and zones with minimum probability of strikes. These are the regions to be avoided when placing mission critical navigational and control systems as well as microelectronic equipment. The severe cloud-to-cloud lightning strikes in which aircraft may get engaged is the most severe threat to navigational, microelectronic, and measurement systems.

2.8.6   A F16 Military Aircraft Flying Between Two Charged Centers

2.8.6.1   F16 Electric charge Model

Figure 2.5 shows the scenario under study where an F16 military aircraft is flying between two oppositely charged electric charge cells (or electric cloud centers). Since the aircraft is positioned inflight, and horizontal and between two electrically charged

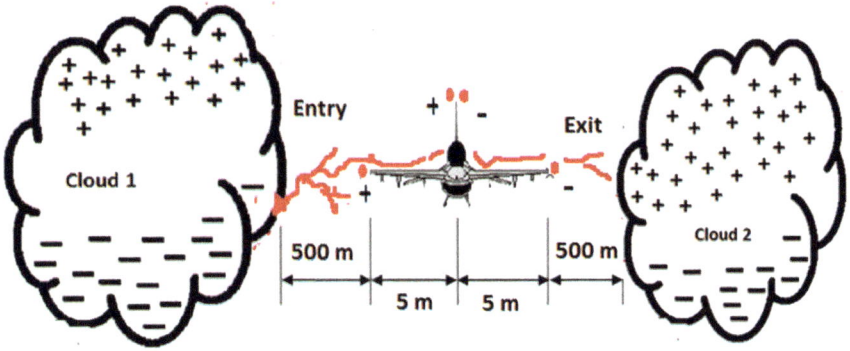

Fig. 2.5   An F16 military aircraft positioned between two charged clouds (Not to scale)

centers, or electric clouds, the dipoles are aligned horizontally with positive electric charges accumulating close to the negatively charged Cloud 1 and the negative charges positioned towards that of the positively charged Cloud 2. The tip of the aircraft wings is at a distance of 500 m from the charged clouds (on both sides). The F16 military aircraft is at an altitude of 1000 m just at level with the two 1000 m altitude cloud cells.

In Fig. 2.5, the electric cloud diameters are taken to be 200 m for Cloud 1 and 200 m for Cloud 2. The voltages are assumed to be -50 MV and $+50$ MV for Cloud 1 and Cloud 2 respectively. The F16 military aircraft is assumed to be at level between the two electric cloud cells of positive (Cloud 2) and negative (Cloud 1) electric charges. The cloud electric charges are calculated using spherical Gaussian surfaces. The capacitance and electric charge of Cloud 1 are calculated to be 1.11212 $\times 10^{-8}$ Farads and -0.556 Coulombs respectively. Similarly, the capacitance and the electric charge of Cloud 2 are calculated to be 1.11212 $\times 10^{-8}$ Farads and $+0.556$ Coulombs. From the horizontal placements of the dipoles as shown in Fig. 2.5, the positive monopoles are aligned horizontally towards the negative electric cloud cell (Cloud 1). Similarly, the negative monopoles are aligned horizontally towards the positive electric cloud cell (Cloud 2). The electric charges on the F16 military aircraft are then calculated from the dipole moment and the aircraft voltage. Finally, the electric fields are determined from the calculated dipole charges. Tables 2.1, 2.2, 2.3, 2.4, 2.5, 2.6 show the results for aircraft voltage V_A and the charges for dipoles on an aircraft surface with aircraft at various altitudes and separation distances from the two charged cloud cells.

2.8.6.2   Analysis of the F16 Military Fighter Aircraft Results

The results in Table 2.6 through Table 2.8 show the electric potential, the electric charges, and the electric fields for the F16 military aircraft flying between two separate electric cloud cells of opposite polarities (positive and negative voltages polarities).

Table 2.6 shows the results when the F16 military aircraft is flying between two electric cloud cells of -50 MV and + 50 MV. The separation distances from the aircraft wings to the two charged cloud cells are each 500 m from the two wings tips as shown in Figs. 2.5 and 2.6. The aircraft potential is 7.742×10^6 V. The electric fields are large at the extremities, the rudder tip, the wing tips, mid fuselage, and the nose boom (a slender metal extension projecting from the nose of the aircraft). The values of the fields are in the range of 10^7–10^9 V/m. The values are extremely high. In practice, these high values would never be reached as the breakdown would have occurred at 400 kV/m for ionization of air to occur. That is, the electric field would have reached a breakdown value before the aircraft reached the altitude to trigger lightning flash. Thus, flying between the separation distance of 500 m between the two electrically charged cloud cell centers is extremely dangerous for an F16 military.

Similarly, from Table 2.7, the potential for an F16 military aircraft at 1000 m altitude with the wing tips at a distance of 5000 m from the two charged cloud cells is 7.316×10^4 V. The high electric field occurred at the rudder tip and the nose boom. The values exceed the breakdown electric field of 4×10^5 V/m (or even lower) at higher altitudes. In practice, the breakdown down electric field would have occurred before reaching these two high values shown in Table 3.7. That is, the breakdown electric field would have probably occurred at electric fields exceeding the high altitude breakdown electric field of400 kV/m. Thus, for an F16 military aircraft within close proximity of a charged electric cloud cell at the potential of -50 MV with a separation distance of 5000 m, the electric field can build up towards the breakdown value.

2.9 Electrostatic Fields of Pre-lightning Thundercloud Environment

The results in Tables 2.1, 2.2, 2.3, 2.4, 2.5, 2.6, 2.7 were compared with the current practice in zoning of aircraft surfaces and geometrical shapes. The zones identify areas of high probability of lightning attachment depending on the electrostatic field enhancement due to electric charge build up in these areas. Moreover there are other zones to which the probability of lightning attachment being swept from an original attachment point is high. The values obtained may be used to identify and more accurately classify the zones during the aircraft design stages, including lightning protection design. The electric field enhanced regions include the rudder, the stabilizers, the radome, and the wing tips for an airbus A380. Similarly, for an F16 military aircraft flying horizontally between two charged electric cloud cells, the accurately identified zones of high electric fields reaching breakdown are mainly the rudder and the nose boom.

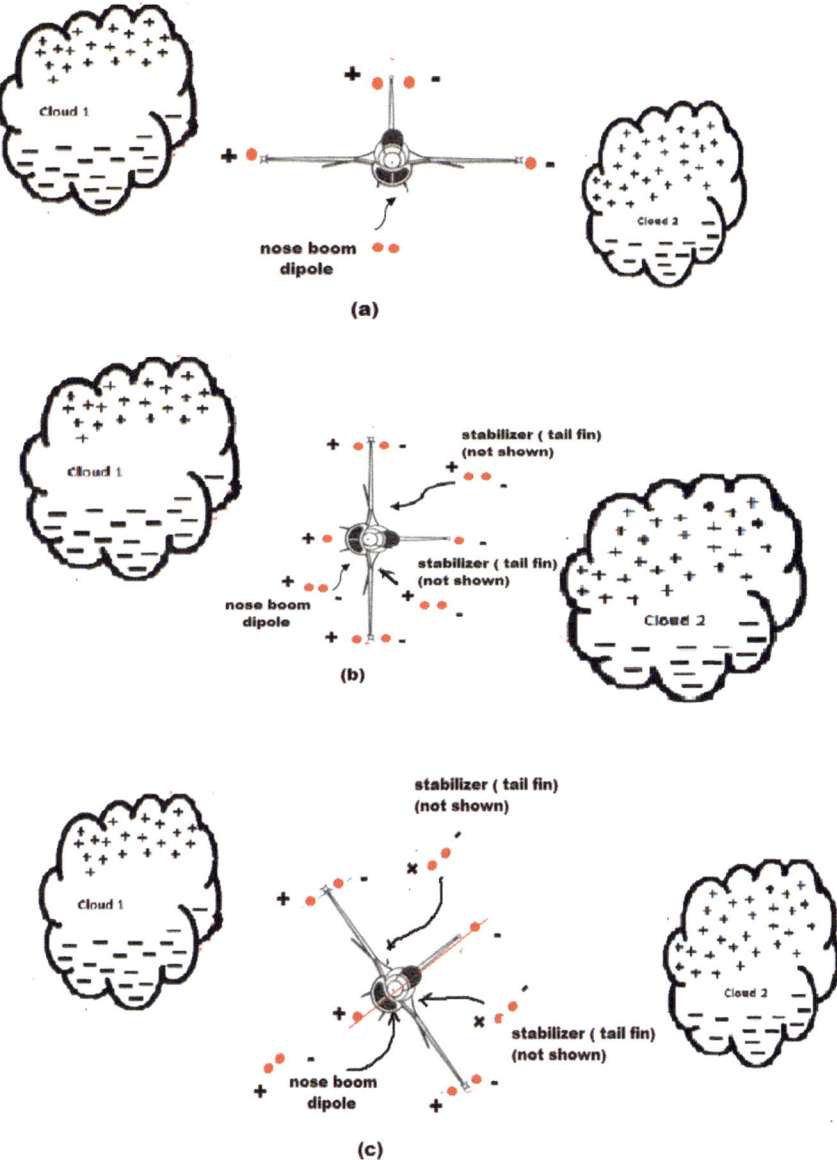

Fig. 2.6 Different roll angle orientations of the F16 military aircraft and dipole alignments **a** horizontal position, **b** vertical position, and **c** positioned at an angle

Table 2.3 Electric charges and electric fields for an airbus A380 at 800 m altitude but 1000 m away from a charged cloud at 1000 m altitude

Computed Voltage V_A: -2.821×10^6 V			
Dipole location and number	Dipole charge (C/m)	Electric field (V/m)	[a]Comment
Rudder tip	1.668×10^{-3}	7.566×10^4	
Mid-fuselage	2.603×10^{-6}	2.121×10^4	
Radome	2.178×10^{-5}	2.176×10^6	Nearing breakdown E-field
Mid-left wing	8.067×10^{-6}	7.536×10^4	
Left wing engine	1.818×10^{-5}	1.611×10^4	
Tip left wing	4.799×10^{-4}	3.32×10^6	Above Breakdown E-field
Mid-right wing	4.025×10^{-5}	3.625×10^5	
Right wing engine	1.652×10^{-5}	1.558×10^4	
Tip right wing	6.179×10^{-4}	4.274×10^6	Above Breakdown E-field
Left horizontal stabilizer tip	1.155×10^{-3}	1.153×10^8	Above Breakdown E-field (high probability of entry point)
Right horizontal stabilizer tip	1.114×10^{-3}	1.113×10^8	Above Breakdown E-field

[a]At high altitude with reduced air density, the breakdown electric field may be as high as 400 kV/m

Table 2.4 Electric charges and electric fields for an airbus A380 at 800 m altitude but 5000 m away from a charged cloud at 1000 m altitude

Computed Voltage V_A: -6.12×10^4 V		
Dipole location and number	Dipole charge (C)	Dipole electric field (V/m)
Rudder tip	1.36×10^{-5}	669.296
Mid-fuselage	2.048×10^{-8}	314.979
Radome	1.552×10^{-7}	1.551×10^4
Mid-left wing	6.684×10^{-8}	676.392
Left wing engine	1.396×10^{-7}	294.147
Tip left wing	3.684×10^{-6}	2.549×10^4
Mid-right wing	3.238×10^{-7}	2.928×10^3
Right wing engine	1.357×10^{-7}	292.391
Tip right wing	4.795×10^{-6}	3.317×10^4
Left horizontal stabilizer tip	9.401×10^{-6}	[a]9.39×10^5
Right horizontal stabilizer tip	9.072×10^{-6}	[a]9.061×10^5

[a]Exceeds breakdown field of 400 kV/m

Table 2.5 Electric charges and electric fields for an airbus A380 at 800 m altitude but 50,000 m from a charged cloud at 1000 m altitude

Computed Voltage V_A: -64.31 V

Dipole location and number	Dipole charge (C/m) WHY PER M?	Electric field (V/m)
Rudder tip	1.603×10^{-9}	2.822
Mid-fuselage	2.374×10^{-12}	2.826
Radome	1.717×10^{-11}	3.309
Mid-left wing	7.927×10^{-12}	2.827
Left wing engine	1.601×10^{-11}	2.827
Tip left wing	4.215×10^{-10}	4.059
Mid-right wing	3.791×10^{-11}	2.846
Right wing engine	1.434×10^{-11}	2.827
Tip right wing	5.517×10^{-10}	4.748
Left stabilizer tip	1.107×10^{-9}	110.60
Right stabilizer tip	1.068×10^{-9}	106.75

Table 2.6 Dipole charges and electric fields for F16 military aircraft at 1000 m in altitude at level with two charged cloud cells of 500 m apart

Cloud altitude: 1000 m
F16 military aircraft altitude: 1000 m and 500 m away from the two charged clouds
Computed Voltage V_A: 7.742×10^6 V

Dipole location	Dipole charge (C)	Dipole electric field (V/m)
Rudder tip	0.04196	5.175×10^9 (above breakdown field)
Fin	4.48925×10^{-4}	5.28×10^7 (above breakdown field)
Mid-fuselage back	7.3322×10^{-4}	3.211×10^7 (above breakdown field)
Wing	3.641×10^{-3}	1.4×10^7 (above breakdown field)
Mid-fuselage front	1.4639×10^{-3}	6.741×10^7 (above breakdown field)
Nose boom	0.07924	1.781×10^{10} (above breakdown field) High probability of strike point

Fly by wire and non-metallic body aircraft designers are interested in the enhanced electric field areas of the aircraft body in order to divide it into zones of protection. The highly probable zones are zones with threatening electric field enhancements and probable lightning strikes and zones with less enhancements of electric fields have minimum probability of strikes. The high strike risk regions are to be avoided

Table 2.7 Dipole charges and electric fields for F16 military aircraft at 1000 m altitude at level with two charged cloud cells and 5000 m from the center of two charged cloud cells

Cloud altitude: 1000 m
F16 military aircraft altitude: 1000 m but 5000 m from cloud charge centers
Computed voltage V_A: 7.316×10^4 V

Dipole location	Dipole charge (C)	Dipole electric field (V/m)
Rudder tip	3.965×10^{-4}	4.89×10^7 (above breakdown field)
Fin	4.242×10^{-6}	4.998×10^5 (above breakdown field)
Mid-fuselage back	6.929×10^{-6}	3.048×10^5
Wing	3.441×10^{-5}	1.353×10^5
Mid-fuselage front	1.3833×10^{-5}	6.377×10^5 (above breakdown field)
Nose boom	7.488×10^{-4}	1.683×10^8 (above breakdown field) High probability of strike point

when placing mission critical navigational and control systems as well as microelectronic equipment. The severe cloud-to-cloud lightning strikes in which an aircraft becomes a part of the lightning flash path is the most severe threat to navigational, microelectronic, and measurement systems.

Finally, presentation of the 3D dipole modeling of electric charges and electric field computations agrees with the norm that electric fields are highest at the extremities or sharp edges. These are situations where high electric fields exist on the surface of a body when the monopole charges that make up these dipoles on the surface have the least separation distances, and large electrically induced charges will appear on the surface of the aircraft. However, if the aircraft body is oriented differently, as when the military aircraft changes its roll angles, this will alter the dipole orientation, and thus the electric field distributions over the aircraft surface. In the results for Fig. 2.6, the F16 military aircraft is horizontally positioned between two charged cloud cells of opposite polarities. The electric field is highest at the nose boom and the tip of the rudder (vertical stabilizer). However, if the F16 military aircraft possesses the flexibility to maneuver its position inflight in different aerobatic orientations as shown in Fig. 2.6b and c, the dipole separation distance would vary. In Fig. 2.6b, the wing tips, the horizontal stabilizers, and the nose boom would have the highest electric fields. For Fig. 2.6c the wing tips, the horizontal stabilizers, and the nose boom would have the highest enhanced electric field induced. The electrostatics computation method we have presented enables the computation of electric potential, electric charge and electric fields at any one of these positions.

2.10 Electrostatic Computation and Evaluation: A Computer-Based Tool

This chapter is presented a reliable approach to calculating the prestrike electrostatic charges and electric fields build up on an aircraft using the 3D dipole method. The method may be used for any structure, whether airborne or ground, to determine a critical understanding of the thundercloud electric threat posed and how to design and to protect to minimize damaging engagement with the pre-lightning electrostatic fields and initiating a lightning flash. The computational tool based on the equations presented in this chapter was developed to handle the thundercloud electrostatic environment, and to evaluate the electric charges induced on the surface of an aircraft and the electric fields produced over any structure by the thundercloud electric charge centers. Using the knowledge of both the electrostatic charges induced and the electrostatic fields generated, it is possible to form zones over the aircraft body to indicate areas of high risk to lightning strike. These are the areas in which microelectronic, navigation, and instrumentation equipment will be subject to high electrostatic stress possibly leading to local electrostatic discharges (ESD). Regions of aircraft structure which may be classified as high-risk zones include the radome, the mid-left and mid-right wing areas as well as the aircraft wing edges in the case of airbus A380. For the F16 military aircraft, the orientation of the aircraft while maneuvering highlights the changing regions of threat including the wing tips, the horizontal stabilizers, and the nose boom depending on the orientation of the aircraft with respect to ground. It was found that even when the aircraft is flying well away from the thundercloud, some zones may still experience large electrostatic stress. The technique reported herein may also be used to study how each zone may dynamically change in experiencing electrostatic stress as the aircraft changes its pitch or roll angles, or when a swept lightning stroke moves over the body of the aircraft.

2.11 Personal Lightning Safety

For purposes of personal safety the following are prudential rules to adopt and follow:

1. When there are lightning flashes in the neighborhood, avoid outdoor activities. Stay indoors. Postpone outdoor events when there is thunderstorm activity in the neighborhood.
2. When you are having a group activity, and you should hear thunder, ask everyone either to go indoors or get into vehicles. Do so immediately. Stay in until 30 min after hearing the last thunder clap.
3. Do not use wired telephones during thunderstorms. Avoid taking showers when there is a thunderstorm. Water is a good conductor, in case the house gets struck by lightning.
4. If you are caught outdoors in a thunderstorm as a group, spread out so that you reduce the chance of multiple causalities if lightning should strike the group.

5.   Stay away from tall, isolated objects such as trees, towers, or utility poles. Stay
     away from metal conductors such as metal fences and wires.
6.   Tents, front porches, picnic shelters, or any covering without metallic covering,
     wiring, or plumbing are not safe.
7.   If someone is struck, give CPR if you are trained. Call emergency. If available,
     use an Automatic External Defibrillator.
8.   Lightning usually strikes outside the area of heavy rainfall and can strike up to
     30 km from where there is rainfall. Lightning strikes occur before rainfall, as
     well as after.
9.   Do everything to avoid becoming a victim to lightning. It can cause permanent
     disabilities to the body and mind.

Bibliography

Abdelal, G.F.: Lightning strike protection system. In: Multifunctional Composites, pp.25–57 (2015).
   http://barbero.cadec-online.com/Multifunctional/seeinsidech2mfc.pdf (2015)
Brooks, H.E.: Severe thunderstorms and climate change. J. Atmosp. Res. (2012). https://doi.org/
   10.1016/j.atmosres.2012.04.002(2012)
Chemartin, L., et al.: Direct effects of lightning on aircraft structure: analysis of the thermal,
   electrical and mechanical constraints. J. Aerosp. Lightn. Hazards Aircraft Launch. **AL05-
   09**(5), 1–15 (2012). http://www.aerospacelab-journal.org/sites/www.aerospacelab-journal.org/
   files/AL05-09_0.pdf
Deng, Y.: Carbon fiber electronic interconnects, PhD dissertation, Department of Mechanical
   Engineering, University of Maryland (2007). http://drum.lib.umd.edu/bitstream/handle/1903/
   6997/umi-umd-4508.pdf;jsessionid=390C062928AF333BD46AAC4DC30D228A?sequence=
   1(2007)
Deo, R.B., Starnes, Jr., J.H., Holzwarth, R.C.: Low-cost composite materials and structures for
   aircraft applications. In: A Paper presented at the RTO AVT Specialists' Meeting held in Loen,
   Norway, 7–11 May 2001, and Published in RTO-MP-069(II) (2001). http://docshare01.docshare.
   tips/files/5085/50851653.pdf
Desch, S.J., Borucki, W.J., , Russell, C.T., Bar-Nun, A.: Progress in planetary lightning. Rep. Progr.
   Phys. **65**(6), 955–997 (2002). https://doi.org/10.1088/0034-4885/65/6/202
Djurdjevic, J.: Lightning Performance Evaluation of Quasi-Hemispherical Air Terminations versus
   a Franklin Rod, M.Sc., Faculty of Engineering and Built Environment, University of the
   Witwatersrand, Johannesburg, R.S.A, pp. 25–30 (2011)
Dwyer, J.R., Uman, M.A.: The physics of lightning. Phys. Rep. **534**(4), 147–242 (2014). Else-
   vier Science Direct (2014). http://www.lightning.ece.ufl.edu/PDF/Published%20Journals/2014/
   Dwyer%20&%20Uman%20a.pdf
Environmental Conditions and Test Procedures for Airborne Equipment, U.S. Department of Trans-
   port Federal Aviation Administration Advisory Circular, AC No: 21–16G, RTCA Document DO-
   160 versions D, E, F, and G (2011). https://www.faa.gov/documentLibrary/media/Advisory_Cir
   cular/AC%2021-16G.pdf (2011)
Fisher, J., Hoole, P.R.P., Pirapaharan, K., Hoole, S.R.H.: Applying a 3D dipole model for lightning
   electrodynamics of low-flying aircraft. IETE J. Res. **61**(2), 91–98 (2015)
Fisher, J.: Lightning-Aircraft Electrodynamics. Ph.D. Thesis, PNG University of Technology (2016)
Fisher, F.A., Plumer, J.A.: Lightning protection of aircraft, Report NASA RP 1008, National Aeronautics and Space Administration, U.S. Government Printing Office. Washington. DC 20402
   (1977). https://ntrs.nasa.gov/archive/nasa/casi.ntrs.nasa.gov/19780003081.pdf

Fisher, F.A., Plumer, J.A., Perala, R.A.: Aircraft lightning protection handbook Report No. DOT/FAA/CT-89/22, U.S. Department of Transportation Handbook Federal Aviation Administration Technical Centre, Atlantic City (1989)

Fisher, J., Hoole, P.R.P., Pirapaharan, K., Hoole, S.R.H.: Parameters of cloud to cloud and intracloud lightning strikes to CFC and metallic aircraft structures. In: Proceedings of the International Symposium on Fundamentals of Engineering, June 2016, To appear IEEE Xplore Digital Library (2016). https://doi.org/10.1109/ISFEE.2016.7803216

Formenton, M., Panegrossi, G., Casella, D., Dietrich, S., Mugnai, A., San'o, P., Di Paola, F., Betz, H.-D., Price, C., Yair, Y.: Using a cloud electrification model to study relationships between lightning activity and cloud microphysical structure. Natl. Hazards Earth Syst. Sci. **13**, 1085–1104 (2013). https://doi.org/10.5194/nhess-13-1085-2013

Genareau, K., Wardman, J.B., Wilson, T. M., McNutt, S.R., Izbekov, P.: Lightning-induced volcanic spherules. Geol. Soc. Am. **43**(4), 319–322 (2015). http://geology.gsapubs.org/content/43/4/319.full.pdf

Golding, W. L.: Lightning strikes on aircraft: how the airlines are coping. J. Aviat./Aerosp. Educat. Res. **15**(1), 41–50 (2005)

Ha, M.S., Kwon, O.Y., Choi, H.S.: Improved electrical conductivity of CFRP by conductivity silver nano-particles coating for lightning protection. In: The 17th International on Composites Materials Proceedings, Edinburgh, Scotland (2009). http://www.iccm-central.org/Proceedings/ICCM17proceedings/Themes/Industry/AEROSPACE%20APPLICATIONS/INT%20-%20AEROSPACE%20APPLICATIONS/IA2.3%20Ha.pdf

Hoffmann, J., Haugh, J.: Classification of cloud types based on form or shape of the clouds, altitude and height and height of clouds. https://www.lcps.org/cms/lib/VA01000195/Centricity/Domain/11412/3%20Classification%20of%20Cloud%20Types.pdf

Holle, R.L.: The visual identification of lightning-producing thunderstorm clouds. In: 23rd International Lightning Detection Conference, Tucson, USA (2014). Retrieved from: http://www.vaisala.com/Vaisala%20Documents/Scientific%20papers/2014%20ILDC%20ILMC/ILMC-Thursday/Holle-Visual%20Identification%20of%20Thunderstorms-2014-ILDC-ILMC.pdf

Hoole, P.R.P., Pirapaharan, K., Hoole, S.R.H.: Electromagnetics Engineering Handbook. WIT Press, UK (2013).

Hoole, S.R.H., Hoole, P.R.P.: A Modern Short Course in Engineering Electromagnetics. Oxford University Press, USA (1996)

Hoole, P.R.P., Thirukumaran, S., Ramiah, H., Kanesan, J., Hoole, S.R.H.: Ground to cloud lightning flash currents and electric fields: interaction with aircraft and production of Ionosphere Sprites. J. Comput. Eng. Article ID 869452, 5 (2014). https://doi.org/10.1155/2014/869452

Hutchins, M.L., Holzworth, R.H., Virts, K.S., Wallace, J.M., Heckman, S.: Radiated VLF energy differences of land and oceanic lightning. Geophys. Res. Lett. **40**, 2390–2394 (2013). https://doi.org/10.1002/grl.50406,(2013)

Jebson, S.: Thunderstorm, The National Meteorological Library and Archive Fact Sheet No. 2 (2007). http://skybrary.aero/bookshelf/books/447.pdf

Industry Document to Support Aircraft Lightning Protection Certifications, U.S. Department of Transport Federal Aviation Administration Advisory Circular, AC No: 20-155A, 2013. Jessa, T., (2016). What are cumulonimbus clouds? Universe Today, http://www.universetoday.com/41646/cumulonimbus-cloud/

Kendall, C., Black, E., Larsen, W.E., Rasch, N.O.: Aircraft generated electromagnetic interference on future electronic systems, Federal Aviation Administration Report No: DOT/FAA/CT-83/49 (1983). Retrieved from http://www.tc.faa.gov/its/worldpac/techrpt/ct83-49.pdf. https://www.faa.gov/documentLibrary/media/Advisory_Circular/AC_20-155A.pdf (1983)

King, S.: Investigation of climate effects on the global atmospheric electrical circuit using surface potential gradient data, M.Sc. dissertation, Department of Meteorology, University of Reading, U.K (2004). http://www.met.rdg.ac.uk/mscdissertations/Investigation%20of%20climate%20effects%20on%20the%20global%20atmospheric%20electrical%20circuit%20using%20surface%20potential%20gradient%20data.pdf (2004)

Kumar, V.V., Jakob, C., Protat, A., May, P.T., Davis, L.: The four cumulus cloud modes and their progression during rainfall events: A C-band polarimetric radar perspective. J. Geophys. Res.: Atmosp. **118**(15) (2013). Retrieved from: https://onlinelibrary.wiley.com. https://doi.org/10.1002/jgrd.50640/pdf

Lal, D.M., Ghude, S.D., Singh, J., Tiwari, S.: Relationship between size of cloud ice and lightning in the tropics. Adv. Meteorol. 2014. Article ID **471864**, 1–7 (2014). https://doi.org/10.1155/2014/471864(2014)

Lalande, P., Delannoy, A.: Numerical methods for zoning computation. J. Aerosp. Lab Numer. Methods Zoning Comput. **ALO5-08**(5), 1–10 (2012). http://www.aerospacelab-journal.org/sites/www.aerospacelab-journal.org/files/AL05-08_0.pdf

Larsson, A., Lalande, P., Bondiou-Clergerie, A., Delannoy, A.: The lightning swept stroke along an aircraft in flight. Part I: thermodynamic and electric properties of lightning arc channels. J. Phys. D Appl. Phys. **33**(15), 1866–1875 (2000). https://doi.org/10.1088/0022-3727/33/15/317(2000

Martin, B.M.: Electromagnetic Effects and Composite Structure. Wichita State University, National Institute of Aviation Research (2015)

Mezuman, K., Price, C., Galanti, E.: On the spatial and temporal distribution of global thunderstorm cells. Environ. Res. Lett. **9**(12), 124023, 1–9 (2014). https://doi.org/10.1088/1748-9326/9/12/124023

Milewski, M.: Lightning Return-Stroke Transmission Line Modelling Based on the Derivative of Heidler Function and CN Tower Data, Ph.D. dissertation, Electrical and Computer Engineering, Ryerson University, Canada (2009). http://digital.library.ryerson.ca/islandora/object/RULA:511

Miranda, F.J., Pinto Jr., O., Saba, M.M.F.: Occurrence of characteristic pulses in positive ground lightning in Brazil. In: 19th International Lightning Conference and 1st International Conference on Lightning Meteorology, Tucson, U.S.A (2006). http://www.vaisala.com/Vaisala%20Documents/Scientific%20papers/Occurrence_of_characteristic_pulses_in_positive_ground_lightning_in_Brazil.pdf

Morgan, M., Hardwick, C.J., Haigh, S.J., Meakins, A.J.: The interaction of lightning with aircraft and the challenges of lightning testing. Aerosp. Lab J. **ALO5-11**(5), 1–10 (2012). http://www.aerospacelab-journal.og/sites/www.aerospacelab-journal.org/files/AL05-complete-issue.pdf

Mrazova, M.: Advanced composite materials of the future in aerospace industry. INCAS Bull. **5**(3), 139–150 (2013). https://doi.org/10.13111/2066-8201.2013.5.3.14

Padhya, A.R.: Manufacturing and certification of composite primary structures for civil and military aircrafts. In: A presentation at ICAS Biennial Workshop on Advanced Materials and Manufacturing – Certification and Operational Challenges, Stockholm, Sweden (2011). http://www.icas.org/media/pdf/Workshops/2011/ICAS%20Workshop%20presentation%2006%20Upadhya.pdf

Peters, M., Leyens, C.: Materials science and engineering. J. Aerosp. Space Mater. 3 (2009). Encyclopedia of Life Support Systems (EOLSS): http://www.eolss.net/sample-chapters/c05/e6-36-05-03.pdf

Petrov, N.I., Haddad, A., Petrova, H., Griffiths, H., Walters, R.T.: Study of effects of lightning strikes on aircraft. In: Agrawal, R., (ed.), Recent Advances in Aircraft Technology, pp. 523–544, Feb. (2012). ISBN978-953-51-0150-5, InTech

Poelman, D.R.: On the Science of Lightning: An Overview, Royal Meteorological Institute of Belgium, Publication Scientifique et Technique, No. 56, (2010)

Prabhakaran, R.: Lightning strike on metal and composite aircraft and their mitigation. In: International Conference on Intelligent Design and Analysis of Engineering Products, Systems, and Computation, Coimbatore, India, pp. 208–221 (2010)

Price, C., Asfur, M.: Inferred long term trends in lightning activity over Africa. Earth Planets Space **58**, 1197–1201 (2006)

Price C., Rind, D.: Possible implications of global climate change on global lightning distributions and frequencies. J. Geophys. Res. **99**, 10823–10831 (1994). https://doi.org/10.1029/94JD00019

Price, C.: Thunderstorm, lightning, and climate change. In; 29th International Conference on Lightning Protection, 23rd – 26th June 2008, Uppsala, Sweden (2008)

Protection Against Lightning – Part 1 General Principles, IEC International Standard, IEC 62305-1, 2 edn., 2010–12 (2012)

Punekar G.S., Kandasamy, C.: Indirect Effects of lightning discharges. Serbian J. Electr. Eng. **8**(3), 245–262 (2011). http://www.journal.ftn.kg.ac.rs/Vol_8-3/02-Punekar-Chandrasekaran.pdf

Rakov, V.A., Huffines, G.R.: Return-stroke multiplicity of negative cloud-to-ground lightning flashes. J. Am. Metorol. Soc. **2**, JAM8805, 1455–1462 (2003). https://doi.org/10.1175/1520-0450(2003)042<1455:RMONCL>2.0.CO;2

Rakov, V.A.: A review of positive and bipolar lightning discharges. J. Am. Meteorol. Soc. 767–776 (2003). http://www.lightning.ece.ufl.edu/PDF/i1520-0477-084-06-0767.pdf

Rakov,V.A.: Lightning phenomenology and parameters important for lightning protection. In: IX International Symposium on Lightning Protection (2007). http://ws9.iee.usp.br/sipdax/papersix/sessao12/12.1.pdf

Rakov, V.A.: Fundamentals of lightning. In: International Symposium on Lightning Protection, Kathmandu, Nepal (2011). http://lightningsafety.com/nlsi_info/Fundamentals-of-Lightning-Rakov.pdf

Rakov, V.A.: The Physics of lightning. Surv. Geophys. **34**, 701–729 (2013). https://doi.org/10.1007/s10712-013-9230-6

Rupke, E.: Lightning Direct Effects Handbook. Report Reference Number AGATE-WP3, 1-031027-043-Design Guidelines Work Package Title: WBS3.0 Integrated Design, and Manufacturing (2002). http://www.niar.twsu.edu/agate/documents/Lightning/WP3.1-031027-043.pdf

Saunders, C.: Charge separation mechanisms in clouds. Space Sci. Rev. **137**, 335–353 (2007)

Schoene, J.D.: Direct and nearby lightning strike interaction with test power distribution lines, Ph.D. Dissertation, Department of Electrical and Computer Engineering, University of Florida (2007). http://etd.fcla.edu/UF/UFE0017522/schoene_j.pdf

Severson, V., Murray, P., Heeter, J.: Establishing cause and effect relationships between lightning flash data and airplane lightning strike damage. In: 23rd International Lightning Detection Conference and 5th International Lightning Meteorology Conference, Tucson (2014)

Simpson, J., Dennis, A.S.: Cumulus Clouds and Their Modifications, NOOA technical Memorandum Report No: ERL-OD-14, U.S Department of Commerce, Environmental Research Laboratories (1972)

Smith, F.: The use of CFC in aerospace: past, present, and future challenges. In: A Presentation Document by Avalon Consultancy Services Ltd (2013). https://avaloncsl.files.wordpress.com/2013/01/avalon-the-use-of-composites-in-aerospace-s.pdf

Soula, S.: Electrical environment in a storm cloud. J. Aerosp. Lab **ALO5-01**(5), 1–10 (2012). http://www.aerospacelab-journal.org/sites/www.aerospacelab-journal.org/files/AL05-01_0.pdf

Uman, M.A., Rakov, R.A.: The interaction of lightning with aerospace vehicles. In: Progress in Aerospace Sciences, vol. 39, pp. 61–81. Elsevier Sciences Ltd (2003). http://www.pas.rochester.edu/~cline/FLSC/Lightning%20Report.pdf

Williams, E.R.: The electrifications of thunderstorms. Sci. Am. (1988). http://www.atmo.arizona.edu/students/courselinks/spring07/atmo589/articles/SciAmNov88.pdf

Chapter 3
Lightning Protection of Domestic, Commercial, and Transport Systems

Abstract This chapter presents a few examples of lightning protection systems for a variety of structures and systems. Particularly, the lightning protection of houses is described. The selection of lightning protections devices, the ratings for different applications, and the graded system protection are discussed for the protection of internal systems of a house. The external protection of the house is also presented. Both basic lightning protection as well as enhanced lightning protection are treated, besides the lightning protection of the following: boats, photovoltaic (PV) installations, frequency converters, network services, wind turbines, and historic buildings.

3.1 General

There are two lightning overvoltage modes. These are the common mode (line-to-neutral or line-to-line surge voltages) and normal mode (line-to-ground or neutral-to-ground surge voltages). Lightning over voltages mostly appears in common mode and enters the internal system at the origin of electrical installation. There are three types of protective wiring. These are the TT, TN, and IT systems. The first of the two letters indicate the earthing conditions of the supplying power source. If the first letter is T, which stands for direct earthing of one point of the power source. If the first letter is I, it indicates the insertion of all live parts from earth or connection of one point of the power source through an impedance. The second letter of the two letters indicates the earthing conditions of the bodies of equipment. If the second letter is T then the body of the equipment is directly earthed. If the second letter is N then the body of the equipment is directly connected to the operational earth electrode, that is the earthing of the power source. A third letter may be added, where S indicates that the neutral and protective conductors are laid separately from each other. If the third letter is C, it indicates that the neutral and protective conductors are combined into one conductor.

The TT system to protect persons is one of the simplest. Here, the exposed conductive parts are earthed and RCDs (residual current devices) are used. Where

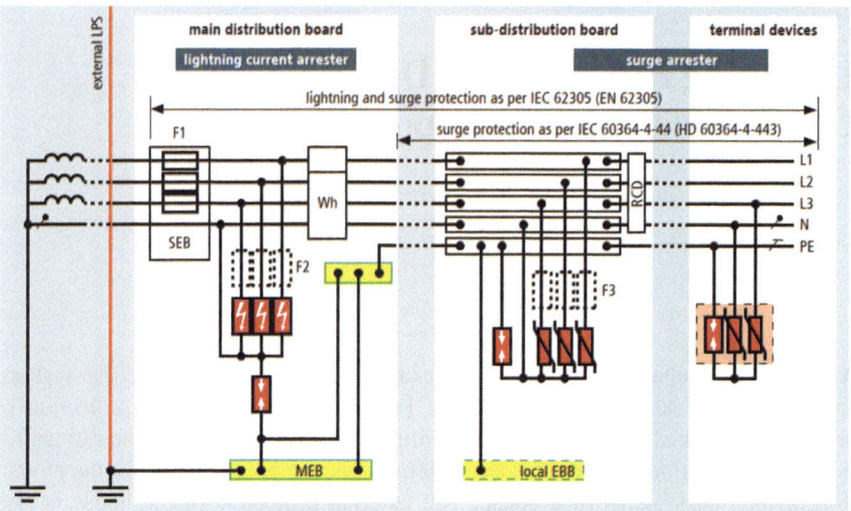

Fig. 3.1 The lightning protection of a TT system

the exposed conductive parts are earthed at a number of points, then each circuit connected to a given earthed electrode must have an RCD installed. Figure 3.1 shows the placements of protective devices in a TT system.

In the TN system, it is essential to have both interconnection and earthing of exposed conductive parts and the neutral. The first fault is interrupted using circuit breakers or fuses as overcurrent protection. In the IT system, the conductive parts are interconnected and earthed. An insulation monitoring device (IMD) is used to indicate the first fault. The second fault is interrupted by overcurrent protectors such as circuit breakers and fuses. The residual current over voltages appears in the TT mode and affects only the sensitive equipment. In the TT mode, when the neutral on the distribution side is linked to a low resistance value earthing system (a few ohms in an installation with earthing electrode resistance of tens of ohms), a phase to neutral protection must be used at each level of the installation. Cascaded protection is used when each level of the installation requires overcurrent protection with ratings appropriate to that level. Several voltage protectors are used in cascade. Spark gap-based protection components with varistors and diodes that limit voltages to compatible levels are used in cascaded protection. For terminal protection close to equipment, proximity voltage surge protectors are used.

When selecting voltage surge protectors, class 1 (Type 1) is compulsory at the origin of the installations and secondary buildings when a lightning conductor is present. In industrial installations and office complexes, high-capacity protection devices such as voltage surge protectors are used when no lightning conductor is used. An increased capacity protection device such as a voltage surge protector is used for smaller installations. At the origin of an installation, an increased capacity protection device or standard main protection device must be coordinated with the main protective device. For very sensitive equipment, a proximity protection device such as a voltage surge protector must be used.

3.2 Lightning Protection of Houses

3.2.1 An Overview

3.2.1.1 Damage from Lightning

A direct lightning strike to a house, rare though it be, poses the greatest threat to the house structure and the contents of the house. Moreover, lightning will be able to damage equipment connected to cables a mile (1.6 km) or more from the location of the strike because the initial lightning impulse is very strong, of the order of 20 kA peak and 1 μs rise time. There are four ways in which that a lightning strike can damage residential equipment. These are strike to power or communication lines, strike to or near equipment outside the house, strike to nearby object, and direct strike to the house structure. These occur in decreasing order of frequency. The four ways by which lightning may pose a threat and damage houses and the electrical and electronic equipment and systems of the house, in detail, are as follows.

(1) Direct lightning strike to power lines and telecommunication and cable television wiring. This is the most common form of lightning threat and damage to houses and home systems. When the network is placed at a higher elevation, it is readily struck by lightning, and the lightning surges travel through the power or communication network into the house. Another route is when lightning strikes the ground close to the house. When this happens, lightning surge currents may travel through underground conductors and piping into the house.

(2) The second most common route is through strikes to neighboring systems that collect lightning surges and may pass them onto the household systems. Examples of these are external electric gate control systems, ground lightning systems, satellite dishes, and security systems. Since all these systems are powered or connected to systems inside the house, lightning surges find a ready route to travel into the home electronic and electrical systems and equipment and cause damage or gradual electric rust.

(3) Pulsed lightning electromagnetic fields, LEMPs, radiated by close by objects including trees and tall structures struck by lightning may travel into the house and induce high voltages and surges in the internal systems. This is so even when the point of strike is not directly connected to home systems and equipment.

Fig. 3.2 Lightning protection of a single-family house without external lightning protection. Credit: DEHN, with permission

(4)   The worst case, though rare, is when lightning directly strikes a house. If the house is not provided with a Lightning Protective System (LPS), such a direct strike can cause immense damage to the house structure and to electrical and electronic systems inside it. The LPS system, both external and internal, is presented in the sections to follow.

Figure 3.2 shows the lightning protection of a single-family house without an external lightning protection. Special attention must be given to protect the electronic equipment and systems in increasing use in homes. Referring to Fig. 3.2, the protection arrangements, with reference to the numbers indicated in Fig. 3.2, are as follows:

CHECK CHANGES IN 1–8

1.   At the main distribution board. Combined arrester mounted at the entrance of the building. Alternative devices may be used to protect against over-voltages caused by LEMP.
2.   For Internet/Telephone Broadband system. Protective device to protect the telecommunication systems at the entrance of the building. Alternatively, other devices may be used to protect broadband connections to PC.
3.   For photovoltaic (PV) systems, 3A and 3B in Fig. 3.2 is presented a protection system for PV systems. If the cable length between the PV and the inverter is longer than 10 m, a separate protector is needed at the roof.

4. For sub-distribution board or terminal devices: Protects downstream distribution boards against over voltages from inductive coupling are in order. Alternative devices are available for three-phase terminal devices, for sensitive electric blinds, and for placing directly at the terminal devices.
5. For office/home office/Ethernet devices, computer workstations, and DSL routers.
6. For TV/Satellite systems: For antenna input of TV sets and satellite systems, urge arresters with F sockets, for antenna splitters and multi-switches are recommended.
7. For home automation/heating/air-conditioning/ventilation: Lightning arresters for sensitive equipment are used. These are installed directly at the electronic components of the air conditioners, heat pumps, etc.
8. For smart home: The protective equipment is mounted in the bus terminal slot of KNX components.

External protection systems, such as an air-termination system, are used to protect the house against the effects of lightning directly striking the house, which may result in fire if unprotected. The air-termination is connected to the earth-termination system using a down conductor. Different lightning protection arrangements to that shown in Fig. 3.2 are available for the following: single-family house with external lightning protection, multiple family house with no external lightning protection, and multiple family house with external lightning protection.

3.2.1.2   Basic Protection Against Lightning

In order to conduct lightning currents safely to ground, and prevent them flowing into electronic and electrical systems and equipment, it is pertinent to provide a good grounding system. In order to ensure this and to prevent potential drops developing to cause excess current flows, it is required that a well-grounded electrode (ground electrode) should be connected to metallic structural parts, metallic pipes, cable sheaths, telephone lines, and broadband connections. A well-protected system will have both good bonding and effective grounding.

There are three main features in the basic grounding and protection systems. These are as follows:

(1) One main grounding point with a soundly earthed electrode will serve as the central point to which all lightning currents must be conducted to be dissipated into the ground. All grounding electrodes must be connected to a central ground in order to avoid potential drops between different grounded electrodes. Any potential drop will cause disastrous circulating currents of large magnitudes, which are injected back into the electric system inside the house, often in a reverse direction. This must be avoided.
(2) Whatever lightning surges that seek to enter the building from external electrical, communication, and entertainment networks must be captured and eliminated at the entry point or systematically, stage by stage, eliminated by using

multiple points of surge protection along its route into the household terminals. Electric power cable sheaths as well as external antenna cable sheaths must be connected to the common grounding system of the house. At the point of entry to the building, a special Network Interface Device (NID) is used to eliminate these surges at the point of entry.

(3) By connecting all metal pipes (e.g. water pipes) and building structural metal parts to the common ground point, lightning surges on metal pipes and building metal parts, as well as any danger of potential rise, may be avoided.

These protections greatly reduce shock or electrocution risk to people inside the house and reduce the possibility of a fire caused by lightning. However, they are insufficient to prevent damage to electrical and electronic equipment.

3.2.1.3 Enhanced Protection Against Lightning

To increase protection in high-lightning areas, the following additional systems are important:

(1) Lightning protection system (LPS).
(2) Surge protectors on the AC power wiring.
(3) Additional surge protectors on signal wiring.
(4) Adjacent to electronic and power equipment, a special protection device.

Where the probability of direct lightning strike is high, one or more lightning rods (air terminals) are placed on top of the building and bonded together and connected to the building ground electrode system. The lightning current is safely conducted from the lightning rods to the building's ground system through down conductors. Surge protectors cut down the voltage level of lightning surges before they enter into system and will need to be handled by point-of-use protectors at the equipment. The lightning currents coming on external wiring are conducted to the ground conductor through the use of surge protectors.

3.2.2 Choosing Service Entrance Surge Protectors (SPDs)

Large surge voltages and currents are experienced by the surge protectors used by the AC mains terminal box typically at the entrance to a house. Indirect penetration of lightning surges occurs through utility and electric distribution systems which pick up the lightning surge from lightning strikes to close by soil, electric power lines, and buildings. Less severe than lightning surges are switching surges caused by switching on or off of heavy electric equipment, load, or machines.

When selecting the ratings of SPD surge protectors, the following conditions must be met: (a) capability to prevent damage to service equipment at the power supply entry point to the house and heavy equipment such as air conditioners and other appliances directly connected to entry circuits; (b) provision gradually to cut down surge voltages, using SPDs at critical points, as the surge travels down the connections cables from the main entry into terminal box to various indoor circuits and equipment; and (c) assurance that the electric wiring and other equipment are not damaged by using SPDs at appropriate points.

Lightning surges are carried by two lines, which include the following two modes: line-to-neutral (L-N) or line-to-line (L-L) Normal Mode; or line-to-ground (L-G) or neutral-to-ground (N-G) Common Mode. At the power line entry to the building and right after a transformer, the L-G and L-N modes need protection. Inside the buildings, the L-N, L-G, and N-G modes need protection.

3.2.3 Surge Current Rating

At the service entry point, typically, a 10 kA (8/20 μs waveform) is defined as the largest surge that can be expected (Note: An 8/20 μs waveform rises to its peak in 8 μs and falls to half that value in 20 μs). SPDs with current ratings of 10–70 kA per phase (for residential systems) and in the range of 40–300 kA (for industrial systems) are available. SPDs are typically tested for 8/20, 4/10, 10/350, or 10/1000 μs impulses. Typically, SPDs with surge current rating of 20–70 kA (8/20 μs) per phase are used for residential or light commercial areas. In order to obtain good lifetime and reliability in areas where lightning incidence is high, SPDs rated in the range of 40–120 kA are used. Those SPDs installed along the circuit from the main entry point need to be coordinated, as the surge voltage is expected to drop significantly as the surge passes through the SPDs. The reduction of the surge voltage, in addition to clipping by the SPD, also depends on the impedance of the circuit between two adjacent SPDs along the stream. When installing SPDs, special attention must be paid to secure low impedance grounding, minimum length leads, twisted SPD wires (to reduce loop impedance) and avoidance of 90-degree bends, which increase lead strength. Over current protection should be provided at the service entrance without an internal fuse. Plug in, point-of-use, or supplementary protectors are used to cover over voltages due to open neutral conductors, utility regulator failures and high voltage power cross.

Telephone cables, CATV/antenna/satellite coaxial cables, and broadband cables that carry power should be protected at their entry point to the building. Bonding of the cable sheath to the building ground where the cable enters the building for coaxial connections is important, and direct connection to the building or power panel ground will deal with the large currents. All ground rods should be connected to the building ground. It is highly recommended that the phone protector and cable are mounted at the entry point, right next to the AC protector/ground. This is to ensure that all points of the grounding system are at the same potential. Where large lightning currents flow through the house grounding system, ground potential rise (GPR) results.

3.2.4 Ground Potential Rise

Voltage drops of the order of 10 kV can develop between two points of a cable sheath, which is grounded for lightning strike toward one end of the cable. This, in turn, can give rise to large potential drops between equipment inside the building, each having its own grounding point connected to the building ground. To prevent the internal GPR from causing damage to equipment, it is important to provide surge protection to all the incoming lines and interconnecting lines between different equipment ports; for instance, two television sets inside the building. Moreover, the surge protection devices of all the utilities (including CATV, telephone, power) should be bonded together and where possible enter the building close to each other. Similarly, equipment outside the building (e.g. emergency generator, spa, well pumps, pool heaters) also experience GPR due to the fact that they are referenced to two grounds. They are referenced to the building ground (the line and neutral connected building ground) and local, equipment ground through the concrete slab on which they may be mounted. For a 20 kA lightning strike and a 30 Ω ground, the potential rise may be of the order of 400 kV. The building ground only experiences a delayed voltage rise due to the cable impedance. This could cause large voltage drops, say of the order of 30 kV between the pad ground and the building ground. In the case of a generator or motor, with the coil referenced to the building ground, there could be an electric flashover between the motor coil and the motor frame referenced to the pad ground. The electric motor/generator insulation could be destroyed. In order to protect such equipment, a surge protector should be installed at the equipment site to bond between all line wires, neutral and ground.

3.2.4.1 Multi-Port Point-of-Use (Plug-In) Protectors

The multi-port point-of-use surge protectors consist of an AC protector and signal-like protectors. They are installed at the equipment connecting to both AC and signal lines. These protectors have a lower effective surge limiting voltage than protectors at the panel, and protect against sustained AC over voltage. These having a lower surge limiting voltage, they protect against very small surge voltages that may slip past the primary signal protector. By bonding the grounds for all protectors, the intersystem voltages are minimized.

3.2.4.2 AC Protection Circuits

Since fast electromechanical relays are available that are fast enough for AC protection circuits. These AC protection systems normally contain circuit breakers, indicator lights, multiple varistors and fuses, besides other voltage-limiting components, and capacitors and inductors, which are used to remove radio frequency interference

(RFI). In case of over voltage, the electronic protection systems quickly disconnect surges when detected. These are called varistors, which provide protection against rapidly changing surges. The current it needs to withstand is close to 10 kA. The plug-in protectors are expected to withstand currents of the order to 20–500 A. Surge arresters with high surge current should be installed at the service entrance.

3.2.5 Signal Protectors

A wide variety of protectors are used depending on the signal service; namely, satellite, computer, phone, and video links. The protectors must allow the signal to go through without significant modifications, while the over voltages must be limited to the safe value of the equipment. They must have a capacity to absorb surge energy without getting damaged. The protectors must be properly coordinated with the main protector at the building entry point. The current flowing into the protector is limited by placing a resistor in series with the signal input line.

3.2.6 Inter-System Bonding

It is preferred that the multi-port protector and its components are mounted in a single unit and the signal protector ground, and the AC grounds should be directly connected. Where needed, hum bars, low voltage MOVs (10–50 V), and ground isolators should be used to handle special problems that may be created by 50 Hz circulating currents.

3.2.7 Special Purpose Protectors

Special purpose protectors are used for power over Ethernet (POW) connections, dog fences, and transceivers for inter-building wireless connections. In general, to protect electronic systems in houses, it is required that proper grounding and bonding be provided, especially at the service entrance. Other installations required are: AC panel and primary signal surge protection at or near the service entrance, and multi-port plug-in protectors near the equipment to be protected.

Both the satellite dish and the coaxial cable sheath must be bonded to the building ground with at least a #10 wire. The satellite RF cable should come into the building near the service entrance to shorten the bond if possible. A separate ground rod is not an adequate substitute for the ground rod. The NID (network interface device, which contains the primary phone protector, and is normally supplied by the telephone

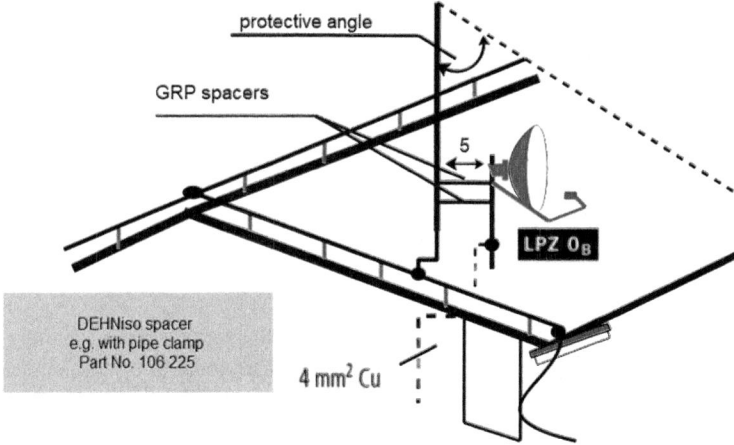

Fig. 3.3 Protection of the microwave dish Antenna (Credit: DEHN)

utility) should be examined, and it should be ensured that its ground terminal is bonded to the building ground. The NID should also be inspected to ensure that there is actually a protector present. The CATV grounding block normally installed at the building entrance should be checked to confirm that it is bonded to the building ground.

Four common types of connections that violate protection integrity are: equipment plugged into an AC outlet, which is not part of the multi-port protector; unprotected rooftop antenna or other signal input connection, bypassing the multi-port protector; downstream signal connections and any ground, unintentional or unintended, to any piece of equipment, bypassing the surge protection.

The rooftop microwave antenna may be protected as shown in Fig. 3.3. It needs to be ensured that the protective angle of the air-termination system covers the antenna as shown in Fig. 3.3. For more details, Cohen (2005) may be consulted.

3.3 Boats

Lightning strikes to ships and boats are higher in number closer to the equator, with higher probability of a lightning strike being closer to shore than at sea. Lightning flash density is higher close to shore than at sea. A typical leisure boat is fixed with electrical wires running down from the top of the mast to the interior of the ship. These include connections to radio antennas, anemometer, and navigation lights. The electrical cables run into various instruments placed inside the interior. There are also underwater sensors such as the depth sounder and log. If the lightning current from a lightning strike to the mast flows down and damages these underwater wiring

Fig. 3.4 The external lightning protection of the boat. The air-termination system is grounded by dangling the earth electrodes into the seawater. (Numbers in the figure: 1—universal earthing clamp; 2—multipole earthing cable; 3—earthing tongs; 4—braided copper strip.) (Credit: DEHN, with permission.)

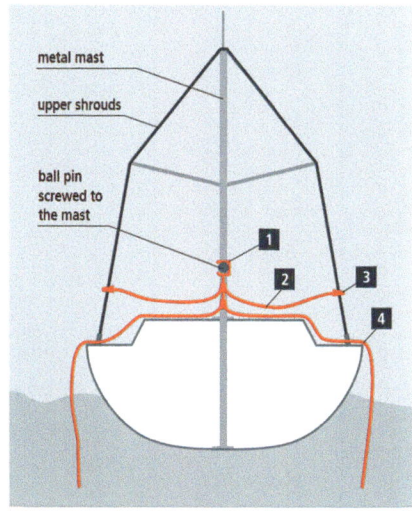

systems, it can result in water seeping into the boat. The lightning leader that strikes the top of the mast is conducted into the water through earthing conductors, with the earthing conductors hanging into the water. This is shown in Fig. 3.4. Mobile lightning protection is used with the metal mast, since mobile protection is much cheaper. The earthing connections are two copper braids let into the water to a depth of 1.5 m. The lower part of the mast has a ball pin for the ease of mounting the mobile lightning protection system.

When the boat is harbored, it is supplied from a power supply on the shore. The power supply system has to be protected from corrosion, and the shoreside earth of the power supply system must not be connected to the earthed part of the boat. An isolation transformer protects people inside the boat from shoreside power supply surges when the boat is harbored. Figure 3.5 shows the protection system employed for the internal electric system of a yacht.

In non-metal boats, additional lightning protection measures need to be provided since the body of the boat cannot be depended on to conduct lightning currents into the water. AN air-termination rod of 12 mm thickness is extended above the wooden mast with the air-termination protruding 300 mm above the mast. A copper conductor of 70 mm² cross-section connects the air-termination rod to earth plates in the outer area of the boat. In large boats, different earth plates are used for power supply grounding in the seawater and lightning earthing. The two plates for the power supply and lightning earthing are kept at a sufficient distance from each other to prevent lightning flashovers between the earth plates. Copper conductors of 16 mm² cross-section are used to connect the mast, shroud, stays, and chain plates to the earth plate. The conductor must be connected using screwing, riveting, and welding.

Equipotential bonding is accomplished connecting all metal parts and electronics to equipotential bonding and the earth-termination system of the power supply. The equipotential bonding helps protect people from touch voltage and sparking. During

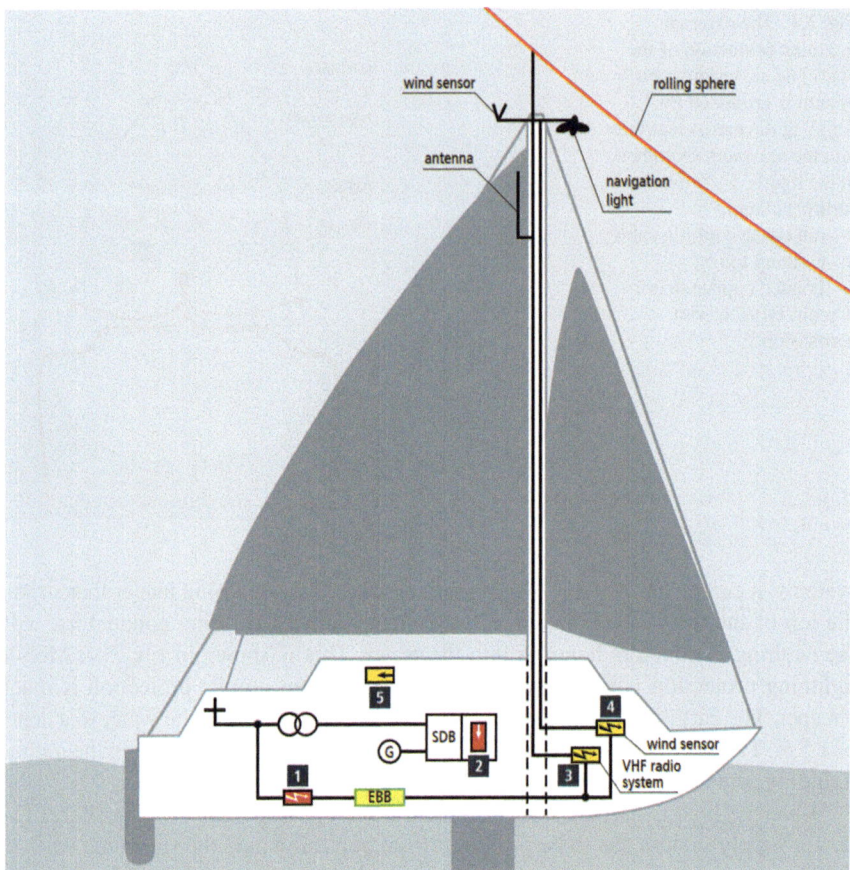

Fig. 3.5 The internal lightning protection of a boat. (Numbers in the figure: Protection for 1—power system; 2—sub-distribution board; 3—VHF radio systems; 4—wind sensor for the navigation system; 5—power supply systems for the navigation system.) (Credit: DEHN, with permission.)

thunderstorm, activity people should not stay outside. Potential differences are generated between the boat metallic parts, wet parts, and the wet skin, which may result in electric flashovers. People should not touch the metal objects, shrouds, or rods. It is important that the lightning protection system should be regularly checked.

3.4 Photovoltaic (PV) Systems

With the year-long availability of solar energy in most countries, it is cheap and clean to use solar radiation to generate electricity. Photovoltaic cells are placed on top of buildings, or larger ground solar farms are constructed by a large collection of solar

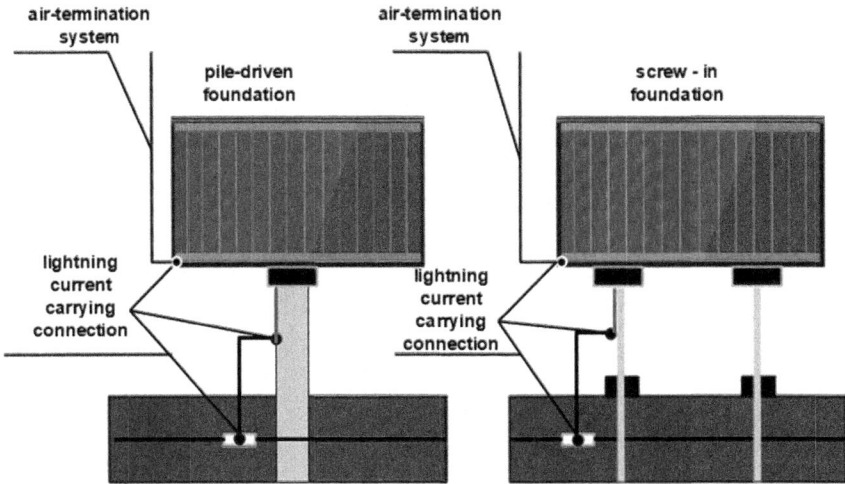

Fig. 3.6 The air-termination protection of photovoltaic (PV) panels (Credit: Adapted from DEHN)

panels connected together in a field. The PV panels are placed outside in the open to expose it to sunlight, either on top of tall buildings or out in open fields. In both cases, whether on top of buildings or in the open fields, they are exposed to lightning strikes. The electric cables connected to the PV panels are taken into the buildings to DC to AC converter electronics and other electronic apparatus and control systems. Figure 3.6 shows the air-termination protection used for the exposed PV panels for both screw-in and pile-driven foundations.

Lightning surges that travel from the PV panels to electronic inverters, electronic apparatuses, and instrument and electrical systems are all exposed to lightning voltage surges that come into the building from the outdoor PV panel system to the systems inside the building. It is important that large voltage loops be avoided to prevent lightning electromagnetic pulses inducing voltage surges in electric loops. The PV systems should be earthed at the point of their installation, with the PV panels mounted on metal mountings. Moreover, air-terminations should be placed adjacent to the PV systems at a distance of 1.08 m (why this odd-looking number?), each with a 10 mm diameter. If the air-terminations are placed too close to the PV panels, shadows cast by the air-termination conductors will reduce the efficiency of the PV electric output. Large PV electricity generators could have currents up to 1000 A and need lightning arresters that combine lightning arresters and surge arresters. (Are these not the same?) PV systems with micro converters should have additional protection systems. The PV systems placed outside in a field require SPDs with high current ratings. Fuses are used to protect the installation from reverse currents. Figure 3.7 shows the general arrangement used for the protection of incoming lines from a PV site on rooftop (Fig. 3.7a). The protection for an open field PV installation, outside the building, with the lightning protection systems, is as shown in Fig. 3.7b.

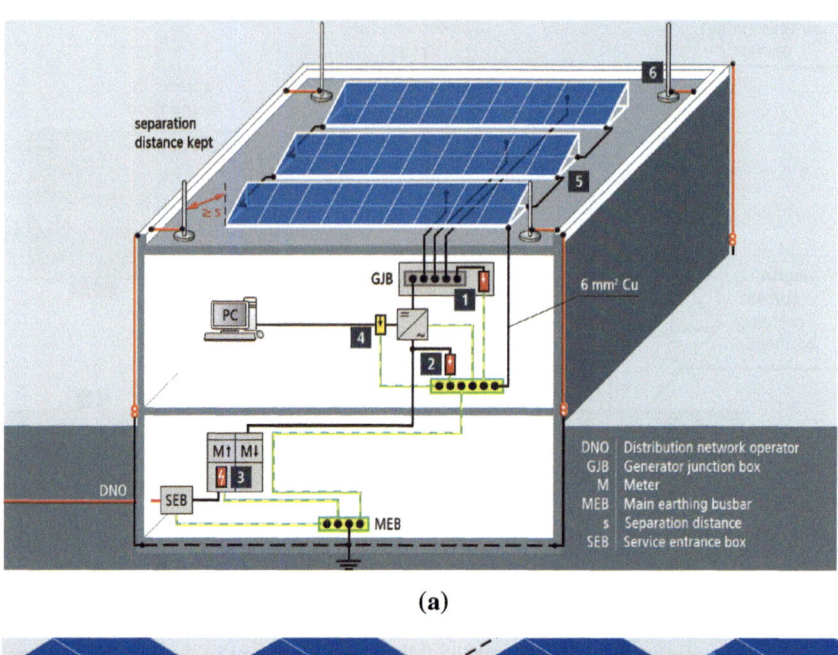

(a)

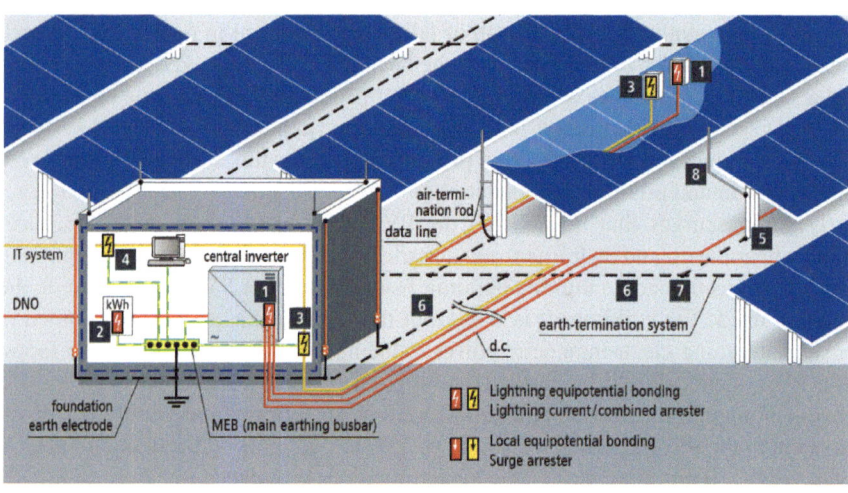

(b)

Fig. 3.7 a Lightning protection of a rooftop mounted photovoltaic system with external protection. (Numbers on the figure: SPDs for 1—the inverter DC input; 2—the inverter AC input; 3—low voltage system; 4—data interface; 5—functional equipotential bonding; 6—air-termination systems.) **b** Lightning protection of DC lines from the PV lines and data lines to the computer from a PV electric generator installed in a field (Numbers on the figure: Protection for 1—the DC input of the inverter; 2—AC side of the grid connection; 3—data interface; 4—remote maintenance ISDN or DSL, earth-termination systems; 5—equipotential bonding; 6—earthing conductor; 7—connection element; 8—air-termination system.) (Credit: DEHN, with permission)

3.5 Frequency Converter Protection

The frequency converter contains a rectifier, DC link, an inverter, and control electronics. These need to be protected from lightning surges. The rectifier generates a pulsating DC voltage. The DC link has residual current protective devices (RCDs), which may experience problems caused by short-time fault currents of the frequency converter, which are high enough to trip the RCD circuit breakers. These RCDs have a discharge capacity of 3 kA for an 8/20 µs waveform. The tripping current is about 30 mA. The inverter has a pulsed output voltage with the pulsed frequency depending on the pulse frequency of the control electronics of the pulsed width modulation (PWM) circuit. A peak pulse is generated with each peak voltage on the fundamental wave. Higher frequencies are used to get a better sinusoid, but these can cause electric field interference. Hence, the electric motor cable must be shielded and earthed at both ends; that is at the frequency converter end and at the motor end. The connections must be made with large area contacts. To reduce voltage drops, earth-terminations that are intermeshed need to be used in order to prevent equalizing currents through the shields. All communication and computer interfaces must be protected by surge protection devices. A suitable protection system for the frequency converter is shown in Fig. 3.8.

Over voltages from system operation such as switching are less severe. But switching surges are more frequent than lightning surges. Although switching surges have lower energy levels than lightning surges, they can still cause large damage. Radiated electromagnetic pulses from inductive and capacitive switching surges radiate in the range of 1–5 MHz. Repetitive starting of welding stations, high-pressure washers, heaters can cause damage as well as age electronic equipment. Filters are used to control high-frequency interference. Voltage rises should be kept below tolerated values of voltage surge protection. Over-voltages should be avoided between protection circuits and exposed metal conductor parts. Equipotential bonding systems

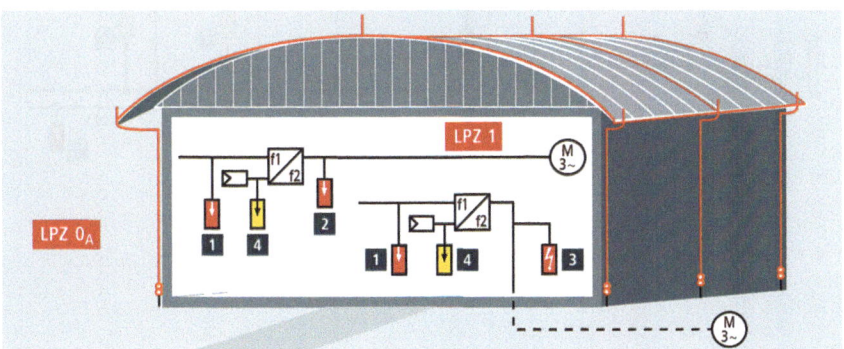

Fig. 3.8 Lightning and surge protection of a frequency converter. (Numbers on the figure: SPDs 1 to 4.) Credit: DEHN, used with permission

should be used. Induction effects due to electric fields should be minimized. Appropriate wiring method and correct location of equipment are essential. In protecting both networks and internal equipment, common fuses and circuit breakers are too slow in operation. Hence, voltage surge protectors should be used for active protection. The protection system must be designed and installed carefully and effectively, particularly including care over the position of the protectors and connections made. Passive protection, seals, equipotentiality, earthing system, and separation of circuits must all be done with proper care.

3.6 Networks and Interactive Services

A lightning and surge protection system for an M-bus (meter-bus) is shown in Fig. 3.9. It is necessary to ensure that antennas on rooftops are earthed. The equipotential bonding of LPS cable networks and shields must be established. The earth-terminations may use one of the following four techniques: a foundation earth electrode, two horizontal earth electrode strips of 2.5 m length and at 60° to each other, a

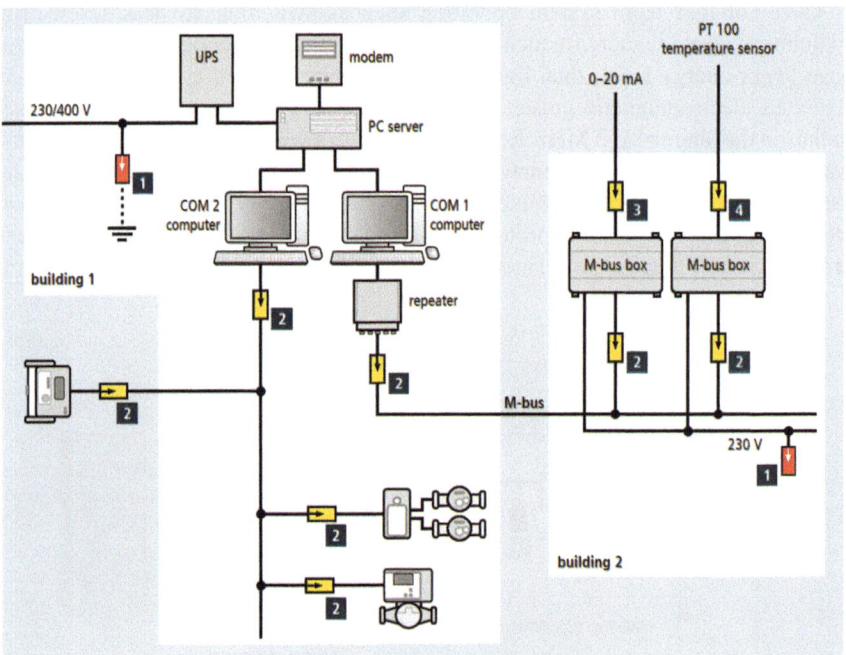

Fig. 3.9 Use of data surge protectors and power surge protectors for M-bus (meter-bus) system without external protection. (Numbers in figure: 1—SPDs for the voltage supply; 2 to 4—SPDs for signal interfaces.) Credit: DEHN, with permission

single 2.5 m long earth electrode or two vertical electrodes of 1.5 m depth and separated from each other by 3 m. The earth-termination must be connected to the main earthing busbar (MEB). The connections must use a 50 mm^2 copper conductor, or 90 mm^2 galvanized steel or a flat strip of 30×35 mm giving a cross-section of 105 mm^2. The cable network must be part of the equipotential bonding. Discharge currents are injected into the network from the device. All cables entering the building must be connected to protective equipment bonding. The lightning equipotential bonding inside the building is done through 4 mm^2 copper conductors. In order to avoid sparking, surge protection devices need to be installed between the inner and outer conductors. At the head end, surge protection is installed. Preventive measures are undertaken as well to avoid inductive coupling. Antennas that are placed under the roofs should be installed 2 m under the roof and must not protrude more than 1.5 m from the wall. These antennas must be positioned within the protective zone.

3.7 Wind Turbines

Since wind turbines are very much exposed to direct lightning strikes, they need to be well protected. In Europe, for instance, about ten direct lightning strikes to wind turbines may be expected each year.Both upward and downward flashes may be expected from wind turbines that are taller than 60 m. Lightning protection of both the rotor blades and the mechanical drive train must be provided. These must be tested for lightning current withstand. The wind turbine tower that is of tubular shape affords good Faraday cage protection for all installations inside the tower from direct lightning strikes. Concrete towers provide a galvanic cage. Connecting cables must have external shields that are able to carry lightning currents. External bonding must be done at both ends of the cables.

Magnetic shielding must be provided along the cable route. Installation of a metal braid on GRP-coated nacelles (that is, DEFINE), metal tower, metal switchgear cabinet, metal control cabinet, and current-carrying cable shields must be ensured. For the external wind turbine structure, lightning protecting air-termination and down conductors inside rotor blades are provided. A lightning protection system arrangement for the wind turbine is shown in Fig. 3.10. The tower foundation must be used as earth-termination with foundation earth electrode and ring earth electrode. A rolling sphere of 20 m radius should be used to determine the strike points that need to be covered by the LPS. Direct lightning strikes with currents up to 200 kA may be expected to the rotor blades, nacelle, sputter structure, rotor hub, or the tower. This current must be safely discharged to ground. A metallic receptor attached to the tip of GRP blade is used to protect the rotor blades. Down conductors from the receptor to the blade root, and down conductor inside the nacelle and tower must be connected to the ground. Meshed earthed terminations are used to distribute the lightning current around the earth at the base of the tower. For this, corrosion resistant ring earth electrodes are used, preventing step voltages.

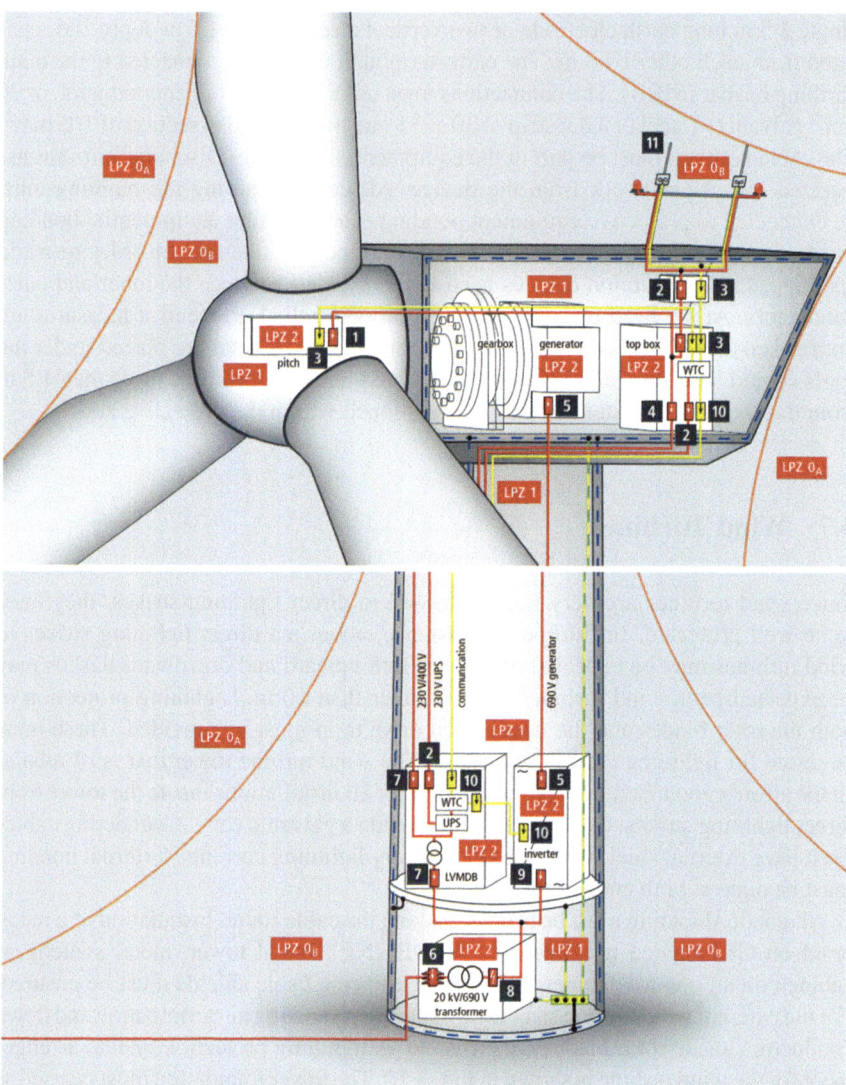

Fig. 3.10 Lightning protection of wind turbine. LPZ—lightning protection zones. Numbers in the figure: areas to be protected: 1—voltage supply of the hub and signal lines between the hub and nacelle; 2—aircraft warning light; 3—signal lines for the weather station and the control cabinet in the nacelle; 4—230/400 voltage supply; 5—protection of the generator; 6—protection of the transformer; 7—protection of tower base voltage supply; 8—main incoming supply; 9—protection of the inverter; 10—protection of the tower base signal lines; 11—protection of the nacelle superstructure.) Credit: DEHN, with permission

Fig. 3.11 Earthing system
for a wind turbine. Credit:
DEHN, with permission

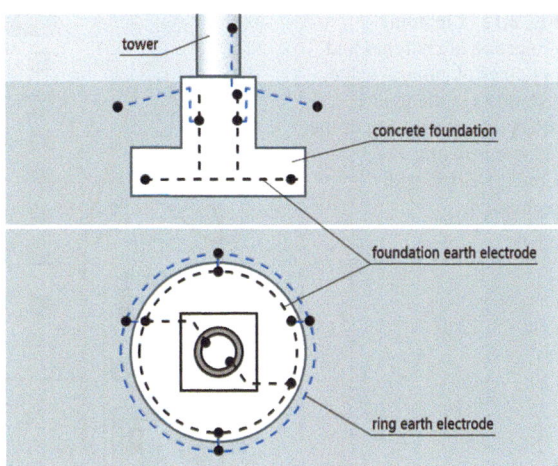

Earthing is an essential element of the LPS. For the wind turbine, the earthing arrangement is shown in Fig. 3.11. The earthing resistance should be 10 Ω. At the earth end, more than three conductors could be arranged in a crow's feet-like geometry and buried 0.5 m under the ground. An alternative is to have the earth rods in a triangular layout. The earthing down conductor should be interconnected with the bonding system of the main equipotential link (Fig. 3.11).

3.8   Historic Buildings

Places of worship like churches and buildings of high cultural value should be equipped with permanent and reliable lightning protection systems. Adequate separation must be maintained between lightning current carrying down conductors and the building electric wiring, especially in the steeple. In order to minimize induced voltage surges due to magnetic coupling, short conductors must be used, loops must be minimized, and where necessary surge protectors installed to protect against surges induced due to the indirect effects of lightning radiated electromagnetic pulses LEMP. Equipotential bonding must be done between all metallic pipes and electric cables. The typical lightning protection of a church is shown in Fig. 3.12.

Fig. 3.12 Lightning protection of churches and buildings of cultural value. (Numbers in the figure: SPDs for each section of the electric network, 1–7.) Credit: DEHN, with permission

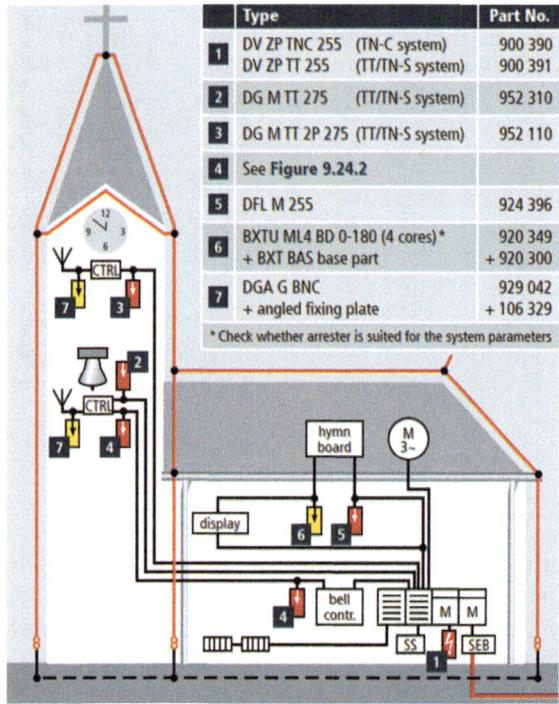

	Type		Part No.
1	DV ZP TNC 255	(TN-C system)	900 390
	DV ZP TT 255	(TT/TN-S system)	900 391
2	DG M TT 275	(TT/TN-S system)	952 310
3	DG M TT 2P 275	(TT/TN-S system)	952 110
4	See **Figure 9.24.2**		
5	DFL M 255		924 396
6	BXTU ML4 BD 0-180 (4 cores) *		920 349
	+ BXT BAS base part		+ 920 300
7	DGA G BNC		929 042
	+ angled fixing plate		+ 106 329

* Check whether arrester is suited for the system parameters

Bibliography

Cohen, R.L.: How to Protect your House and its Contents from Lightning: Surge Protection. IEEE Press (2005)

DEHN, Lightning Protection Guide, 3rd edn (2014)

DEHN, Dynamic Times: DEHN protects residential houses (2020)

Golde, R.H.: Lightning Vol 2: Engineering Applications. Academic Press (1977)

IEC Protection against Lightning Vol 3, Physical damage to Structures and Life Hazard

IEC Protection against Lightning Vol 4: Electrical and Electronic Systems within Structures

NPFA, Standard for the Installation of Lightning Protection System, NPFA-780 (2020)

Chapter 4
Practice of Lightning Protection: *Risk Assessment, External Protection, Internal Protection, Surge Protection, Air Termination, Down Conductor, Earthing, and Shielding*

Abstract In this chapter, we review the risk analysis that needs to be performed to determine the probability of lightning strike to any structure that is to be protected from lightning damage. Moreover, presented herein is how different aspects of the structure will change the lightning damage risk. Details of protection of electrical and electronic installations inside a structure are presented, including the choice and placement of surge protection devices to prevent lightning damage to internal electronic and electrical equipment and installations. The practice of zoning for the design of lightning protection is described. The chapter also reviews external protection, including an air-termination system, bonding, and down conductor to a suitable earthing termination system and the earthing system. The chapter closes with a discussion of shielding from lightning electromagnetic pulse (LEMP) radiation.

4.1 Introduction

The components of a lightning protection system (LPS) are the air-termination system, the down-conductor system, earth-termination system, separation distances between conductors and lightning equipotential bonding, and, lastly, the surge protection devices and systems for electrical and electronic installations and devices. The external LPS should intercept a direct lightning strike to the structure, line, or people through air terminations. The down conductor must safely conduct the lightning currents to the ground through good grounding arrangements. The lightning current dissipated into the ground must be safely distributed without causing any underground coupling to other electrical systems or spark overs. The internal LPS must prevent dangerous sparking inside the structure and damage to electrical and electronic equipment and systems. Good electrical bonding is essential, as well as maintaining safe distance of separation between the LPS and internal wiring and devices.

A lightning current often peaks at 20 kA or more. Suppose an external lightning protection system installed with the building structures, which may take 99.9% of the current and dissipate it in the ground. Then the electrical wiring of the building

Fig. 4.1 The Four parts contained in Lightning Protection Standards

(household) takes the remaining 0.1% of 100 kA of lightning current which is 100 A. A 100 A surge can burn the whole electrical wiring system. Hence, the need for having a proper lightning protection system.

A lightning protection guide and the associated design standards are government approved for each country. These usually comprise four parts, as shown in Fig. 4.1. Pt 1 describes the general principles, Pt 2 describes the risk management, and Pt 3 describes the physical damage to structure and life hazards and the final part deals with protection from damage to electrical and electronic systems within a structure. The IEC standards are listed in the reference at the end of this chapter.

General Principles contain information on how to design a Lightning Protection System (LPS) in accordance with the approved standards. Risk Management concentrates more on risk of loss of human life, loss of service to the public, loss of cultural heritage, and economic loss. Protection of structure describes four classes or protection levels of LPS and protection methods used in designing an LPS. The Electronics Systems Protection covers the protection of electrical and electronic systems within the structure. It introduces Lightning Protection Zones (LPZs) for the design and installation of Surge Protection Methods (SPM).

4.2   General Principles of Lightning Protection

The general principles identify lightning damage and loss from the following four main sources: flashes to the structure, flashes near the structure, flashes to the lines connected to the structure, and flashes near the lines connected to the structures. Each of these flashes may result in injury to living beings through an electric shock, physical damage due to lightning current, or damage due to an electromagnetic impulse LEMP. These can result in losses including loss of human life, service to the public, structure with an economic value, or a cultural heritage. In Fig. 4.2, the zoning of lightning protection systems and the use of surge arresters to protect internal electrical and electronic equipment from lightning surges due to the direct and indirect effects of lightning are shown. There are direct lightning strikes to the house structure, traveling voltage surges from exposed power, and telecommunication lines, LEMP radiated from nearby lightning flash and ground potential gradients producing circulating

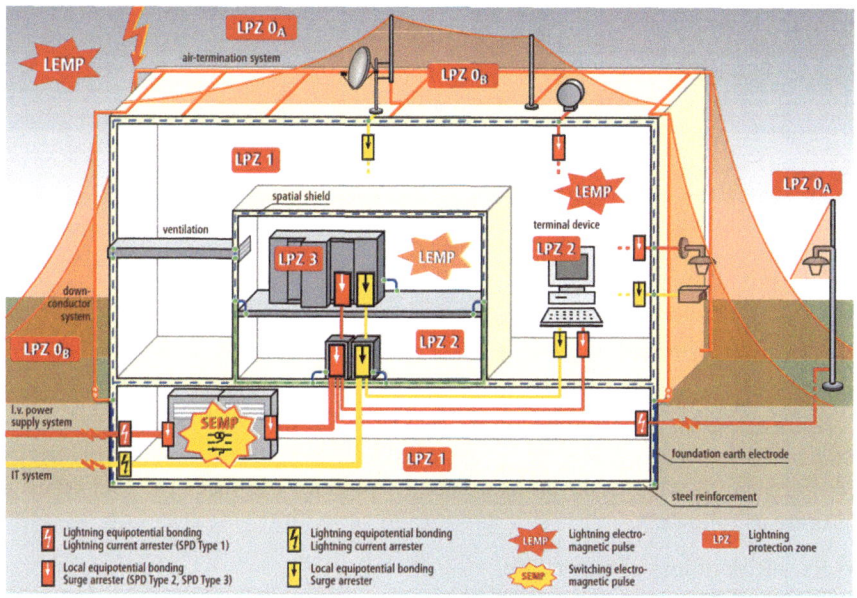

Fig. 4.2 Lighting protection zones (LPZ) and the placement of lightning surge arresters Credit: DEHN, with permission)

currents from the point at which the lightning struck and at which point the potential is temporarily raised to millions of volts.

The ideal protection scheme of any protection system would be to enclose the structure within an earthed and perfectly conducting metal shield box, known as a Faraday cage. Even though this will prevent any penetration of the lightning current and induced electromagnetic fields, it is neither practical nor cost-effective Hence under general principles, standards are set out to define a set of lightning current parameters and how they fall within limits defined as Lightning Protection Levels (LPL).

Four LPLs have been identified based on the maximum and minimum current parameters, as shown in Table 4.1.

The maximum values are used in the design of products such as lightning protection components and surge protection devices. The minimum current value has been used to derive the rolling sphere radius for each level. Depending on these LPLs the corresponding Lightning Protection System (LPS) is classed separately, as described in Table 4.2.

Table 4.1 Lightning current for each LPL based on a 10/350 μs waveform

LPL	I	II	III	IV
Maximum current (kA)	200	150	100	100
Minimum current (kA)	3	5	10	16

Table 4.2 Relationship between lightning protection level and class of LPS

LPL	LPS
I	I
II	II
III	III
IV	IV

4.3 Risk Management

4.3.1 Introduction

A risk assessment is performed to decide on the level of the Lightning Protection System (LPS) needed. As the first stage in risk assessment it is important to identify which of the four losses the structure and its contents can face. The four types of risks are as follows:

- R1—Risk of loss of human life.
- R2—Risk of loss of service to public.
- R3—Risk of loss of cultural heritage.
- R4—Risk of loss of an entity of economic value.

Each primary risk (RN) should be calculated according to data gathered about climate, population, etc. For each of the first three primary risks, a tolerable risk level (RT) can be set. If the actual risk (RN) is smaller or equal to tolerable risk (RT) then protection measures are not needed. If the actual risk is larger than the tolerable risk, then protection measures are needed and deployed.

This process is repeated using new values of the chosen protection measure until the actual risk is less than the tolerable risk. This process decides the choice of Lightning Protection Level (LPL) of a Lightning Protection System (LPS) and the Surge Protective Measures (SPMs) to be used to counter lightning surges and Lightning Electromagnetic Impulse (LEMP). Risk Management software is used to calculate risk of loss due to lightning strikes and transient overvoltages caused by lightning strikes. This software analyzes risks in a few minutes. If done manually it will take days to calculate risks.

4.3.2 Risk Assessment: Basics

Risk assessment is used by the owner of a building, a design engineer, the architect, and safety engineer to determine the risk of damage or injury due to a building being struck by lightning. The key concerns include issues such as continuity of electrical services, safety of a large crowd gathering, height of isolated buildings, density of lightning flash, and buildings with rare cultural heritage, and presence of inflammable and explosive materials. The factors taken into consideration are the type of construction, the environment in which the building stands, human occupancy, contents of the building, and consequences of the building being struck by lightning. The annual threat occurrence N_T is defined by the equation.

$$N_T = N_Y \times A_C \times C_L \times 10^{-6} \text{ events per year,} \tag{4.1}$$

where

N_Y = the yearly number of flashes per km^2.
A_C = the lightning collection area around the building.
C_L = the relative location factor.

If there are, for instance, 25 thunderstorms each year in the area where the building is located, then we have $N_Y = 4$ lightning flashes per km^2 each year. The value of N_Y ranges from 0.25 to 15. For a rectangular building structure, let the rectangular structure have the dimensions L (length) × W (width) and height H. For that rectangular building, the lightning collection area

$$A_C = LW + 6H(L + W) + 9\pi H^2. \tag{4.2}$$

The value of the location factor C_L varies. It is 0.25 when the structure is surrounded by tall structures or trees within a distance 3H, 0.5 for when the structure is surrounded by equal or lesser structures or trees within a distance of 3H from the building, 1 when the structure is isolated up to a distance of 3H, and 2 when the structure stands completely isolated or on top of a hill.

In order to determine the lightning risk to a structure, we compare the lightning frequency N_T to the risk of damage to a structure N_D. We determine the risk of damage to a structure N_D from

$$N_D = 0.0015/S \text{ events per year,} \tag{4.3}$$

where the structural coefficient

$$S = S_1 \times S_2 \times S_3 \times S_4. \tag{4.4}$$

The value of S_1 depends on the type of roof of the structure. S_1 is 0.5 to 2 for a metallic roof, 0.5 for metallic structure, 2 for a combustible structure, 1 to 2.5 when it is a nonmetallic roof, and 2 to 3 for a combustible roof. The value of S_3 depends on the contents inside the structure. The value of S_2 is 0.5 for low value, noncombustible material inside, 1 for standard value, noncombustible contents, 2 for high value, moderately combustible contents, 3 for exceptional value, flammable liquids, digital electronic equipment, and 4 for exceptional and irreplaceable contents such as cultural heritage artifacts. The value of S_3 is occupancy related. It is 0.5 for an unoccupied structure, 1 for normally occupied structure, and 3 for structure where it is difficult to evacuate the people inside. The value of S_4 depends on the requirement or not whether the continuous use of the building is required and whether or not there is any adverse impact on the environment. The value of S_4 is 1 if the continuity of the use of the facility is not required, 5 if the continuity of the use of the structure is required but environmental impact is negligible, 10 if there are serious consequences to the environment, as, for instance, a tree or a substation transformer struck by lightning and catching fire to set fire to other surrounding trees.

Once the measure of risk of damage N_D is determined, it needs to be compared with the lightning frequency N_T. If N_T is less than or equal to N_D then installing lightning protection is optional. However, if N_T is greater than N_D, then the design and installation of lightning protection is recommended.

Lightning flash counters are used to measure lightning flash densities. The CIGRE lightning flash counter is a standardized device, which, by registering the number of lightning flashes within a specified area, enables the density of lightning flashes to ground, per unit area, and per unit time, to be estimated. Long-term average values for the number of lightning flashes per year were obtained for a Pacific Island from a lightning flash counter network. These records are used with local storm observations to obtain the ground flash density N_Y. The average number of thunder days (T) and lightning days (L) are estimated. The values are compared with data from the literature of other regions used in transmission system design. Average values as high as about 20 lightning faults per 100 km per year were measured in the Pacific Region. It is apparent that designs from temperate countries would require adaptation for the higher incidence of lightning in this country, although the lower values of N_Y will compensate to some extent for the higher values of number of lightning flashes.

4.3.3 Advanced Risk Assessment

Let us define some variables
  S_1 = Coefficient for direct lightning strike.
  S_2 = Coefficient for lightning strike near structure.
  S_3 = Coefficient for direct lightning strike to an incoming line.
  S_4 = Coefficient for lightning strike near an incoming line.
  L_1 = Coefficient for loss of human life.
  L_2 = Coefficient for loss of service to public.
  L_3 = Coefficient for loss of cultural heritage.
  L_4 = Coefficient for loss of economic value.
  The risk component

$$R_x = N_x \cdot P_x \cdot L_x, \tag{4.5}$$

where

N_x = Number of dangerous events.

P_x = Probability of damage.

L_x = Loss factor, quantitative evaluation of damage.

The risk component R_x should be less than R_t, the tolerable event.

Collection area of line, $A_L = 40\,L_l$, where L_l is about 1000.

The ground flash density is the lightning strikes per km^2 per year.

$N_G = 0.1\,T_D$, where T_D is thunder days per year.

Direct strikes $N_D = N_G.\,A_D.\,C_B.\,10^{-6}$, where C_D = The location factor, including surrounding

$$A_D = \ LW \ + 2(3H).\,(L+W) + \pi\,(3H)^2. \tag{4.6}$$

For nearby lightning

$$N_M = \ N_G.A_M.10^{-6}, \tag{4.7}$$

with A_M (drawing line) at 500 m.

Lightning strikes lines at an annual rate of

$$N_L = \ N_G.A_l.\ C_I.C_E.C_T.10^{-6}, \tag{4.8}$$

where N_L is the annual number of surges in line section with maximum 1 kV surges. C_I = insulation factor, C_T = line type factor (building density). $A_L = 40\,L_L$, with $L_L = 1000$.

$$N_I = \ N_G.A_I.C_I.C_E.C_T.10^{-6} \tag{4.9}$$

$$A_I = 4000\,L_L. \tag{4.10}$$

Let us work out the probabilities of damage. For direct lightning strikes, let P_A = The probability of physical damage (for example, from fire, explosion, mechanical, chemical reactions). Let P_C = the probability of failure of electric or electronic systems due to a direct lightning strike to a building or structure. If lightning strikes the ground near the structure, let P_M = the probability of failure of electrical/electronic systems. In the case of a direct strike to an incoming line (bringing electric power into a building), let P_U = the probability of injury, P_r = the probability of physical damage, and P_W = the probability of failure of electrical/electronic systems.

The probability of damage in case of a direct lightning strike is given by

$$P_A = \ P_{TA}.\ P_B. \tag{4.11}$$

P_{TA}, the probability that the lightning strike will cause shock, is 1 for no protection, 10^{-1} for warning notice, 10^{-2} electric insulation with 3 mm XLPE (cross-linked polyethylene) down conductors, 10^{-2} effective potential control in the ground, and 0 for physical instructions or building framework as down conductor. P_B, the probability of damage to a physical structure that is hit is 1 for no coordinated SPD, with air termination it is 0.05, with lightning protection system it is 0.1 for class III to IV systems, 0.1 for class II, 0,02 for class I system with continuous metal down conductor it is 0.01. In the case where coordinated SPDs (surge protective devices) are installed, the probability of damage caused by direct lightning is

$$P_C = P_{SPD}.C_{LD}, \qquad (4.12)$$

which is the probability that lightning strike will cause damage to the electrical or electronic systems.

Let LPL be the Lightning Protection Level, and then the probability of damage P_{SPD}, with coordination with lightning protection level (LPL) is 1 with no coordinated SPD, 0,05 (with class III to IV LPS), 0.02 (with class II LPS), and 0.01 with class I LPS. C_{LD} is either 1 or 0. For instance, it is 1 for shielded buried cable as the external line and 0 when there is no external line. C_{LI} is 1 for an unshielded external line, 0.2 for a power line for multi-grounded shield line (a shield line grounded at multiple points), and 0 for a multi-grounded shielded underground cable as the external line.

Probabilities of damage in case of nearby lightning strike is

$$P_M = P_{SPD}. P_{MS}, \qquad (4.13)$$

where

$$P_{MS} = (K_{s1}.K_{s2}.K_{s3}.K_{s4})^2 \qquad (4.14)$$

and K_{s1} is the shielding effectiveness of structure, K_{S2} is the shielding effectiveness of internal shields of structure at boundaries, K_{S3} is the shield effectiveness for shields of internal cables, and K_{S4} is the rated impulse voltage withstand voltage of protected system.

The probability of damage due to direct lightning strike to a power line is given by

$$P_U = P_{TU}. P_{EB}. P_{LD}, P_{LD}, \qquad (4.15)$$

where P_{TU} is the probability of touch voltage protection warning notice (1 or 0.1, with physical restrictions from touching, 0), P_{EB} is the lightning equipotential bonding induced protection (1 with no SPD, 0.05 for LPS II, IV; 0.02 for LPS II; 0.01 with LPS 0.01; with surge protection devices it is 0.005 to 0.001). P_{LD} is the probability that the internal system will fail, and coefficient C_{LD} is considering earthing, shielding and insulation conditions of the line. We have for the probability of physical damage

$$P_V = P_{EB} \cdot P_{LD} \cdot C_{LD}. \tag{4.16}$$

The probability of internal system failure is

$$P_W = P_{SPD} \cdot P_{LD} \cdot C_{LD}. \tag{4.17}$$

And the probability of damage in case of indirect lightning strikes to line is

$$P_Z = P_{SPD} \cdot P_{LI} \cdot C_{LI}. \tag{4.18}$$

P_{LD} is 1 for impulse withstand voltages of 1 kV, 5 kV, or 6 kV with shielded overhead line without bonding of shield to the same equipment bonding bar; it is 0.6 (for 1 kV withstand voltage), 0.2 (for 5 kV withstand voltage), or 0.02 (for 6 kV withstand voltage) for a shielded overhead line or buried bonded cable bonded to the same equipotential bonding bar as equipment with shield resistance less than 1 Ω/km. It is closer to 1 if the shield resistance is in the range of 5 to 20 Ω/km. The values of P_{LI} are 1 for 1 kV withstand voltage, 0.3 for 2 kV withstand voltage, or 0.1 for 6 kV withstand voltage for power lines. It is much lower, that is, 0, 2 (2.5 kV) and 0.04 (6 kV) for telecommunication lines. The coefficient C_{LI}, depending on the type of external line, takes values of 1, 0.2, or 0. The risk factor varies according to whether it is agricultural land (10^{-2}), marble or ceramic (10^{-2}), gravel (10^{-4}) or asphalt or wood (10^{-5}).

The tolerable risk P_T/year is dependent on what is being considered, whether loss of human life or injury (tolerable value is 10^{-5}), service to public (10^{-3}), cultural heritage (10^{-4}), or economic values (10^{-3}).

4.4 Inspection of Lightning Protection System

The Lightning Protection System (LPS) should be regularly inspected over the entire phase of its design, installation, acceptance, and maintenance stages. Both measurements and visual inspection need to be made. The LPS used in critical systems and situations should be completely inspected annually. Class I and Class II LPSs should be annually inspected and a complete inspection made every 2 years. Class II and IV LPSs should be visually inspected once a year and a complete inspection performed every 4 years. Reports should be prepared giving information on structures, the LPS, fundamental inspection activities, the results of inspection, and the inspector. Maintenance should be carried out to prevent loss of LPS quality and the effects of direct lightning strikes and any damage to the LPS. Inspection of all conductors and components of the LPS system should be performed. Continuity of installation should be tested. The earth resistance at earth terminations should be measured. The

SPDs should be visually inspected. Fixings of conductors and components should be tested. It should be ensured that the effectiveness of the LPS remains unchanged.

4.5 Internal Lightning Protection

4.5.1 Surge Protection Measures

By careful design of the LPS, earth bonding of metallic services such as water and gas and cabling routes, structures, and screened rooms, the internal electrical and electronic systems can be protected from lightning surges. Proper installation of surge protective devices will ensure the proper operation of equipment and will protect them from damage. These are known as surge protection measures. Initially, according to the standards used, it needs to be determined whether structural and/or LEMP protection is required. Once the need is decided upon, the proper selection and location of Surge Protection Devices (SPDs) need to be done.

- Coordinated SPDs
  Coordinated SPDs have to work together to protect equipment. The lightning current SPD at the entrance of the service should handle most of the surge energy, sufficiently relieving the downstream overvoltage SPDs to control the overvoltage. The overvoltage SPDs as well as equipment to be protected can be damaged due to poor coordination.
- Enhanced SPDs
  Standard SPDs may only protect against common mode surges (between live conductors and earth), providing effective protection against outright damage but not against downtime due to system disruption. Enhanced SPDs provide lower let through voltage protection against surges in both common mode and differential mode (between live conductors). They also provide additional protection over bonding and shielding measures.

The Surge Protection Device is connected in parallel on the power supply circuit of the loads that it is to protect, as shown in Fig. 4.3.

According to the characteristics of the current wave or voltage wave, SPDs can be divided into three types, namely, Type1 SPD, Type 2 SPD, and Type 3 SPD.

- Type 1 SPD
  The Type 1 SPD is characterized by a 10/350 μs current wave. This type is recommended for the specific case of service sector and industrial buildings to be protected by a lightning protection system. It protects electrical installations against direct lightning strokes. It can discharge the back current from lightning spreading from the earth conductor to the network conductors.

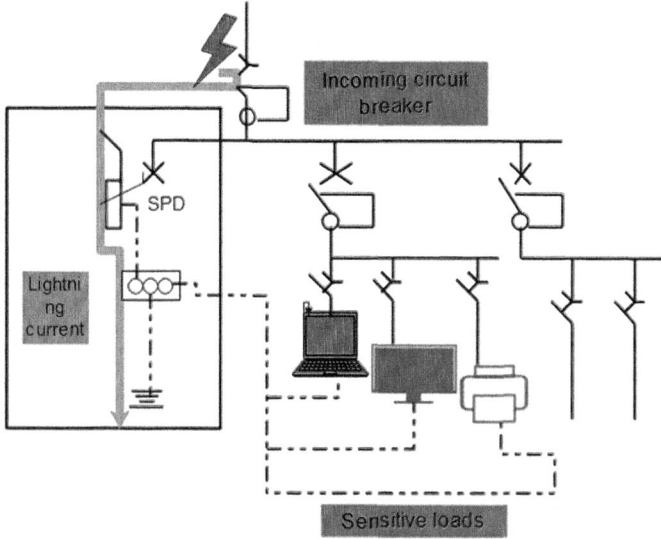

Fig. 4.3 SPD used in protection system. Adapted from DEHN

- The Type 2 SPD

  The Type 2 SPD is characterized by a 8/20 μs current wave. It is the main protection system for all low-voltage electrical installations. They are installed in each electrical switch board; it prevents spread of overvoltages in the electrical installations and protects the loads.

- The Type 3 SPD

  The Type 3 SPD is characterized by a combination of voltage waves (1.2/50 μs) and current waves (8/20 μs). These SPDs have low discharge capacity. They are installed as support to the Type 2 SPD and in the vicinity of sensitive loads.

Summary

1. A lightning protection system consists of an external and an internal lightning protection system. According to the International Electrotechnical Commission (IEC), the components that make up the lightning protection system are an air-termination system, a down-conductor system, separation distances, and lightning equipotential bonding. The separation distance between an external lightning protection system and metal structures is important to minimizing the probability of partial lightning current from entering the internal structures. The main purpose of a lightning protection system is to protect buildings from fire and persons from injury or death in the event of overcurrent due to lightning. The function of an external lightning protection system is to intercept lightning strikes via an air-termination system, to safely conduct lightning current to ground via a down-conductor system and distribute lightning current in the ground via an earth-termination system. On the other hand, the main function

of internal lightning protection system is to prevent dangerous sparking inside buildings by using equipotential bonding or maintaining a certain separation distance between components of the lightning protection system and conductive elements inside the structure. Lightning equipotential bonding reduces potential differences, between internal devices and between conducting parts, by connecting all conductive parts directly by conductors or through surge protective devices.

2. The selection for the appropriate design of the protection system and measures to be taken are calculated during risk assessment where the source and type of damage are evaluated to predict the severity and type of damage due to lightning. Four major sources of damage are flashes to the structure (S1), flashes near the structure (S2), flashes to a service line (S3), and flashes near a service line (S4). The types of damage that may be inflicted are injury due to step and touch voltages (D1), physical damage such as fire and explosion, mechanical damage or chemical release due to lightning current effects including sparking (D2), and failure of internal systems due to Lighting Electromagnetic Impulse (LEMP) (D3). Lightning Protection Standards define a set of parameters where protection measures should be taken to reduce damage due to a lightning strike. There are four classes of lightning protection levels (LPL) which are labeled as I, II, III, and IV. Each of these is determined using a set of construction rules. The LPLs are directly proportional to the class of Lightning Protection System (LPS). The higher the LPL, the higher the class of LPS required. LPL I has a maximum current of 200 kA and minimum current of 3 kA; LPL II has maximum current of 150 kA and minimum current of 5 kA; LPL III has maximum current of 100 kA and minimum current of 10 kA; LPL IV has a maximum current of 100 kA and minimum current of 16 kA. The above parameters are based on a 10/350 μs waveform. Other than lightning protection levels, lightning protection zones (LPZ) were introduced to determine protection measures to prevent LEMP in a building. LPZs are divided into external and internal zones. External zones LPZ 0_A are zones with risk of a direct lightning strike and LPZ 0_B are zones with risk of partial lightning current. LPZ 1, LPZ 2, and LPZ 3 are internal zones where the higher the number of the zone, the lower is the risk of electromagnetic effects.

4.5.2   Lightning Protection Zones

4.5.2.1   General

Electrical and electronic systems can be damaged if overvoltages occur near the sensitive parts of the components. Usually, overvoltage happens in areas of residential and functional buildings due to lightning discharge. Hence, protection is essential to prevent the owner wasting lots of money by repairing or replacing the damaged

Fig. 4.4 Lightning protection zones. Adapted from DEHN

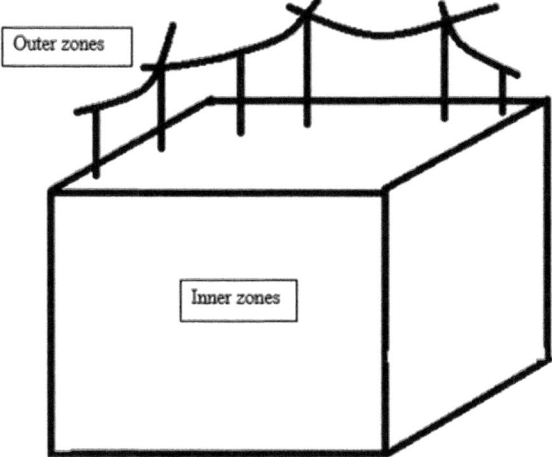

components. Nowadays, the operator also sets very high demands regarding the availability and reliability of these systems. In lightning protection zone (LPZ) principles, the inner zones and outer zones are identified as shown in Fig. 4.4.

Before designing the protection systems, information on the computer system in use, the electrical installation, and the earth-termination system must be collected and collectively evaluated for a comprehensive overall protection system. Protection of electrical and electronic devices can be classified into several parts because of the threat of direct lightning strikes and lightning electromagnetic field. The principle of lightning protection zones (LPZ) is implemented to get rid of incoming surges resulting from lightning electromagnetic pulses (LEMP). According to the LPZ principle, the building structure that needs to be protected must be divided into outer zones and inner zones based on the risk level. By using this flexible concept, the protection of the structure is maximized and the minimum damage and loss of service costs can be achieved. Regarding the inner zones and outer zones, these are classified as follows.

4.5.2.2 Outer Zones

LPZ0—This refers to zones where the threat results from un-attenuated lightning electromagnetic field and where the internal systems may be subjected to the full or partial lightning current. LPZ 0 is subdivided into $LPZ0_A$ and $LPZ0_B$.

LPZ0A—This is a zone where the threat is due to direct lightning strikes and the full lightning electromagnetic field. The internal systems may be subjected to the full lightning current.

LPZ0B—This is the zone that is protected from direct lightning strikes, but the threat is due to the full lightning electromagnetic field. The internal systems may be subjected to partial lightning currents.

4.5.2.3   Inner Zones

LPZ1—This is a zone where the impulse currents are limited by current distribution and isolating interfaces or by Surge Protection Devices (SPDs) at the zone boundaries. Spatial shielding may attenuate the lightning electromagnetic field.

LPZ2—This zone is where the impulse currents are limited by current distribution and isolating interfaces or by additional Surge Protection Devices (SPDs) at the zone boundaries. Additional spatial shielding may be used to further attenuate the lightning electromagnetic field.

The dielectric strength of the electrical and electronic systems to be protected plays an important role in determining the requirements for the inner zones. Equipotential bonding needs to be established at the boundary of each inner zone for all incoming metal parts and supply lines either directly or by means of suitable SPDs. These zone boundaries are formed according to the shielding measures used. Before commencing on the design of the protection systems, information such as the computer system, the electrical installation, and the earth-termination system must be collected and centrally evaluated for a comprehensive overall protection system.

4.5.3   SPM Management

The owner and operator usually emphasize optimum protection of the electronic systems with a minimum of expenses. However, this can be only achieved if the electronic devices and systems are designed together with the building and before its construction undergoes changes. The costs of the LEMP protection measures freshly installed for an existing, old structure are higher than for new structures. By choosing the LPZs appropriately when existing installations are used or upgraded, much cost can be reduced.

- The SPM should be planned by a lightning protection specialist having sound knowledge of Electromagnetic Compatibility (EMC).
- There ought to be close coordination between building and LEMP experts (e.g. civil and electrical engineers).

The SPM management plan is shown in Table 4.3.

- The final risk must be assessed and it must be proven that the residual risk is less than the tolerable risk.

The interconnection of all metal components by equipotential bonding inside the structure forms a low-inductance equipotential bonding network which is a three-dimensional meshed network. An ideal equipotential network is around 5 m × 5 m in size as it is able to reduce the electromagnetic field in an LPZ by a factor of 2 or by 6 dB. Electronic devices and systems are integrated in the equipotential bonding network by short connections. Hence, a sufficient number of equipotential bonding bars must be allocated in the structure as all the bars must be connected to the

Table 4.3 SPM management plan for new buildings and for comprehensive changes to the construction or use of buildings according to IEC 62305-4 (EN62305-4)

Step	Aim	Action to be taken by (if relevant)
Initial risk analysis	Assess the necessity for LEMP protection measures. If necessary, an appropriate LEMP Protection Measures System (LMPS) must be chosen based on a risk assessment	• Lightning protection specialist • Owner
Final risk analysis	The cost-benefit ratio of the protection measures chosen should be optimized again by a risk assessment. The following must be determined: • Lightning protection level (LPL) and lightning parameters • LPZs and their boundaries	• Lightning protection specialist • Owner
Design of the LEMP Protection Measures System (LPMS)	Definition of the LMPS: • Spatial shielding measures • Equipotential bonding networks • Earth-termination systems • Conductor routing and shielding • Shielding of incoming supply lines • SPD system	• Lightning protection specialist • Owner • Architect • Designer of internal systems • Designer of relevant installations
Design of the LPMS	• General drawings and descriptions • Preparation of tender lists • Detailed drawings and schedules for installation	• Engineering office or equivalent
Installation and inspection of the LMPS	• Quality of the installation • Documentation • Possible revision of the detailed drawings	• Lightning protection specialist • Installer of the LMPS • Engineering office • Supervisor
Acceptance of the LMPS	• Inspection and documentation of the system	• Independent lightning protection expert • Supervisor
Periodic inspections	• Ensuring an appropriate LMPS	• Lightning protection specialist • Supervisor

equipotential bonding network. Protective conductors and cable shields of data lines are integrated in the equipotential bonding according to specifications of the manufacturer, in a meshed or star configuration. It is important that all metal components of the electronic system must be sufficiently insulated against the equipotential bonding network when using a star configuration. Due to this matter, star configurations are typically limited to small applications or locally confined systems. The star configuration is connected to the equipotential bonding network at a single earthing reference point (ERP) and all lines must enter the structure at a single point. On the other hand, metal components of electronic systems do not need to be insulated against the equipotential network when using the meshed configuration. The difference between star and mesh configurations is that all components in the mesh configuration are integrated in the equipotential bonding network at as many equipotential bonding points as possible. As a result of that, the meshed configuration is extensive and is an open system with many lines between individual devices. An added advantage of the meshed configuration is that the system can enter the structure at different points unlike the star configuration where the system is only allowed to enter at a single point. For more complex systems, the star and meshed configurations are combined to benefit from the advantages of both systems.

4.6   Equipotential Bonding for Metal Installations

4.6.1   Prologue

Equipotential bonding provides protection by eliminating the potential difference between different devices or systems and thus prevents circulating lightning currents. There are two types of equipotential bonding, which are protective equipotential bonding and supplementary protective equipotential bonding. All buildings must be equipped with a protective equipotential bonding system as specified by the standards. Supplementary protective equipotential bonding is used when the conditions for disconnecting the supply cannot be met or for special installations or locations.

4.6.2   Equipotential Bonding for Metal Installations at the Boundary of $LPZ0_A$ and $LPZ1$

All metallic electrical lines or systems passing through the boundary between the electromagnetic compatibility (EMC) lightning protection zones must be integrated in the equipotential system. Measures must be taken to reduce the radiated electromagnetic field at this boundary. Lightning equipotential bonding must be implemented along with protective equipotential bonding for electrical and electronic lines at this boundary. A good practice is to implement the equipotential bonding as close as possible to the points where lines and metal installations enter the building and the lines should be as short as possible to lower line impedance.

4.6.3 Equipotential Bonding for Metal Installations at Boundary of LPZ 1 and LPZ 2

This is similar to the equipotential bonding at boundary of $LPZ0_A$ and LPZ1, where the equipotential bonding system must be installed as near as possible to where the lines and metal installations enter the zone of transition between LPZ1 and LPZ2. Applying ring equipotential bonding allows low-impedance connection of the system in this zone.

4.6.4 Protective Equipotential Bonding

There are a few individual conductive parts that must be directly connected in the protective equipotential bonding system such as the protective bonding conductor, metal foundation or lightning protection earth electrode, central heating system, metal water supply pipe, any conductive parts of a building structure such as lift rails, steel frame, ventilation or air conditioning ducts, metal drain pipe, internal gas pipe, earthing conductor for antennas and telecommunication systems, protective conductor of electrical installations, metal shields of electrical and electronic conductors, metal sheaths of power cables up to 1000 V, and earth-termination systems of power installations exceeding 1 kV.

Extraneous conductive parts are conductive parts that do not form part of an electrical installation, but are capable of introducing a potential called the earth potential. Conductive floors and walls are also classified as extraneous conductive parts as long as they are capable of introducing an electric potential. There are parts that must be connected indirectly via isolating spark gaps to the protective equipotential bonding system such as installations with cathodic corrosion protection and stray current protection measures, earth-termination systems of power installations exceeding 1 kV, traction system earth in case of AC or DC railways, and signal earth for laboratories if it is separated from the protective conductors.

Lightning equipotential bonding is an extension of protective equipotential bonding. Both equipotential bonding systems have to be connected with the main earthing busbar of the earth-termination system, as shown in Fig. 4.5. Lightning equipotential bonding provides safe integration of conductors entering the equipotential bonding system in the event of lightning strikes at the protection system or the entering conductors. Figure 4.5 illustrates a basic diagram of lightning equipotential bonding.

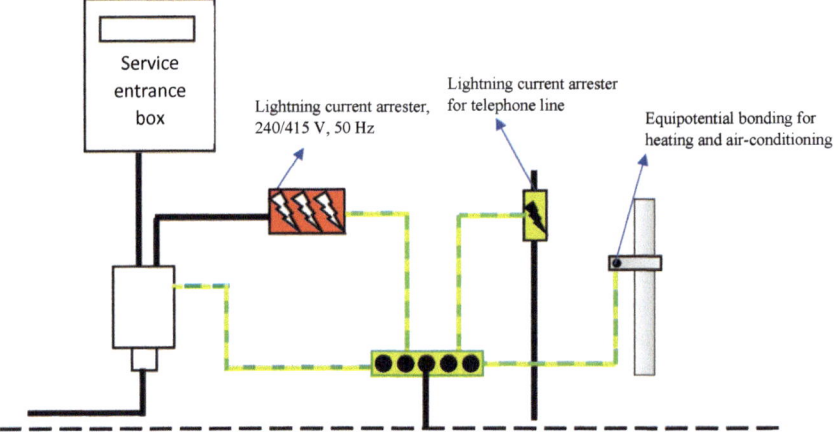

Foundation earth electrode

Fig. 4.5 Lightning equipotential bonding (Credit: DEHN. With permission)

4.6.5 **Earth-Termination System for Equipotential Bonding**

Low-voltage electrical installations require certain low earth resistances. The foundation earth electrode is capable of providing good earth resistances if installed effectively and it is a complement of the equipotential bonding system by improving earthing effectiveness.

4.6.6 **Protective Bonding Conductors**

Equipotential bonding conductors are labeled as protective conductors which are green or yellow as long as they are for protective purposes. Equipotential bonding conductors do not carry operating currents from the main supply and can either be bare or insulated. The minimum cross section of protective bonding conductors which are to be connected to the main earthing busbar is 6 mm^2 for copper, 16 mm^2 for aluminum, and 50 mm^2 for steel. The minimum cross section for earthing conductors of antennas is 16 mm^2 for copper, 25 mm^2 for aluminum, and 50 mm^2 for steel. The larger cross-sectional area required is due to higher frequency antenna currents needing a larger surface to flow in order to reduce the conductor resistance.

4.6.7 Equipotential Bonding Bars

Equipotential bonding bars must be able to clamp all connecting cables and cross sections so that they will have high contact stability and must be able to safely carry currents in addition to be sufficiently corrosion resistance. In order to get high contact stability, the equipotential bonding connections must be able to provide good and permanent contact.

4.6.8 Integrating Pipes in Equipotential Bonding System

Earthing pipe clams corresponding to the diameters of the pipes are used to integrate pipes in the equipotential bonding system. Typically, stainless steel earthing pipe clamps with tensioning straps are used because it offers greater flexibility to clamp pipes of different materials and also allows through-wiring.

4.6.9 Testing and Monitoring Equipotential Bonding System

A resistance value of less than 1 Ω is sufficient for equipotential bonding connections. The test equipment with 200 mA test current must be used in a continuity test.

4.6.10 Supplementary Protective Equipotential Bonding

If conditions for disconnection cannot be met, supplementary protective equipotential bonding is required to interconnect all accessible parts including stationary equipment and to connect all individual conductive parts to keep touch voltage to its minimum. For installations of Information Technology (IT) systems with insulation monitoring, supplementary protective equipotential bonding must be used. This type of bonding is also used if environmental conditions in special installations present a risk. These include areas containing bath or shower; basins of swimming pools; and other water basins, agriculture, and horticulture premises. The minimum cross sections required for supplementary protective bonding copper conductor are 2.5 mm^2 for protected installations and 4 mm^2 for unprotected installations. As compared to protective equipotential bonding, the cross section of conductors in supplementary protective equipotential bonding is smaller as it can be limited to a particular area.

4.6.11   Minimum Cross Section for Equipotential Bonding Conductors

Cross sections of conductors for lightning protection must be designed so that it is capable of handling high stress since it is carrying lightning currents. Hence, it must have larger cross sections. All classes of Lightning Protection System (LPS) have the following minimum cross sections depending on type of material: 16 mm² for copper, 25 mm² for aluminum, and 50 mm² for steel. The minimum cross sections for conductors that connect internal metal installations to the equipotential bonding bar can be smaller since only partial lightning currents of reduced amplitudes flow through these conductors. The cross section for copper is 6 mm², aluminum is 10 mm², and steel is 16 mm².

4.6.12   Equipotential Bonding for Power Supply Systems

Feeder cables of low-voltage installations need to be integrated in the equipotential bonding system. A unique feature of this system is that connections to the equipotential system is only possible via sufficient Surge Protection Devices (SPDs). As done for equipotential bonding for other metal installations, the bonding for feeder cables of low-voltage installations should be fixed directly at the entry point.

4.6.13   Equipotential Bonding for Power Supply Systems at the Boundary of LPZ0$_A$ and LPZ1

All electrical power and data lines entering the building transitions from LPZ0$_A$ to LPZ1 must be included in the equipotential bonding system. The boundary of LPZ0$_A$/LPZ1 as shown in Fig. 4.6 is assumed to be the boundary of the building if the installations are supplied by low-voltage systems.

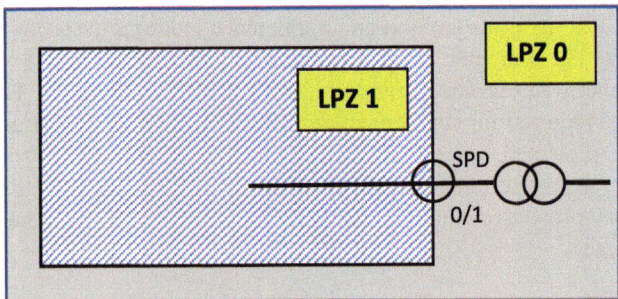

Fig. 4.6 Boundary of LPZ 0$_A$/LPZ 1 with transformer outside the structure. Adapted from DEHN

However, LPZ0$_A$ is extended up to the secondary side of the transformer for installations which are directly supplied by the medium-voltage system as shown in Fig. 4.7. The lightning equipotential bonding is installed at the 240/415 V side of the transformer. It is advisable to install surge protective devices on the high-voltage side of the transformer to prevent damages. In addition, it is recommended that additional shielding measures at the incoming medium-voltage line be used to prevent lightning currents in LPZ0 from flowing into the system in LPZ1. Implementation of lightning equipotential bonding for all incoming metal, electrical power, and data lines at a central point prevents equalizing currents between various equipotential bonding points. Low-frequency equalizing currents are dangerous because they can be superimposed on power frequency current flow. This may cause cable fires in extreme cases. If such an arrangement is not possible, a ring equipotential bonding bar should be used instead. The discharge capacity of the lightning current arrester used in this zone is classified as a Type 1 surge protective device and must be capable of handling the stress from the lightning current. The level of lightning protection chosen is based on risk assessment. If risk assessment or information on lightning current distribution at the transition from LPZ 0$_A$ to LPZ 1 is unavailable, the class of lightning protection system with the highest requirements is used (Level 1). The minimum lightning-carrying capacity of Type 1 surge protective device is 75 kA/m.

The lightning current distribution varies depending on the installation conditions. If there are several parallel load systems, the stress on the building which is hit by lightning will increase. The resulting earth resistance of a low-voltage system consisting of several connected buildings and a transformer is lower compared to the single earth resistance of a single building hit by lightning. Other than that, the current will not be evenly distributed between the low-voltage installation and the earth-termination system. Hence, the Type 1 surge protective devices in the low-voltage system discharge a significantly larger amount of current compared to the earth-termination system.

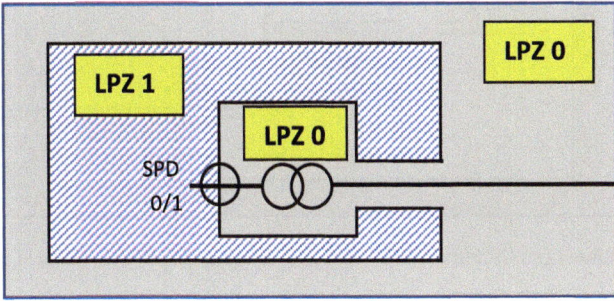

Fig. 4.7 LPZ 0 integrated in LPZ 1 with transformer inside the structure. Adapted from DEHN

4.6.14 Equipotential Bonding for Power Supply Systems at the Boundary of LPZ0$_A$ and LPZ2

Transition from LPZ0$_A$ to LPZ2 is usually found in compact installations. The layout of such transition is shown in Fig. 4.8. This type of transition places a high surge voltage stress on surge protective devices. Low voltage protection level and high limitation of the interference energy conducted by the arrester are the foundations for safe energy handling demands on surge protective devices in LPZ2 or with surge-limiting protective components of input circuits in equipment. Spark-gap-based combined arresters with voltage protection level less than 1.5 kV provide optimum protection of terminal devices and are suitable even for sensitive equipment with rated impulse withstand voltage of 1.5 kV. It also ensures safe operation of equipment and systems in LPZ2. Thus, it offers the advantage of combined lightning equipotential bonding and coordinated protection of terminal devices of a Type 1, Type 2, and Type 3 arresters in just a single device.

Both lightning protection zones are adjacent to each other for transitions from LPZ0 to LPZ2. Hence, it is very important and necessary to provide for a high degree of shielding at the zone boundaries. It is best to maintain the area of adjoining lightning protection zones LPZ0 and LPZ2 as small as possible. Optimally, LPZ2 should be enhanced with an extra zone shield which is installed individually away from the lightning current-carrying zone shield at the zone boundary of LPZ0, if the structure allows it. This is so that LPZ1 covers a majority of the installation as shown in Fig. 4.8. The implementation of this method not only decreases the magnitude of electromagnetic field in LPZ2 but also eliminates the need for constant shielding of all lines and systems found in LPZ2.

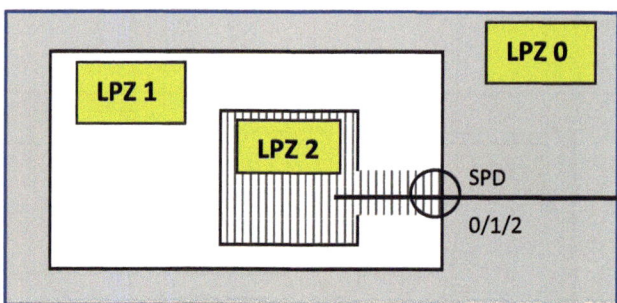

Fig. 4.8 Transition from LPZ 0A to LPZ 2 which is integrated in LPZ 1. Adapted from DEHN

4.6.15 *Equipotential Bonding for Power Supply Systems at the Boundary of LPZ1–LPZ2*

To limit surge voltage and decrease magnitude of electric field for transitions from LPZ1 to LPZ2 and higher transitions, the electrical power supply and data lines integrated in the equipotential bonding system at each LPZ transition have to be parallel to all metal systems. Shielding the rooms and devices, such as hospital operating rooms and monitoring devices, will further attenuate the electromagnetic effect. The main function of placing surge protective devices at the transitions from LPZ1 to LPZ2 is to further decrease the residual current from surge protective devices in the external zones. It must be capable of reducing induced surges affecting the lines and also surges generated within the LPZ. The discharge capacity of the SPDs used in this situation is Type 2 and it should be able to discharge at least 5 kA (8/20 μs) per phase without any damage. The 8/20 μs wave characterizes current waves from an indirect lightning strike. The surge protective devices can either be assigned to a device for device protection or to form infrastructural basis of a building for proper operation of a device or system depending on the location and type of protective measures taken. Hence, for LPZ transitions from LPZ 1 to LPZ 2 and higher transitions, different types of surge protective devices can be used.

4.7 Equipotential Bonding for Information Technology (IT) Systems

4.7.1 *Introduction*

It is required in lightning equipotential bonding that all metal conductive parts at the entrance point into the building to be integrated in the equipotential bonding system. This is to reduce the impedance to its minimum. Examples of parts in Information Technology (IT) systems are antenna lines, telecommunication lines with metal conductors, and optical fiber installations containing metal elements. The arresters and shield terminals must be chosen according to the expected lightning current parameters. The following additional steps are recommended to minimize induction loops within buildings: cables and metal pipes should enter the building at the same location, power and data lines should be laid spatially close but shielded. Lastly, unnecessarily long cables should be avoided by laying lines directly.

4.7.2 Equipotential Bonding for IT Systems at the Boundary of $LPZ0_A$ and LPZ1

Lightning current arresters with adequate discharge capacity must be placed as close as possible to entry points into the building to protect information technology lines. Typically, transition from $LPZ0_A$ to LPZ1 requires a discharge capacity up to 2.5 kA (10/350 μs) per core of information technology lines. However, this method is not used to rate discharge capacity for installations containing multiple information technology lines. The partial lightning current to be expected for the information technology cable is calculated. This is followed by determining the impulse current per core by dividing the expected lightning current with the number of single cores in the cable. Therefore, partial lightning current stress per core is lower in multi-core cables compared to cables with only single cores. In this situation, surge protective devices specified with discharge current of 10/350 μs are used. Other than that, surge protective devices with impulse current discharge capacity of up to 20 kA (8/20 μs) are compatible if equipotential bonding is set up for lines at the transition from $LPZ0_B$ to LPZ1. This is because there would be no galvanically coupled partial lightning currents flowing through.

4.7.3 Equipotential Bonding for IT Systems at the Boundary of $LPZ0_A$ and LPZ2

A majority of interference energy from lightning current is discharged by the lightning current arrester from LPZ0 to LPZ1 to protect systems and devices in the building. However, it is also a common occurrence that the level of residual interference from the lightning current arrester is still too high for the terminal devices. To remedy this problem, additional surge protective devices are installed at the transition from LPZ1 to LPZ2 to restrict interference so that residual voltage level is adjusted to dielectric strength of the terminal device. This is so that the electric field will not exceed the level in which the devices are designed to handle and to ensure that the devices are not damaged.

If equipotential bonding is implemented from LPZ0 to LPZ2, partial lightning current of single cores and shields must be determined using the method similar in boundary of $LPZ0_A$ to LPZ1 and the place of installation is chosen after calculating risks and taking into account the layout of the structure. The requirements of the surge protection device to be installed at this point of transition and the requirements on the wiring cause this transition change to be in the latter part of the system. A combined arrester coordinated with the terminal device must be used because such arresters have exceptionally high discharge capacity and low residual interference level and is particularly suited to protect the terminal device. In addition to that, the outgoing line from the protective device to the terminal device is shielded and both ends of the cable shield are included in the equipotential bonding system to prevent inflow of interference.

The installation of combined arresters is recommended in the following situations: the terminal devices are close to the entry point of cables into the building, low-impedance equipotential bonding between protective device and terminal device, line from protective device to terminal device is always shielded and earthed at both ends, and if cost-effective solution is needed. The use of lightning current arresters and surge arresters is recommended if cable distances between the protective device and the terminal device are long, inflow of interference is to be expected, the surge protective devices used for power supply and information technology systems are earthed via different equipotential bonding bars, use of unshielded lines, and existence of high interference in LPZ1.

4.7.4 Equipotential Bonding for IT Systems at the Boundary of LPZ 1 and LPZ 2 and Higher

Additional protective measures are taken at LPZ transitions in buildings to reduce interference level in IT systems. Most terminal devices are installed in LPZ2. Therefore, the protection measures at this zone must ensure that any residual interference is below the nominal value which the terminal device is able to cope with. This can be achieved by installing surge protective devices in areas near the terminal devices, integrating cable shields in the equipotential bonding system, connecting low-impedance equipotential bonding system of the surge protective device for information technology systems with the terminal device, surge protective device for power supply systems, coordinating energy flow of upstream surge protective device with the terminal device, maintaining a distance of at least 130 mm between telecommunication lines and gas discharge lamps, placing distribution board and data distributor in different cabinets, making sure that low-voltage and telecommunication lines cross at 90°, and crossing the cable along the shortest possible route.

4.8 Protection of Antenna Systems

Antenna systems are typically mounted in exposed locations for convenience of radio communication. Hence, it is more likely to be affected by lightning currents and surges if there is a direct lightning strike. Parts of antenna that are connected to an antenna feeder, but cannot be directly connected to the equipotential bonding system, should be protected by lightning current-carrying arresters. It can be assumed that 50% of the direct lightning current flows away via the shields of all antenna lines. If an antenna system is dimensioned for lightning currents up to 100 kA (lightning protection level LPL III), the lightning current splits so that 50 kA flows through the earthing conductor and 50 kA flows through the shields of all antenna cables.

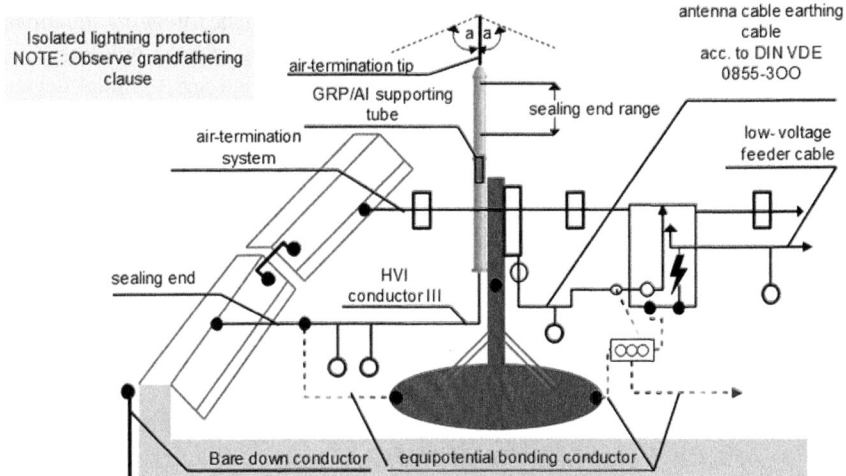

Fig. 4.9 Protection of roof top communication antenna (Credit: DEHN)

Antenna systems which cannot carry lightning currents must be equipped with air-termination systems. The factors that must be taken into account when choosing a suitable cable is the lightning current shared by the antenna line with the down conductor. The required dielectric strength of the cable is determined from the transfer impedance, length of antenna line, and the amplitude of lightning current. Antenna systems on buildings can be protected by air-termination rods, elevated wires, or spanned cables as stated in the lightning protection standard where a certain separation distance must be maintained for each of the methods above, as shown in Fig. 4.9. The main function of electrical isolation of lightning protection system from conductive parts of buildings and isolation of lightning protection system from electrical lines in buildings is to prevent partial lightning currents from entering the control and power supply. Therefore, such isolation is able to prevent electrical and electronic devices from being affected or destroyed by lightning currents.

4.9   Protection of Optical Fiber Installations

Optical fiber installations with metal elements are typically divided into the following categories, namely, cables with metal core but with metal sheath or metal supporting elements, cables with metal elements in the core and with metal sheath and lastly, and cables with metal elements in the core but without metal sheath. The minimum peak value of lightning current must be determined for all types of cables since it has adverse effects on the transmission of optical fiber cables. For such situations, cables capable of carrying lightning currents must be chosen and the metal parts must be connected to equipotential bonding directly or through a surge protection device.

The metal sheath is connected by shield terminals at the entrance point into the building while the metal core is connected by earthing clamp, for example, installing a protective conductor terminal near the splice box. An indirect connection is required via a spark gap to prevent equalizing currents.

4.10 Telecommunication Lines

Telecommunication lines with metal conductors usually consist of cables with balanced or coaxial stranding elements and can be divided into various types. The types are cables without additional metal elements, cables with metal sheath and metal supporting elements, and cables with metal sheath and additional lightning protection reinforcement. The individual cables must be integrated in the equipotential bonding system according to the following methods:

(i)   Unshielded cables are connected by surge protective devices which are able to carry partial lightning currents. Partial lightning current per core is obtained by dividing the partial lightning current of the cable by the number of single cores.

(ii) If the cable is shielded and the shield is capable of carrying lightning currents, the lightning current flows through the shield. However, capacitive or inductive interferences can reach the cores and hence it is necessary to use surge arresters. The requirements for this type of cables are that the shield at both cable ends must be connected to the main equipotential bonding system such that it can carry lightning currents. The lightning protection zone concept must be used in the building where the cable ends. The active cores must be connected in the same lightning protection zone, typically LPZ1. LPZ1 is an inner zone protected against direct lightning strikes and is defined as a zone where impulse currents are limited by current distribution and isolating surfaces or by surge protective devices. Spatial shielding may be used to decrease the intensity of lightning electromagnetic field. Unshielded cables in a metal pipe are treated as a cable with a lightning carrying cable shield.

(iii) If the cable shield does not carry lightning currents then the procedure of integration into the equipotential bonding system is similar to a signal core in an unshielded cable if the shield is connected at both ends. The partial lightning current per core is calculated by the partial lightning current of the cable divided by the number of single cores added with a shield. However, if the shield is not connected at both ends, it is treated as if it were not there and the partial lightning current per core is obtained by dividing the partial lightning current of the cable with the number of single cores.

Figure 4.10 shows typical lightning and surge protection for a telecommunication system.

If the exact core load cannot be determined, the appropriate threat parameters given in protection standards must be used. For telecommunications lines, the maximum

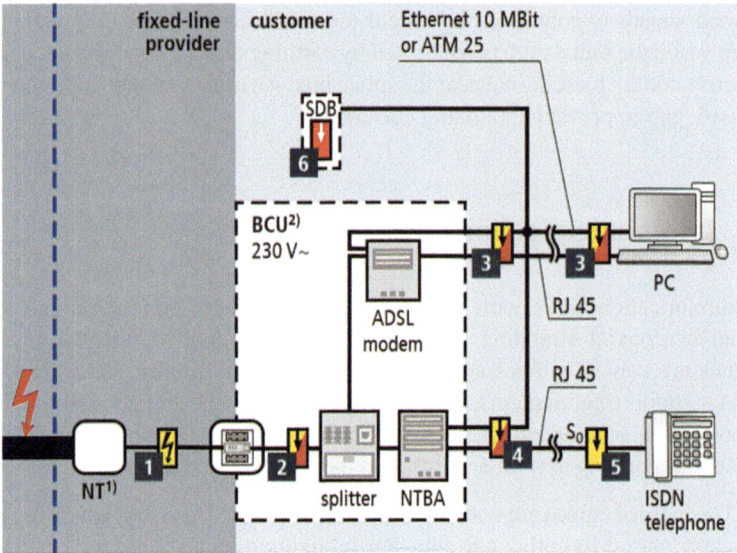

Fig. 4.10 Lightning and surge protection of telecommunications system: ISDN connection with ADSL. Credit: DEHN, with permission

lightning current load per cable core is an impulse of 2.5 kA (10/350 μs). A 10/350 μs current wave indicates a direct lightning stroke with a 10 μs rise time in which the magnitude of current reaches its peak within 10 μs and the impulse has a 350 μs voltage surge duration. The surge protective devices must be able to withstand the lightning current and have a discharge path to the equipotential bonding system.

The advantages of having lightning protection zones are minimal coupling of surge voltages into other cable systems because dangerous lightning currents are directly attenuated at the entry point of the building and at the transition points between zones. It also reduces equipment malfunction due to magnetic field.

4.11   Choosing Internal Lightning Protection System: Type of Surge Protection Devices (SPDs)

Surge protection devices are essential in the protection system. There are three classes of surge protection devices. It is important to choose the appropriate surge arrester because each type is designed for different situations. The Type 1 lightning current arrester is capable of discharging powerful lightning currents and is installed in the main switchboard or at the entry to the building and should be incorporated in lightning protection system, for example, when lightning rods or meshed cages are installed. Type 2 surge arresters are used in main and sub-distributors. It discharges currents from indirect lightning strikes, protects from inductive and conductive overvoltages, and also switching transients. The Type 3 surge arrester has very low discharge capacity. Surge voltages typically occur between the phase to neutral cable. The Type 3 surge arrester is used to protect against inductive coupling and switching surges in device power circuits. It is a supplementary surge protective device used in surge arrester Types 1+2+3 combinations where there is a lightning protection system or in Types 2+3 combination when there is no lightning protection system. Choosing a suitable lightning protection system involves risk assessment to determine which areas are at the most risk or at least risk. A rule of thumb is to always install a Type 2 surge arrestor and if the distance between the surge arrestor and the equipment to be protected is greater than 10 m, a Type 2 or Type 3 arrestor is added because wave reflection starts increasing from 10 m. Surge wave reflections can double the voltage at 30 m. Therefore, it is necessary to install a surge protection device if the distance to the equipment exceeds 10 m. Although surge arrestors do not trip, it is important to protect them to work optimally and prolong its lifespan. There are a few situations that damage the surge protection device. One such situation is thermal runaway which is caused by constant excessive current which does not exceed the device specification, but it eventually leads to slow destruction of internal components. This is sometimes called electronic rust. In this situation, a thermal fuse in the surge protection device disconnects it. Next, short circuit occurs because of a fault at power frequency system at 50 Hz electrical distribution network or due to the current exceeding the maximum current flow capacity. To protect the surge protection device, an external or integrated short circuit protection device such as a fuse or circuit breakers is installed to disconnect the surge protection device.

When designing the protection system for a building, it is important to determine the quantity and type of surge protection device, maximum discharge current, and short circuit current at the point of installation. Table 4.4 summarizes the location and type of surge protection device to be installed.

Table 4.4 Summary of type of SPD to be used

Distance between sensitive equipment from lightning protection system in main switchboard	Lightning rod on the building or within 50 m of the building	
	No	Yes
D < 30 m	One Type 2 SPD in main switchboard	One Type 1+2 SPD in main switchboard
D > 30 m	One Type 2 SPD in main switchboard, one Type 2 or Type 3 SPD in enclosure near to sensitive equipment	One Type 1+2 SPD in main switchboard, one Type 2 or Type 3 SPD in enclosure near to sensitive equipment

4.12   External Lightning Protection

A lightning protection system is a system that protects buildings from direct lightning strikes, injected lightning current as well as from potential fire. The function of external lightning protection system is to prevent direct lightning strikes to buildings via an air-termination system. Moreover, the external protection systems conduct the lightning current to the ground safely via a down-conductor system and distributes it in the ground via an earth-termination system.

When there is a lightning strike there is a possibility of an explosion at a structure under construction. Hence, it is essential to design either an isolated or non-isolated protection system depending on the material with which the structure is constructed. If the structure is built with a combustible material, there is a high chance of explosion due to a lightning strike and this requires the design and installation of an isolated lightning protection system.

An external LPS consists of

- An air-termination system.
- Down-conductor system.
- Earth-termination system..

Air rods may be used for air termination. These are small, vertical protrusions designed to act as the "terminal" for a lightning discharge. Most are topped with a tall, pointed needle or a smooth, polished sphere. Alternatively, they may be in the form of catenary (suspended conductors) or meshed conductor network. These should be installed at corner, exposed points, and edges of the structure. The places where these air-termination systems should be positioned are determined by one of the following methods.

i.   The Rolling Sphere Method (RSM)

This is a method which identifies the areas of the structure that need protection, taking into account the possibility of side strikes to the structure. This method uses rolling spheres to identify the areas that are vulnerable and that require air termination. This is illustrated in Fig. 4.11.

Rolling sphere method is based on an electro-geometric model. For a cloud-to-ground lightning, a downward leader grows from cloud towards the earth. As the downward leader gets close to the earth, when it is at about ten to hundreds of meters from the earthed structure, upward leaders start to grow towards the head of downward leader. The intersecting point at which the downward leader and the upward leader meet is the point of lightning strike. The closest distance between starting point of the upward leader and the head of downward leader is known as the final striking distance, which corresponds to the radius of the rolling sphere. The proportionality between final striking distance (radius of the rolling sphere) and peak value of lightning current I is given by

$$r = 10 \cdot I^{0.65} \tag{4.19}$$

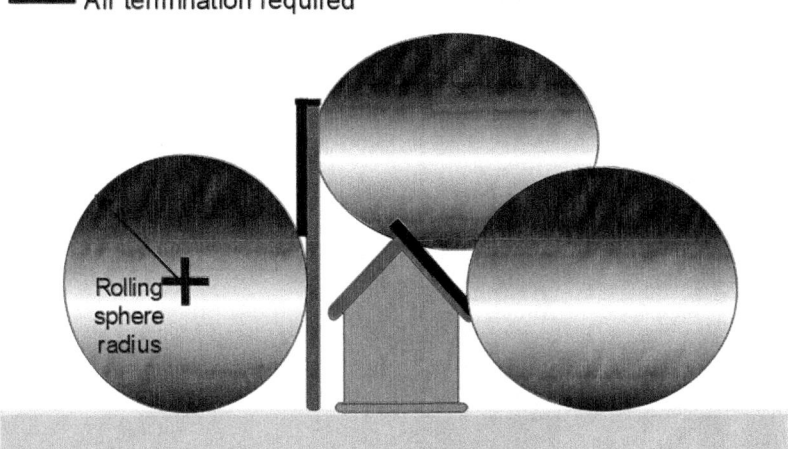

Fig. 4.11 Application of the rolling sphere method

where

   r is the radius of the rolling sphere in m and I is the peak value of the lightning current in kA.

   In order to use the rolling sphere method, a scale model of the structure to be protected is required. The rolling sphere is rolled around the scale model, and the points where the circumference of the sphere touches the model may be vulnerable to lightning strikes. The relation between the lightning protection level (LPL) and the radius of the rolling sphere is shown in Table 4.5. The area where the sphere does not touch is less vulnerable to lightning strikes, as shown in Fig. 4.11.

ii.   **The Protective Angle Method (PAM)**

Table 4.5 Relationship between lightning protection level, interception probability, final striking distance, and minimum peak value of lightning current

Lightning protection level (LPL)	Probabilities for the limits of the lightning current parameters		Radius of the rolling sphere (final striking distance), r in m	Minimum peak value of lightning current, I in kA
	Minimum value	Maximum value		
I (maximum risk)	0.99	0.99	20	3
II	0.97	0.98	30	5
III	0.91	0.95	45	10
IV (minimum risk)	0.84	0.95	60	16

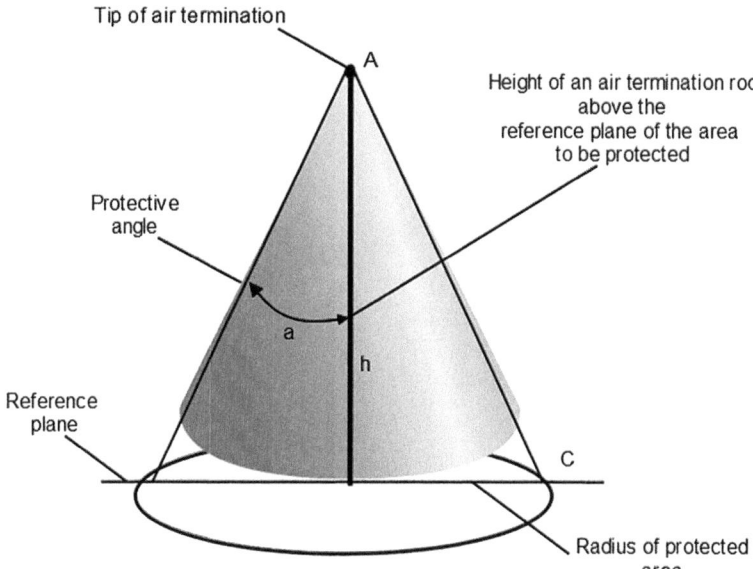

Fig. 4.12 The protective angle method for a single air rod. Adapted from DEHN

The protective angle is defined by the angle created between the tip of the vertical rod used for air termination and the line projected down to the surface on which the rod sits. The cone from the rod is called a cone of protection. The protective angle differs depending on the class of LPS. But in most cases it is 45°.This method is best suited with simple shaped buildings and is valid only up to a height equal to the rolling sphere radius for the corresponding LPL. The PAM is shown in Figure 4.12.

iii.   **The Mesh Method (MM)**

Meshed air-termination system can be used regardless of the height of structures and shape of the roof. By using the outer edges and ridge of the structure serving as an air-termination system, the individual meshes can be placed at any desired points. The air-termination conductor on the outer edges of the building must be placed as close as possible to the edges. Using the rolling sphere method on meshed conductor network, the mesh must be mounted at certain distance above the roof plane, to make sure that the rolling sphere does not touch the roof plane, as shown in Fig. 4.13. Table 4.6 shows the typical mesh sizes.

• Down conductors:

Down conductor is the direct route from the air-termination system to the earth-termination system. The earth-termination system can be ground rods which are long, thick, and heavy buried deep into the earth around a protected structure. The down-conductor cables (which carry lightning currents from the air-termination rods to the ground) are run along the top and around the edges of roofs, then down one

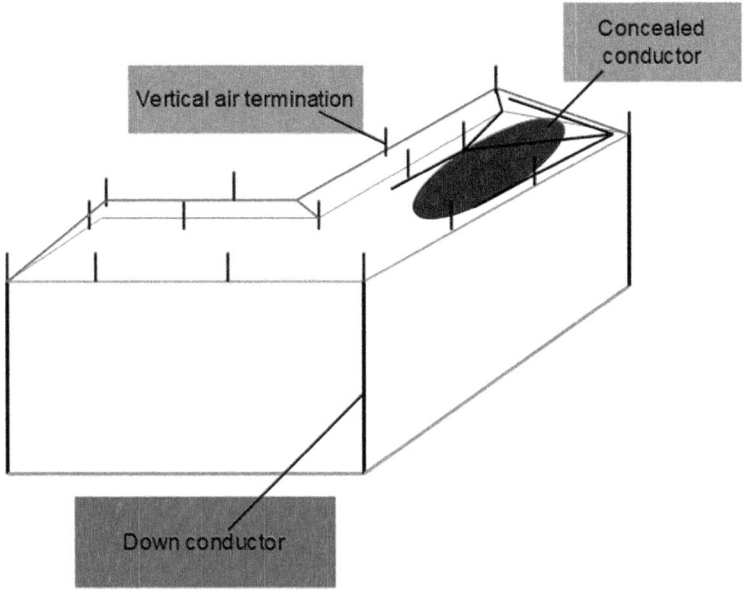

Fig. 4.13   Concealed air-termination network. Adapted from DEHN

Table 4.6 Mesh size based on lightning protection level	Lightning Protection Level (LPL)	Mesh size
	I	5×5 m
	II	10×10 m
	III	15×15 m
	IV	20×20 m

or more corners of a building to the ground rod. They are connected to these rods to complete a safe path for a lightning discharge around a structure.

It is advisable to use metal parts on or within the structure to be incorporated into the Lightning Protection System (LPS). Examples of these are when the internal reinforcing bars are connected to external down conductors through clamps or using the reinforcing bars as down conductors. But when doing so the continuity from the air termination must be tested.

• Earth-termination system:

The earth-termination system is vital for the dispersion of lightning current safely and effectively to the ground. A good earthing connection should possess the following characteristics:

1. Low electrical resistance between the electrode and the earth.
2. Good corrosion resistance.

The standards state that the earthing resistance should be 10 Ω or less for a good and efficient lightning protection.

There are three types of basic earth electrode arrangements which are given below:

- Arrangement A: Horizontal or vertical electrodes connected to each down conductor fixed on the outside of the structure.
- Arrangement B: This is a fully connected ring earth electrode that is suited around the periphery of the structure and is in contact with the surrounding soil.
- Foundation earth electrode: These are the conductors that are installed in the concrete foundation of the structure as described above.

Adding a protection system doesn't prevent a strike, but gives it a better, safer path to ground. The air terminals, cables, and ground rods work together to carry the large lightning currents away from the structure, preventing fire and most appliance damage:

4.13   Air-Termination Systems

4.13.1   *Isolated and Non-isolated Air-Termination Systems*

Isolated air-termination system protects the buildings from a direct lightning strike by using air-termination rods and mast with cables spanned over it. During the installation of air-termination systems, the separation distance between air-termination system and the buildings must be fixed. Isolated air-termination system is usually installed for structures with roof that are covered with flammable materials and structures located in hazardous area. Glass Reinforced Plastic (GRP) air-termination system is frequently used for buildings with roof-mounted system, such as heat exchanger and ventilation system. An isolated air-termination system consists of air-termination rods and air-termination conductors. A single air-termination rod is able to protect a small roof-mounted structure, such as a small roof-top fan. Self-supporting air-termination rods up to the height of 14 m are installed by using tripod stand that is fixed on a concrete base. Additional supports are required for higher air-termination rods so that they can withstand the wind load. Usually, they are widely used for lightning protection of Photovoltaic (PV) solar systems and antennas. Air-termination conductors are usually installed above the structure that need to be protected. These conductors create a tent-shaped protected region at the side and a cone-shaped protected region at the ends. The protective angle varies according to the class of LPL and the height of the air-termination system above the structure that needs to be protected. The dimensions of the air-termination conductors are determined by using the rolling sphere method (RSM).

Non-isolated air-termination system can be installed in two ways, depending on the type of the roof material. If the roof is made of non-flammable material, the conductors of air-termination system are installed directly on the surface of the buildings. However, if the roof is made of highly flammable materials, the flammable parts of the roof must be kept at a certain distance from the conductors of the air-termination system to ensure there is no direct contact between the flammable parts of the roof and air-termination system.

4.13.2 Air-Termination System for Buildings with Different Types of Roof

Every type of roof must install its unique, suitable design of air termination in order to maximize the efficiency of the external lightning protection system. There are eight different types of buildings that have their own specific designation:

1. Gable roofs.
2. Flat roofs.
3. Metal roofs.
4. Thatched roofs.
5. Accessible roofs.
6. Green roofs.
7. Steeples and churches.
8. Wind turbines.

4.13.3 Air-Termination System for Building with Gable Roofs

Buildings with gable roof are usually installed with meshed network of air-termination system. The individual meshes are placed by using ridge, outer edges and other metal parts as a part of the air-termination system. Normally, a metal gutter is used for closing the meshed network of air-termination system on the surface of the roof. If the gutter is connected in an electrically conductive way, a gutter clamp is mounted at the cross point between the gutter and air-termination system.

Non-conductive roof-mounted structures are sufficient to protect the roof against lightning strikes provided they do not exceed the final striking distance of 0.5 m from the plane of the meshed network air-termination system. If exceeded, it is required that these structures are connected to the nearest air-termination conductor and equipped with air-termination system. On the other hand, metal roof-mounted structures with non-conductive connections do not need to be equipped with air-termination system as long as the roof-mounted structures are less than 0.3 m from roof level, have a maximum enclosed area of $1m^2$, and with a length of less than 2 m.

Air-termination rod must be installed for a chimney so that the whole chimney is under the lightning strikes protection region. The dimension of the air-termination rod is determined by using the Protective Angle Method (PAM). If the chimney is made of bricks, the air-termination rod can be mounted directly on the chimney. However, if there is metal pipe within the chimney, the chimney must be equipped with isolated air-termination system and installed with air-termination rods using spacers. The metal pipe is then connected to the equipotential bonding system.

A similar method is used to protect parabolic antennas. If there is a direct lightning strike to the antenna, the lightning current will enter the building through coaxial cable, causing damages and interference. This can be avoided by equipping the antenna with isolated air-termination system.

4.13.4   Air-Termination System for Buildings with Flat Roofs

The meshed network of air-termination system is installed for buildings with flat roofs according to the mesh size and LPL as shown in Table 4.5. The roof parapet which acts as the natural component of the air-termination system is connected with air-termination conductors. The length of materials of the roof parapet changes according to the changes in temperature. Therefore, the individual segments are connected by using bridging braids, brackets, or cables to ensure that they are always interconnected and electrically conductive when they are changing length due to changes in temperature. Unfortunately, the material used can be melted when struck by lightning. Thus, an air-termination tip is installed using the rolling sphere method. Roof sheeting will move across the roof surface during windy condition if they are not fixed properly on the roof surface. A common way of fixing the air-termination conductor safely is by using the roof conductor holder with strips. The roof conductor and strips have to be placed next to a roof sheeting joint at a distance of around 1 m. If the slope of the roof is less than 5°, every second roof conductor holder is fixed, while if the slope of the roof is more than 5°, every roof conductor holder must be fixed. However, some roof conductor holder is not suitable for use if the angle of the roof slope is more than 10°.

4.13.5   Air-Termination System for Buildings with Metal Roofs

When lightning strikes a metal roof without any protection system, it will leave a hole and damage the metal roof. Therefore, an external lightning protection system with lightning current-carrying wire and clamps are installed on the metal roof to avoid this kind of damages. A separate air-termination system with many air-termination tips is installed on the metal roof to ensure the rolling sphere does not touch the metal roof. The recommended height of air-termination tip is shown in Table 4.7.

There are numerous types of air-termination conductor holders available for metal roof, such as round standing seam, standing seam, and trapezoidal. The conductor in the air-termination conductor holder located at the highest point of the metal roof must be fixed, while the conductors in other air-termination conductor holders are routed loosely because of the changes in length with changes in temperature.

Table 4.7 Lightning protection for metal roofs	Distance of the horizontal conductors, in m	Height of the air-termination tip, in m
	3	0.15
	4	0.25
	5	0.35
	6	0.45

4.13.6 Air-Termination System for Buildings with Thatched Roofs

Buildings with thatched roofs are usually installed with external lightning protection systems of Lightning Protection Level (LPL) Class III. Air-termination conductors on buildings with thatched roof have to be fastened with insulating material in order to allow them to move freely. Note that some distance should be maintained around the eaves. The exact distance between each down conductor can be calculated according to the separation distance specified in the lightning protection standard. Generally, ridge conductors must have span with 15 m width and 10 m length of down conductors without any other supports. Anchor bolts and washers are used to connect span stakes to the roof structure. The metal parts around the roof surface such as antennas, metal sheet, and wind vanes have to be protected by isolated air-termination system. Air-termination rods and air-termination conductors must be installed on the building in order to increase the efficiency of lightning protection system. If the thatched roof is located near to a metal roofing material, non-electrically conductive roofing material is inserted between the metal roofing material and thatched roof.

A new possibility to install an isolated lightning protection system is by the use of insulated down conductors. This type of lightning protection system is widely installed in historical farmhouses. The rolling sphere method is used when designing the air-termination system to determine the protected region from lightning strikes. A GRP supporting tube is used to elevate the air-termination system and support the insulated down conductors.

4.13.7 Air-Termination System for Buildings with Inaccessible Roofs

It is impossible to mount air-termination conductors on inaccessible roof. However, the air-termination conductors can be installed in the joints between the decks. The air-termination studs are then fixed at the intersections of the meshed network of air-termination system as the point of lightning strike. The rolling sphere and protective angle methods are used to determine the dimension of the air-termination system when designing an external lightning protection system. These air-termination systems consist of air-termination rods and these rods are fixed to the parapet.

4.13.8   Air-Termination System for Buildings with Green Roofs

Meshed air-termination system is installed for buildings with green roofs. A meshed air-termination system is usually installed on the surface of the green roof for easier inspection. The common wire material used for air-termination system of green roof is stainless steel.

4.13.9   Air-Termination System for Steeples and Churches

A lightning protection system of class III LPL is required for steeples and churches. Steeples with the height of less than 20 m must be equipped with a down conductor. This down conductor has to be connected to an external lightning protection system of the nave if the steeple is joined together with the nave. Steeples with the height that is higher than 20 m must be equipped with two or more down conductors and one of the down conductors must be connected to the external lightning protection of the nave. The down conductors of the steeples have to be routed along the outer surface of the steeple to the ground because the installation inside the steeple is not allowed.

Some of the modern churches are made of reinforced concrete. The reinforced steel can be used as a natural component of down conductors provided that it has permanent, electrically conductive connection. Lightning equipotential bonding or surge protection of the electrical equipment, for instance, power installation, telephone, and loudspeaker system, is installed at the entrance of the building, while for the bell controller in the steeple, surge protection is installed at the control system.

4.13.10   Air-Termination Rods Subjected to Wind Loads

Self-supporting air-termination rods are installed on the roof of the building. They experience mechanical stress due to wind speeds. Therefore, isolated air-termination rods must meet the requirement regarding their mechanical stability. The local wind conditions and the height of the buildings have to be taken into account when calculating the wind load stress.

In order to design self-supporting air-termination rods which are able to withstand required wind load stress, the tilt resistance, bending resistance of the air-termination rods, and the fixed separation distance between the protected structures must be determined. The stability of the air-termination rods is calculated by considering the following: the area of the air-termination rods exposed to wind, the area of the braces exposed to wind, the weight of the air-termination rods and braces, the weight of the post, and tilt lever of the post. Since the wind load stress will exert bending stress on the air-termination rod, the break resistance of the air-termination rod has to be determined. The calculation to determine bending stress of the air-termination rod must include the following information: Finite Element Method (FEM) calculation model, characteristics of the material used (density, elasticity, cross-sectional value), and wind loads.

4.13.11   Safety System and Lightning Protection

Industrial buildings with flat roofs are commonly installed with safety rope system. The advantage of using safety rope system is that the operators can walk along the rope by hooking the rope slide or rope guide within the safety rope system. Lightning protection system and rope safety system are two different systems that are installed on the roof of the building. Each of them works independently and therefore they must be installed with their own experts. The rope safety system should be installed within the protected region of the air-termination system to prevent it getting damaged from lightning strikes.

4.14   Down Conductors

4.14.1   Determination of the Number of Down Conductors

A down conductor is an electrically conductive connection between earth-termination system and air-termination system. The function of the down conductor is to conduct the lightning current straight to the earth without causing any damage to the building. There are some factors that need to be paid attention to when mounting the down conductor to minimize or avoid the damage caused by the lightning current when discharging to the earth-termination system:

- The length over which the current flows should be kept as short as possible, preferably vertical and straight without looping.
- Several parallel current paths may exist.

The number of the down conductors required is determined using the perimeter of the projection from the external edges of the roof to the ground surface. The distance between each consecutive down conductors is categorized depending on the class of LPL as shown in Table 4.8. The exact number of down conductors required can only be obtained through calculation of the separation distance. The separation distance can be reduced through balancing the distribution of the lightning current by interconnecting down conductors at the ground level.

Table 4.8 Distance between down conductors based on class of LPL

Class of LPL	Typical distance in m
I	10 m
II	10 m
III	15 m
IV	20 m

4.14.2   Down Conductors for a Non-isolated Lightning Protection System

Down conductors for a non-isolated lightning protection system are usually direct mounted onto the building without separation distance. This is due to the rise of temperature when lightning strikes the external lightning protection system. Another reason why down conductors are mounted directly on the building is because of the non-flammable material used for the wall. If the wall is made of flammable material it must be ensured that the rise of temperature due to lightning current flows is not dangerous.

4.14.2.1   The Installation of Down Conductor

Down conductor is installed with direct continuation from air-termination system and shortest possible vertical straight line connection to the ground directly. Down conductors cannot be installed in the downpipe or gutter because the moisture of the gutter and downpipe will corrode the down conductor. The down conductors are recommended to have a fixed separation distance from windows and doors.

4.14.2.2   Natural Components of Down Conductor

Some of the parts of the structure that may be used as natural components of down conductor are stated as follows:

i.   Metal installations.
ii. Metal framework of structure.
iii. Interconnected reinforcement of the structure.
iv. Precast parts.
v.   Facade elements, ISO (International Standards Organization) standard rails, and metal sub-structures of facade.

4.14.2.3   Internal Down Conductors

Internal down conductors are installed if the edges of the structures are four times greater than the distance of down conductors, depending on the class of LPL. Some of the lightning current may flow through the internal down conductors within the building, which needs to be constrained.

4.15 Earth-Termination System

Earth-termination system is an external lightning protection system that allows energy from lightning strikes to be dissipated quickly into the earth with the usage of earth electrodes. The overall resistance for the whole earth-termination system should be less than 10 Ω. The earth electrodes are categorized according to their installation location.

1. Surface earth electrodes consist of the following:
    Earth electrodes are installed into the ground up to 1m depth.
    Round materials or flat strips are used.
    Common designs are radial, ring, or meshed earth electrodes.
2. Earth rods are earth electrodes that are driven vertically into the deep ground.
3. Foundation earth electrodes: one or more conductors that are combined together and connected to the earth in a large area.
4. Control earth electrode: arrangement and the shape of earth electrodes which serve to control the ground potential.
5. Ring earth electrodes: earth electrodes that are formed in a closed ring.
6. Natural earth electrode: metal parts that are in contact with water or with earth.

The earth electrode consists of three types of resistance-related parameters which are the earth resistivity, ρ_E; the earth resistance, R_A; and conventional earth impedance, $R_{st.}$ The earth resistance R_A can be explained with the aid of a metal sphere that is buried into the ground. The earth resistance R_A includes some of the resistances of the single sphere layer. The resistance of the sphere layer is calculated by using following formula:

$$R = \rho_E \cdot \frac{l}{A}, \qquad (4.20)$$

where
ρ_E is the earth resistivity of the ground,
l is the assumption of the thickness of the sphere layer, and
A is the center surface of the sphere layer.
The earth resistance R_A is then calculated by using the following formula:

$$R_A = \frac{\rho_E \cdot 100}{2\pi \cdot r_k} \cdot \frac{1 + \frac{r_k}{2d}}{2} \qquad (4.21)$$

where

ρ_E is the earth resistivity of the ground in Ωm,

d is the burial depth in cm, and

r_k is the radius of the metal sphere.

The earth resistivity ρ_E can be calculated from the measured resistance R by using the following formula:

$$\rho_E = 2 \cdot \pi \cdot d \cdot R, \tag{4.22}$$

where

d is the probe spacing in m,

R is the measured resistance in Ω, and

ρ_E is the earth resistivity of the ground.

If there are a few earth rods that are installed near to each other in an area, Table 4.9 can be used to calculate the earth resistance from the distance between the electrodes. As shown in Fig. 4.14, the earth resistance of the earth electrodes is frequency dependent.

We may classify two types of earth electrode arrangement for earth-termination system. These are the Type A and Type B arrangements. The arrangement of Type A earth electrodes is the placement of earth rods (vertical earth electrodes) or surface earth electrodes (horizontal radial earth electrode) that are connected to the down conductor. This type of arrangement needs two or more earth electrodes. If different types of earth electrodes (horizontal and vertical electrodes) are used together, the

Table 4.9 Calculating earth resistance

Earth electrode	Approximate formula	Auxiliary
Surface earth electrode (radial earth electrode)	$R_A = \frac{2 \cdot \rho_E}{l}$	–
Earth rod	$R_A = \frac{\rho_E}{l}$	–
Ring earth electrode	$R_A = \frac{2 \cdot \rho_E}{3 \cdot d}$	$d = 1.13 \sqrt[2]{A}$
Meshed earth electrode	$R_A = \frac{\rho_E}{2 \cdot d}$	$d = 1.13 \sqrt[2]{A}$
Earth plate	$R_A = \frac{\rho_E}{4.5 \cdot a}$	–
Hemispherical or foundation earth electrode	$R_A = \frac{\rho_E}{\pi \cdot d}$	$d = 1.57 \sqrt[3]{V}$

R_A is Earth resistance (Ω),

ρ_E is Earth resistivity (Ωm),

l is length of the earth electrode (m),

d is diameter of a ring earth electrode, the area of the equivalent circuit or a hemispherical earth electrode,

A is area (m^2) of the enclosed area of a ring or meshed earth electrode,

a is edge length (m) of a square earth plate

(in case of rectangular plates: a is substituted with $\sqrt[2]{b \cdot c}$, where b and c are the two sides of the rectangle), and

V is volume of a single foundation earth electrode

Fig. 4.14
Frequency-dependent earth
impedance of different
earthing systems. Adapted
from: DEHN

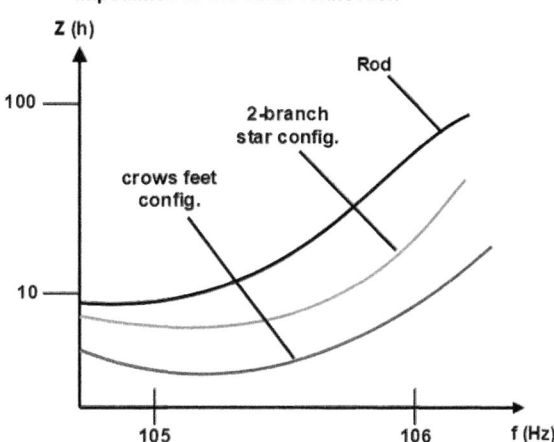

equivalent total length of horizontal and vertical earth electrodes has to be determined. Generally, earth rods are installed vertically deep into natural soil because the deeper the soil layers they are installed into, the lower the earth resistivity, as compared to the areas that are close to the earth surface. The earth electrodes should be placed at least 500 mm below the ground surface. The electrodes must be distributed as evenly around the building as possible to prevent electrical coupling. Earth rods which are made of high-alloy stainless steel are used widely due to its large range of benefits.

Type B earth electrode arrangement is the arrangement with earth electrodes encircling the structures or buildings that need to be protected from lightning strikes. The arrangement is also known as foundation earth electrodes. If the building cannot be encircled in a closed ring arrangement, the ring must be completed by using the conductors inside the building, such as pipework or other electrically conductive metal components. About 80% length of the earth electrodes is driven into the soil so that it can be used as a base to determine the separation distance. The earth electrodes should be driven at least 500 mm below the surface of the ground. The ring earth electrode is recommended in natural soil to ensure that the earth resistance is not affected. The earth electrode chosen should resist corrosion, preferably made of stainless steel. Type B earth electrode arrangement is suitable for installing on rocky ground because it is often the only way to install earth-termination system on a rocky ground with a resulting low resistance. Ideally, Type B earth electrode is always used for: (i) dissipating lightning current from down conductors to the ground, (ii) connecting equipotential bonding of down conductors at the ground, (iii) manipulating the potential in the vicinity of electrically conductive wall of a building, and (iv) buildings with high fire risk or with many electronic equipment.

Some of the systems that need special requirements when installing earth-termination system are: (i) electrical systems with the disconnection requirements of the relevant system configuration, (ii) equipotential bonding, (iii) electronic systems such as data information systems, (iv) antenna earthing, (v) electromagnetic compatibility (EMC) earthing, and (vi) transformer power substation.

4.16 Manufacturer's Test of Lightning Protection Components

Lightning protection components that are made of metal material such as air-termination conductors, air-termination rods, earth electrodes, or clamps, which are exposed to seasonal changes or different weather conditions must undergo artificial conditioning or aging which are tested to ensure their suitability for real-time application. The testing of metal lightning protection components with artificial conditioning or aging can be done in two steps.

1. Salt mist treatment:

     This test forms an artificial saline condition to test the metal lightning protection components to determine whether they can withstand it for a long period of time. The test chamber consists of a salt mist chamber, where the metal components are sprayed with sodium chloride (NaCl) solution three times with 2-hour period at a temperature between 15 and 35 °C and relative humidity of 93% for 20 to 22 h, to ensure their sustainability.

2. Humid sulphurous atmosphere treatment:

     This test forms a condensed humidity condition that is filled with sulphur dioxide. The metal lightning protection components are accessed in seven test cycles. Each cycle has an 8-h heating process at a temperature around 40 °C followed by a 16-h duration of saturated humidity condition, with a total duration of 24 h.

Another test that needs to be done is the testing of connecting components such as clamps, which are used to connect air-termination conductors, down conductors, and earth entries with one another during the installation of external lightning protection system. These clamps must be able to withstand the thermal and electrodynamic forces that are produced by lightning current flow. Table 4.10 shows that the permissible material combinations of air-termination system and down conductors with one another or with other structural parts.

Table 4.10 The possible material combination of air-termination system and down conductors with one another or with other structural parts

	Steel	Aluminum	Copper	StSt (V4A)	Titanium	Tin
Steel (StZn)	Yes	Yes	No	Yes	Yes	Yes
Aluminum	Yes	Yes	No	Yes	Yes	Yes
Copper	No	No	Yes	Yes	No	Yes
StSt (V4A)	Yes	Yes	Yes	Yes	Yes	Yes
Titanium	Yes	Yes	no	Yes	Yes	Yes
Tin	Yes	Yes	yes	Yes	Yes	Yes

4.17 Shielding of electrical and electronic systems against LEMP

4.17.1 Magnetic Field Calculations for Shielding

The primary interference for devices and installations is from the lightning currents and the associated electromagnetic field LEMP. First approximation is used to determine the complex distribution of the magnetic field inside a grid-like shield. Magnetic field coupling of each rod in the grid-like shield (shown in Fig. 4.15) with all other rods including the simulated lightning channel is considered in performing the calculations. To determine either the electromagnetic field of the first return stroke or the subsequent return stroke, the magnetic fields produced by the following are considered:

- Maximum value of the current of the first positive return stroke ($i_{f/max}$).
- First negative return stroke current ($i_{fn/max}$).
- Maximum value of the current of the subsequent return strokes ($i_{s/max}$).

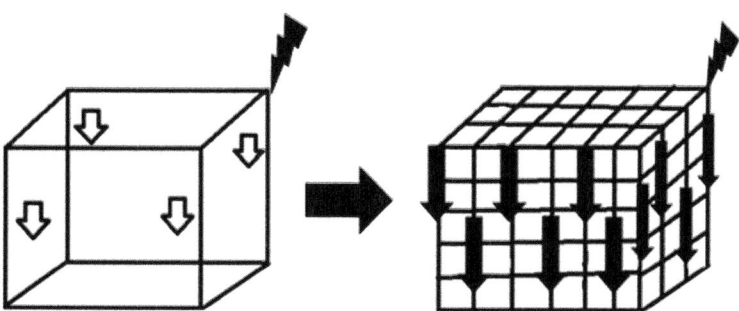

Fig. 4.15 Reduction of the magnetic field using grid-like shields (Adapted from DEHN)

Internal electronic systems may only be installed within a safety volume with a safety distance from the shield of the LPZ. To calculate the safety distances, the following must be considered:

$d_{s/1}$—Safety distance in case of a spatial shield of LPZ1 if lightning current flows into the spatial shield. (The spatial shield of LPZ1 produces a magnetic field due to the currents induced in it by the LEMP.)

$d_{s/2}$—Safety distance in case of spatial shield if no lightning current flows into these spatial shields.

4.17.2 Calculation of the Magnetic Field Strength in Case of A Direct Lightning Strike

In order to attenuate the amount of lightning LEMP radiated energy penetrating into sensitive electronic equipment, we need to form cage-like shields which would attenuate the energy that gets into the shielded area. It is important to ensure that the size of the grid mesh is less than the minimum wavelength of the electromagnetic field it needs to keep out. The main task is to determine the size of the mesh-like cage we need to construct to get a particular shielding factor to keep the equipment safe and stable. To calculate the magnetic field strength in case of a direct lightning strike, the following formula may be used:

$$H_1 = \frac{k_h . I_0 . h_m}{d_w . \sqrt{d_r}} \text{ in A/m,} \tag{4.23}$$

where

d_r is the shortest distance between the point considered and the roof of the shielded LPZ 1 in m;

d_w is the shortest distance between the point considered and the wall of the shielded LPZ 1 in m;

I_0 is the lightning current in LPZ 0A in A;

k_h is the configuration factor, typically $kh = 0.01$ in $1/\sqrt{m}$; and

h_m is the mesh size of the grid-like shield of LPZ 1 in m.

Depending on which lightning stroke is being considered, the current I_0 may be set as one of the following three currents:

$I_{f/max}$ is the maximum value of the first positive stroke current in accordance with the LPL in A;

$I_{fn/max}$ is the maximum value of the first negative stroke current in accordance with the LPL in A; and

$I_{s/max}$ is the maximum value of the subsequent stroke current in accordance with the LPL in A.

Depending on the shielding factor SF required, we have

$$d_{s/1} = \frac{h_m . SF}{10} \; for \; F \geq 10 \; in \; m \tag{4.24}$$

$$d_{s/1} = h_m \; for \; SF \leq 10 \; in \; m \tag{4.25}$$

where
SF is the shielding factor in dB and
h_m is the mesh size of the grid-like shield in m.

4.17.3 To Determine the Magnetic Field in Case of Nearby Lightning Strike

To calculated the magnetic field strength in case of a nearby lightning strike:

$$H_0 = \frac{I_0}{2.\pi.r} in \; A/m, \tag{4.26}$$

where
I_0 is the lightning return stroke current in LPA0A in Aand
r is the distance between the point of strike and the center of the shield volume in m.

From this follows, for the maximum value of the magnetic field in LPZ 0, the shielding factor SF is determined from Table 4.11.

h_m = mesh size [M] (hm $\leq$ 5m); rc = rod radius [m]; the permeability of the shield wires is μ and it approximates to 200.

In Table 4.12, the shielding factors for different materials used and the size of the shield mesh at two different frequencies of the LEMP frequency spectrum are shown. In Table 4.12, w_m is the width of the mesh and r is the distance of the lightning strike from the mesh. Note that where shielding from very high-frequency wireless communication system signals need to be constructed, with frequencies much higher (e.g. 2 GHz, or 2×10^9 Hz) compared to much lower frequencies for lightning (e.g. 5 to 100 MHz or 5×10^6 to 10^8 Hz), mesh sizes should be very much smaller. If the frequency of the signal is f, then the wavelength is $\lambda = c/f$, where c is the velocity of light, $c = 3 \times 10^8$ m/s.

Table 4.11 Determining the shielding factor SF

Material	Shielding factor SF (dB)	
	25kHz (first stroke)	1MHz (subsequent stroke)
Copper or aluminum	20log(8.5/ h_m)	20log(8.5/ h_m)
Steel	$20log \frac{(8.5/h_m)}{\sqrt{1+(18\times10^{-6})/r_c^2}}$	20log(8.5/ h_m)

Table 4.12 Magnetic attenuation of grids in case of a nearby lightning strike

Example steel grid			
w_m (m)	r (m)	dB at 25 kHz	dB at 1 MHz
0.012	0.0010	44	57
0.100	0.0060	37	39
0.200	0.0090	32	33
0.400	0.0125	26	27

The reduction of the magnetic field intensity from H_0 to H_1 in the LPZ 1 depends on the SF and is given by

$$H_{1/max} = \frac{H_{0/max}}{10^{\wedge}(SF/20)} \text{ in A}/\text{M}, \tag{4.27}$$

where
SF is the shielding factor and
H_0 is the magnetic field in LPZ 0 in A/m.

4.17.4 Implementation of the Magnetic Shield Attenuation of Building/Room Shield

To implement the magnetic shield attenuation of the building/room shields, extended metal components are crucial when shielding against the magnetic fields. Generally, a meshed interconnection is used to create an effective electromagnetic shield. The distance between adjacent mesh wires should be less than the lowest wavelength of the incoming signal from which we want to shield the systems inside the shielded cage. The steel reinforcement, when used in building, can be designed into an electromagnetic cage (hole shield). In reality, it is not possible for us to weld or stick together every junction in very large structures. A system typically having a size around 5 m is usually used to install a meshed system of conductors into the reinforcement. This meshed network is connected in an electrically safe way at the cross points.

Reinforcement mats in concrete are suitable for shielding purposes and it is usually laid at a later date when upgrading the existing system. It requires reinforcement mats to be galvanized to protect them from corrosion. The magnetic field inside the structure can be reduced over a wide frequency range by means of reduction loops, which arise as a result of the meshed equipotential bonding network. Three-dimensional meshed equipotential bonding network is formed by the interconnection of all metal components both inside and on the structures. This equipotential bonding network when installed in the lightning-protection zones will reduce the magnetic field by a factor of 2.

4.17.5 Cable Shielding

Cables need to be shielded as well. By shielding the cable, we reduce the effect of interference on the active cores and the interference emitted from the active cores to neighboring systems.

4.17.5.1 Double-Ended Shield Earthing

For good conductivity, shielded cables must be continuously connected along its length and the shields must be earthed at least at both ends. This is because only a shield earthed at both ends is able to reduce inductive and capacitance coupling. Cross-sectional area of the cable shields entering a building needs to be considered as a certain minimum to avoid the risk of the dangerous sparking. Without doing this, the shields are hardly able to carry the lightning current.

Minimum cross section of a cable shield (S_{Cmin}):

$$A_{cmin} = \frac{I_f . \rho_c . L_c . 10^6}{V_w} \left[mm^2 \right], \qquad (4.28)$$

where
ρ_c is shield resistivity;
I_f is lightning current flowing along the shield;
V_w is impulse with stand voltage of the system; and
L_c is cable length.

The shield connection system is typically tested with lightning current up to 10 kA (10/350 μs). For the first approximation, the lightning current of 10 kA can be used as the maximum value. Besides, V_w can be interpreted in different ways. The impulse withstand voltage strength of the cable is decisive.

4.17.5.2 Indirect Single-Ended Shield Earthing

For common operation, cable shields are sometimes earthed at only one end. This protection may only provide a certain attenuation from capacitive interference fields. However, it does not provide any protection against the electromagnetic induction arising from lightning strikes. The reason we sometimes use shields with single-ended earthing is to prevent the flow of low-frequency equalizing currents. In the extended installation like a bus cable, it can often stretch many hundreds of meters between buildings. For the older installations, one part of the earth-termination system may not operate normally if the meshed equipotential bonding network is absent. This will lead the interferences to occur as a result of multiple shield earthing. For a building, resulting potential differences of the different earth-termination system can allow low-frequency equalizing currents and the transients superimposed thereon, to flow.

At the same time, the current cable may burn if current is up to a few amperes. Furthermore, if signal frequency is in the similar frequency range to the interference signal, crosstalk can cause signal interference.

Implemented Electromagnetic Compatibility (EMC) requirements and preventing equalizing current can solve the signal interference by combining direct single-ended and indirect shield earthing. These shields are directly connected to the local equipotential bonding system at a central point such as the control room. The shields are indirectly connected to the earth potential via isolating spark gaps at the far ends of the cable. Basically, the resistance of the spark gap is around 10 GΩ, which means that during the surge-free operation, the current will be prevented from being equalized. If lightning strike occurs, the spark gap will need to ignite and discharges the interference pulse without destruction. This helps to reduce the residual impulse on the active cable cores and the terminal devices are subjected to become less stressed. Furthermore, a gas discharge tube is recommended at one side between the cable shield and the equipotential bonding system to eliminate the interference impulses.

4.17.5.3   *Low-Impedance Shielding Earthing*

A cable shield has to conduct impulse currents up to several kA. The impulse current will flow to the shield, then from the shield to the earth when it is discharging. At the same time, the potential differences between shield and the earth is created by the impedances of the cable shield and the shield terminal. This can be very dangerous since the potential differences formed are able to destroy the insulation of the conductors or connected device. Therefore, quality of the cable shield used needs to be considered and it will affect the number of shield earthings required. Suitably large-area contact terminals with the slipping spring elements are used for shield protection.

4.17.5.4   *Maximum Length of the Shielded Cables*

The interference impulse currents usually flows through the shield resistance, creating a voltage drop on the cable shield. Thus, the length of the cable needs to be controlled because it will determine the permissible transfer impedance for the cable shield. Voltage drop due to the length of the shield cannot be ignored in this case. This is because if the voltage drop becomes higher than the insulation strength of the system, a surge arrester needs to be present.

4.17.5.5   Extension of the LPZs with the Help of Shielded Cables

Surge protectors or arrestors are not needed if the shielded cable is used in between two identical LPZs. This is because the interferences from the surroundings of the shield cable and the meshed equipotential bonding will be suppressed by the shield.

However, this needs to be monitored because adverse situations may arise due to peculiar installation conditions. Potentially adverse situations may arise due to (i) the supply of the terminal devices at a different main low-voltage distribution boards, (ii) the TN-C systems, (iii) high transfer impedance of the cable shield, or (iv) insufficient earthing of the shield. Failure could lead to residual interferences with the signal transferred by the cable core. This type of interferences can be controlled by using a high-quality shielded cable or surge protection devices.

References

Barry, J.D.: Ball Lightning and bead Lightning. Plenum (1980)
Betz, H.D.: Schumann, U., Laroche, P. (eds.) Lightning: Principles, Instruments and Applications. Springer (2009)
Blume, S.: High Voltage Protection of Telecommunication Systems. IEEE Press (2011)
Bazelyan, E.M., Raizer, Y.P.: Lightning Physics and Lightning Protection. IOP Publishing (2000)
Chalmers, J.A.: Atmospheric Electricity. Pergamon, 2nd edn (1984)
Cohen, R.L.: How to Protect your House and its Contents from Lightning: Surge Protection: IEEE Guide for Surge Protection of Equipment Connected to AC Power and Communication Circuits. IEEE Press, New York (2005)
Cooray, V.: An Introduction to Lightning. Springer (2015)
Cooray, V. (ed.): The Lightning Flash. IET (2003)
Cooray, V. (Ed.): Lightning Protection (2010)
Cooray, V. (Ed.): Lightning Electromagnetics (2012)
DEHN.: Lightning Protection Guide, 3rd edn (2014)
DEHN.: Lightning Protection Guide, 2nd Edition (2012)
Federal Aviation Administration.: Aircraft Lightning Protection Handbook (1989)
Hasse, P.: Overvoltage Protection for Low Voltage Systems. IET (2000)
Henshaw, M.L.: Protection against Lightning. BSI (2007)
Golde, R.H.: Lightning Protection. Arnold (1971)
Golde, R.H. (Ed.): Lightning Vol 1: Physics of Lightning. Academic Press (1977)
Golde, R.H. (Ed.): Lightning Vol 2: Engineering Applications. Academic Press (1977)
Legrande, Protection Against Lightning Effects (2009)
IEC Protection against Lightning Vol 1: General Principles
IEC Protection against Lightning Vol 2: Risk Management
IEC Protection against Lightning Vol 3, Physical damage to Structures and Life Hazard
IEC Protection against Lightning Vol 4: Electrical and Electronic Systems within Structures
IEC Protection against Lightning Vol 5: Services
Kramer, J.: Lightning: Nature in Action. Lerner (1992)
NFPA.: Standard for Installation of Lightning Protection Systems (2020)
Rakov, V.A., Uman, M.A.: Lightning: Physics and effects. CUP (2003)
Rakov, V.A.: Fundamentals of Lightning. CUP (2016)
Uman, M.A.: All About Lightning. Dover (1986)
Uman, M.A.: The Lightning Discharge. Academic Press (1987)
Uman, M.A.: The Art and Science of Lightning Protection. CUP (2008)

Chapter 5
Lightning Physics, Modeling, and Radiated Electromagnetic Fields

Abstract This chapter presents a self-consistent, science-based mathematical model for representing the most destructive part of the lightning flash: the lightning return stroke. It presents the validation of representing the lightning return stroke wave as an electromagnetic wave. Following from that, the chapter presents the modeling of the return stroke electromagnetic wave using a distributed electric circuit model, which is an approximation of the electromagnetic phenomena. The electric circuit model is connected to the generation of the lightning electric and magnetic pulses (LEMP) radiated by the lightning return stroke currents. The lightning return stroke model and the calculation of the radiated LEMP provide a self-contained computer-based model to determine the most important parameters required by the lightning protection engineer for designing the protection of both ground and airborne electrical/electronic systems and structures. The computer-based tool is validated by comparing computer-simulated results for cloud to ground lightning flash to ground-based return stroke current and LEMP measurements.

5.1 Introduction: The Need for Computer-Based Testbeds for Lightning Testing

Lightning return stroke currents and voltages can be of the order of 30 kA and 50 MV, respectively, which are much larger than the 2 kA current that may flow over the 500 kV very high-voltage power system lines, cables, and apparatuses such as transformers, circuit breakers, and switches. The electric power system which spans large geographical areas of a country is exposed to lightning flashes and the system components as well as the protective equipment such as circuit breakers need to be able to handle these high voltages and currents. Hence, they need to be tested for their ability to withstand and handle these severe electric stresses caused by lightning currents and voltages. They are normally tested in very expensive high-voltage laboratories piece by piece and parts of the apparatus (e.g. a short length of the power cable or the bushing of a transformer). However, it is impossible to

© The Author(s), under exclusive license to Springer Nature Switzerland AG 2022
P. Hoole and S. Hoole, *Lightning Engineering: Physics, Computer-based Test-bed, Protection of Ground and Airborne Systems*,
https://doi.org/10.1007/978-3-030-94728-6_5

generate the complex lightning waveforms and characteristics accurately in a high-voltage laboratory. Moreover, the whole systems when interconnected together, or a large aircraft in its entirety, cannot be tested under laboratory conditions. Therefore, there is an important need for reliable mathematical models of the lightning flash and its interaction with systems such as the electric power system and apparatuses, which may be used to develop a reliable computer-based test setup of apparatus, structures, and systems.

The need for such models also extends to airborne vehicles and systems as well. An aircraft, for instance, whether military or commercial, is struck by lightning when flying under a 50 MV thunder cloud or close to a thundercloud. Normally pilots are instructed to keep the aircraft about 50 km away from thunderclouds to avoid being struck by lightning or impacted by the lightning-radiated electromagnetic pulses (LEMPs) which can be as severe as nuclear bomb-generated electromagnetic pulses (NEMPs). Although various parts of an aircraft may be tested in expensive, local high-voltage laboratories as with power grid equipment and apparatuses, there are many conditions of the lightning flash that cannot be reproduced in a high-voltage laboratory, including the very peculiar lightning return stroke waveform and the LEMPs generated by not only the return stroke current at the point of attachment to ground, but also at the points of attachment to the aircraft in flight and the LEMPs produced by the traveling return stroke wave. Because of their expense, there are only a few large high-voltage laboratories found in the world, most with voltages limited to a few megavolts. Often the entire aircraft is attached to the lightning return strike channel, and thus to understand the currents and voltages and the forces generated on the aircraft body there is a need for reliable lightning and aircraft models to computer-simulated aircraft-lightning electrodynamics and the engineering parameters of the voltages, currents, and LEMPs produced.

In Fig. 5.1a and b, the lightning flash, with a single lightning channel connecting the thundercloud and the ground in a cloud to ground (CG) flash is shown. The branches are minimal. This is an ideal condition to be considered as a single conductor transmission line carrying the lightning return stroke, The return stroke is taken to be an electromagnetic wave traveling from ground to cloud, when the 50 MV lightning leader coming down from the cloud connects to the zero-volt ground. In Fig. 5.1c, additional electrical activities in the form of intracloud (IC) flashes taking place almost at the same time as the cloud to ground (CG) flash are shown. These additional electrical activities could have an effect on the transmission line parameters such as the distributed capacitances and inductances close to the cloud. Similarly, the lightning flashes inside the cloud can also be modeled by a transmission line, with the parameters worked out for a line parallel to the ground. In Fig. 5.1d, the bend or kink in the upper part of the lightning flash path could cause additional reflections of the return stroke at such bends. A horizontal intracloud or cloud to air flash can be represented by a horizontal, instead of vertical, transmission line.

Moreover, the telecommunications electronics, control equipment, and navigational electronics which are mostly digital are very sensitive to voltages and currents generated on the low-voltage digital and analogue electronic circuits. To protect these critical systems, a clear understanding and knowledge of the rates of rise of

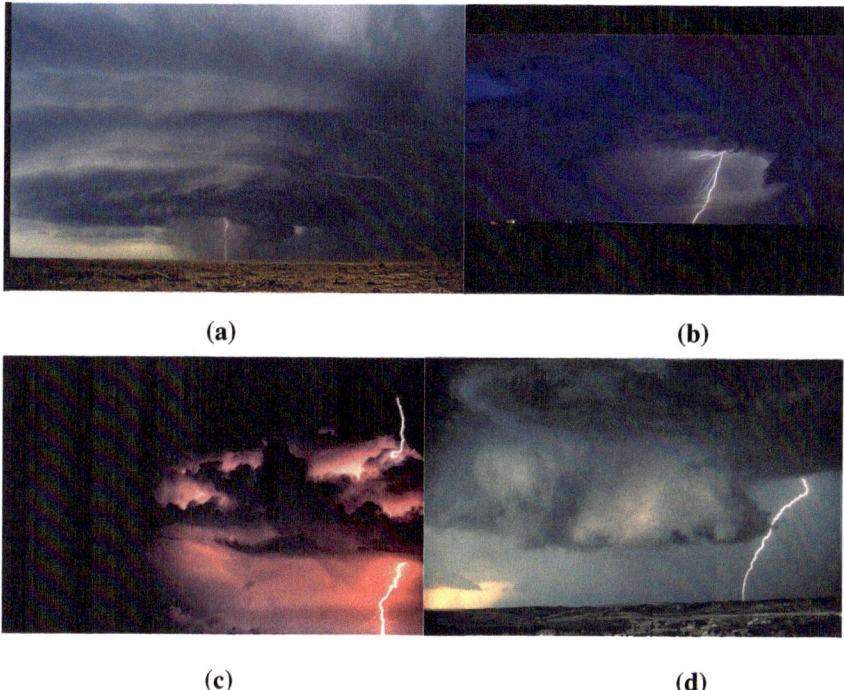

(a)   (b)

(c)   (d)

Fig. 5.1 The lightning flash and the transmission line model. **a** The return stroke as an electromagnetic wave traveling along a straight transmission line connected between the thundercloud and earth. **b** A cloud to ground flash with a straight flash path to the ground. **c** The return stroke as a cloud to earth transmission line, with simultaneous lines of conduction (flashes) inside the cloud. **d** A cloud to ground flash with a kink in the leader close to the cloud, thus requiring the return stroke (the bright flash observed) to travel through a bended transmission like structure (Credit for photographs: NOAA_NSSL, USA. With permission)

currents and voltages, as well as the energy and electric charges of the coupled lightning surges induced need to be known. These may be determined from digital computer-simulated models (using software codes implementing the mathematical models) that calculate realistic transient currents and voltages, long-term continuing currents, and electric charges produced by the return stroke of the lightning flash. The return stroke is the most severe electric threat portion of the lightning flash.

5.2 Lightning Return Stroke

5.2.1 Electromagnetic Wave Nature of the Lightning Return Stroke

Before we discuss the modeling of the lightning return stroke, we will seek to establish here the validity of looking at the lightning return stroke as an electromagnetic wave. The electric circuit model will follow thereon, since the electric circuits are but approximations of electromagnetic wave phenomena, where the capacitance of the electric circuit is associated with the electric fields, the inductance with the magnetic fields, and the resistance of the circuit with the Joule loss of the electromagnetic energy transferred into heat energy. The transmission of electromagnetic waves along a single electrical conducting channel of the vertical or horizontal electric conductor, with no outer conductor, is well established. Fundamentals of electromagnetic theory are used to show the validity of considering the return stroke as a transverse magnetic wave along such a transmission line. Maxwell's equations-based electromagnetic wave equation given below is considered for electromagnetic waves traveling over coaxial cables using a transmission line model.

$$\nabla^2 \underline{E} - \mu\sigma \frac{\partial \underline{E}}{\partial t} - \mu_0\varepsilon_0 \frac{\partial^2 \underline{E}}{\partial t^2} = \nabla\rho / \varepsilon_0, \tag{5.1}$$

where the charge relaxation time in a linear, homogeneous, and isotropic conductor which is given by $\tau = \varepsilon_0/\sigma = 2 \times 10^{-5}$ for $\sigma = 4242 \ \Omega^{-1} \ m^{-1}$. A single conductor transmission line model for the return stroke waves is valid with the assumption that the net free charge within the conductor rapidly vanishes and any excess charge is located on the surface of the conductor. Thus, for a wave propagating in the positive z-direction, the equation is

$$\nabla^2 E_z - \mu_0\varepsilon_0 \frac{\partial^2 E_z}{\partial t^2} - \mu\sigma \frac{\partial E_z}{\partial t} = 0, \tag{5.2}$$

where the cylindrical coordinates r, ϕ, and z are used. Let $E_z = E(R)\exp(i(-\omega t + hz))$, where h is the vertical (z-directed) wave number.
   Defining

$$k^2 = \varepsilon_0\mu_0\omega^2 + i\mu_0\sigma\omega, \tag{5.3}$$

$$R = r\sqrt{k^2 - h^2}, \tag{5.4}$$

and permittivity, for f << $1/2\pi \ (\sigma/\varepsilon_0) = 8 \times 10^{12}$ Hz (with $\sigma = 4242 \ \Omega^{-1} \ m^{-1}$ for an ionized lightning channel), is given as a complex permittivity,

$$\varepsilon_p = \varepsilon_0 + i\sigma/\omega \tag{5.5}$$

$$= i\sigma/\omega = \varepsilon_{pi} \quad \text{inside the lightning channel} \tag{5.6}$$

$$= \varepsilon_0 = \varepsilon_{pe} \quad \text{outside the lightning channel} \tag{5.7}$$

Consider the transverse magnetic wave, where only B_ϕ, E_z, and E_r have nonzero values. From Faraday's law,

$$i\omega B_\varphi = ihE_r - \partial E_z/\partial r. \tag{5.8}$$

From Ampere's law in Maxwell's equations,

$$-\mu_0\varepsilon_p i\omega E_r = -ihB_\varphi. \tag{5.9}$$

From (5.8) and (5.9),

$$B_\varphi = \left[i\varepsilon_p\omega/(\omega^2\mu_0\varepsilon_p - h^2)\right]\partial E_z/\partial r. \tag{5.10}$$

The factor $\exp(i(-\omega t + hz))$ is dropped since it is common to B_ϕ, E_z, and E_r. It is to be noted that $h = \beta - i\alpha$, $ih = \alpha - i\beta$. Once E_z is solved for, B_ϕ and E_r, may be determined from (5.9) and (5.10). Equation (5.1) now becomes Bessel's equation

$$d^2E/dR^2 + (1/R)dE/dR + E = 0. \tag{5.11}$$

For the axially symmetric solution of (5.11), mode $n = 0$, and for E to be finite at the axis of the conductor and everywhere else, the solution for $0 < r < a$ is

$$E = a_0 J_0(R), \tag{5.12}$$

where a_o is a constant and J_0 is Bessel's function of zeroth order. Outside the conductor, remembering that open space surrounds the vertical lightning channel, for complex values of R only H_0^1 Hankels' function of the first kind vanishes as r goes to infinity on the positive imaginary half plane of R. Hence for $a < r < \infty$,

$$E = b_0 H_0^1(R). \tag{5.13}$$

Substituting (5.12) and (5.13) into (5.10),

$$B\phi = ik\sqrt{\left[\mu_0\varepsilon_p/(k^2 - h_2)\right]}a_0[dJ_0(R)/dR] \quad 0 < r \le a \tag{5.14}$$

$$= ik\sqrt{\left[\mu_0\varepsilon_p/(k^2 - h_2)\right]}b_0\left[dH_0^1(R)/dE\right] \quad a < r \le \infty \tag{5.15}$$

The general permittivity has been retained to keep the expressions neat. Both B_ϕ and E_z must satisfy the continuity conditions at the boundary. When r = a, substituting r = a into the two pairs of Eqs. (5.12)–(5.15), two equations are obtained, dividing one by the other and rearranging,

$$
\left[\sqrt{(k_{e^2} - h^2)}/k_e/\sqrt{(\mu_0\varepsilon_{pe})}\right] \times \left\{H_0^1(R_e)/\left[dH_0^1(R_e)/dR\right]\right\}
$$
$$
= \sqrt{(k_{i^2} - h^2)}/k_i\sqrt{(\mu_0\varepsilon_{pe})} \times \left\{J_0(R_i)/\left[dJ_0(R_i)/dR\right]\right\} \tag{5.16}
$$

where the subscripts i and e stand for internal to the conductor and external to the conductor, respectively. For small values of R_e, $H_0(R_e)$ can be approximately represented by

$$
H_0(R_e) = (2i/\Pi)\log_e\left(\eta R_e/2i\right); \quad dH_0^1(R_e)/dR = 2i/\Pi \ R_e \tag{5.17}
$$

where $\eta = 1.781$ the Euler-Mascheroni constant. Hence, re-write (5.16) as

$$
-(2/\eta)^2\left(\eta R_e/2i\right)^2\ln\left[\eta R_e/2i\right]^2 = J_0(R_i)/\left[dJ_0(R_i)/dR\right] \times 2a \ k_e\left(\varepsilon_{pe}/\varepsilon_{pi}\right)^{1/2} \tag{5.18}
$$

The lightning return stroke traveling wave-radiated light intensity measured for the return stroke current peaks by lightning photography. The electromagnetic wave current peaks that are determined by the phase constant of the wave number of Eq. (5.18) match as shown in Table 5.1. Therefore, the electromagnetic wave along a conductor represented by (5.18) correctly captures the lightning return stroke wave traveling along the lightning leader channel, captured by photographs of the return stroke-generated traveling light intensity of the moving return stroke.

Table 5.1 shows that the transverse magnetic wave along an unmagnetized electric plasma (the lightning channel). It shows electric current wave peaks observed through photography of the natural lightning return stroke. The return stroke peaks at 5 µs, and the return stroke modeled as a transverse magnetic wave peaks at 4.9 µs at 400 m above ground. At a height of 1600 m above ground, the lightning return stroke current wave peaks at 20 µs, and the transverse magnetic wave peaks at 18 µs. The close correspondence indicates that the lightning return stroke is an electromagnetic wave,

Table 5.1 Time for current peak at different heights with return stroke light intensity

Height from ground (m)	Time to peak of return stroke light intensity (µs)	Time to peak of transmission line transverse magnetic (TM) waves (µs)
400	5	4.9
800	10	9
1600	20	18
2000	28	25

more specifically, a transverse magnetic (TM) wave that may be more approximately modeled by an LCR distributed transmission line with a voltage and current wave traveling along it.

5.2.2 Lightning Return Stroke Models

Among the various models for lightning return strokes (LRS) that exist, the lossy, distributed transmission line (DLCR) model, which is presented herein, is a dependable, comprehensive, and well-tested model. The model contains inductance (L), capacitance (C), and the heat-loss resistance (R). Recently, many alternative models have been proposed, and the adequacy of the DLCR model (DLCRM) has been questioned because of some shortcomings in the previously reported DLCRM simulation results. This section corrects some of these shortcomings, such as correct representation and computation of the LRS current pulse wavefront, and the special nature of the attachment point at the earth end. In this section where the DLCRM model proposed is a self-consistent model, within the assumptions stated and justified, it is shown that the LRS velocity predicted by the DLCRM model is about 50 to 70 percent less than the velocity of light (e.g. c/3), as expected. The velocity determined from the DLCRM presented here agrees with the measured LRS velocity and captures also the drop in velocity as the LRS moves away from the segments away from the ground.

When considering both the physical principles and observations of the earth flash lightning return stroke (LRS), the DLCRM yields results that are consistent with lightning measurements. The DLCRM may be used to obtain important engineering parameters that are not easily measured; one such example is the very high rate of rise of currents on a sub-microsecond timescale (e.g. 98 kA/μs), whereas the microsecond rate of rise of current may be a tenth of the sub-microsecond values. Relating the computed electric and magnetic fields radiated by the LRS currents obtained from the DLCRM shows the correlation between the LRS current waveforms and the electromagnetic field waveforms at different distances from the LRS channel. Moreover, for unbranched first and subsequent return strokes, the model's electrical parameter values such as inductance (L), capacitance (C), and resistance (R) may be calculated from basic principles, with the assumptions made clearly defined and justified. Among the various models for lightning return strokes, the lossy transmission line model (the DLCRM) remains the most dependable when considering both the physical principles and measurements that provide a consistent and self-contained justification for the LCR model.

Although the frequent lightning flashes are the flashes that occur within a thunder cloud (intra-cloud flashes) the most frequently studied flashes are those that occur between the thunder cloud and ground. These earth flashes are of most interest from an engineering point of view because of their close interaction with power and telecommunication systems, aircraft, and rockets in flight close to a thunderstorm, and the threat they pose to various electronic systems, and to human life in a limited sense.

A single lightning flash between a thunder cloud and earth may last for half a second. This single flash will contain the first return stroke and two to three subsequent return strokes. Each of these strokes may last for about one hundred milliseconds, with an interval between each stroke. Each stroke is made up of a rapidly moving current pulse (electromagnetic pulse) with sub-microsecond rise times and fractional changes. Even when the cloud to ground flash does not directly attach itself to an electronic system or electrically sensitive object (e.g. a rocket), it radiates electromagnetic waves with sub-microsecond changes which may interact destructively with avionics and ground electronic systems. The mathematical modeling and computer simulation of the earth to ground flashes are not only of interest from the perspective of gaining greater knowledge of lightning physics (since they yield parameters that are normally not measurable, such as currents through the channel above the ground), but lightning return stroke simulations may help us also to predict and take protective action against the adverse electrical and thermal effects of a lightning flash on airborne and ground vehicles and systems.

We consider first the general measured characteristics of lightning return stroke currents and radiated electromagnetic fields. Second, the origin of the electrical circuit model of the lightning return stroke, the lumped circuit model, is presented. Third, the transmission line model and the dispersion characteristics, that is, the quasi-transverse electromagnetic (quasi-TEM) wave and the distributed circuit model are considered. Fourth, the accuracy of the numerical solution of the quasi-TEM return stroke wave is tested. Fifth, simulation results of the downward earth flash return stroke, including currents and voltages, are presented. Lastly, an analysis of the LRS currents and the radiated electromagnetic pulses (LEMP), calculated from the DLCRM currents, is compared to measured LRS currents and LEMPs.

5.3   Analysis of Experimental Data of Lightning Return Stroke

5.3.1   Background

Photography, current measurements, and electromagnetic field measurements have been extensively used since the early days of lightning research. Boy's EXPLAIN. WHICH BOY'S?camera in 1926 originated the era of lightning photography. The progressions of both the lightning leader and the return stroke have been photographed. These photographs first showed the stepped nature of the first leader and gave good estimates of the bright tips observed in the stepped leader, the dart leader, and the return strokes well above the ground. Photography has also been extensively used in triggered lightning investigation to obtain the geometry of the return stroke channel and the stroke velocities. Return stroke currents have been measured by measuring the current along a tall conductor struck by lightning. These measurements gave an idea of the return stroke peak current, current rise rate, the action integral, and the current wave forms at the foot of the channel. Electric field changes due to the leader (L—change), return stroke (R—change), and continuing discharge (C—change) have been recorded. These records further reveal short, sharp pulses during intervals between component strokes (J—change) as well as during the flow of continuing currents (M—change). Recent electromagnetic field measurements have sought to measure sub-microsecond changes and fields from positive flashes. Spectroscopic measurements and sound measurements too have been made. There are not many spectroscopic measurements available, and what has been analyzed does not agree well: there are obvious practical difficulties in getting a clean light spectrum of lightning. A lack of correlated measurements does make the understanding of data precarious. Although artificially (rocket) triggered lightning lends itself to correlated measurements, there is still much work to be done to correlate and interpret them in agreement with the physics of lightning. Moreover, the relation between natural lightning and artificially triggered lightning is another area in which more precise work still needs to be done. A review of the important measurements that are pertinent to natural lightning return stroke modeling is given below with comments. A most exhaustive amount of data has been obtained for return stroke currents, return stroke velocity, and the lightning electromagnetic pulse (LEMP).

5.3.2 Lightning Current and Electromagnetic Field Measurements

Although the lightning phenomenon has been observed for centuries (mainly associated with light and fire) it is only in the past 60 years that a massive amount of data has been published on the lightning discharge, the major part of it being confined to earth (cloud-to-ground) flashes. The most notable work has come from Schonland (1930–1950s), Berger (1960s), and Uman and associates at the University of Florida (1970s to date). Schonland's work forms a good foundation. Berger's work is the best and most comprehensive. Uman's output is massive, spanning over 40 years, providing very valuable data on triggered lightning, though still inconclusive and at times controversial in its interpretation and with regard to the empirical return stroke models developed on the basis of the observed electric and magnetic fields. There is indeed a great need for different schools of thought on lightning to come together to work towards an understanding and mathematical model that is not only consistent with the measurement but also with the plasma and electromagnetic principles that underlie the observations. A summary of measured parameters of earth flash return stroke is given in Tables 5.2 and 5.3.

From Table 5.2, it is obvious why the positive strokes, mostly observed in winter thunderstorms, are more severe. The peak current of a positive first stroke can be as high as 250 kA, whereas for a negative stroke it is around 30 kA. However, if we consider that the destructive power of the lightning current is to be associated with the rate of rise of current, the negative stroke is more severe because of the lower rise

Table 5.2 Characteristics of return strokes

	Negative and Positive flashes				
	Negative Downward flashes			Upward flashes	
	Downward Negative strokes	Downward	Upward Negative Strokes		
	Negative first strokes	Subsequent strokes	First strokes	Subsequent strokes	Positive first strokes
Peak current(kA)	30	12	07	08	4.6–250
Maximum current di/dt steepness (kA/μs)	12	40	5	13	0.2–32
Time to crest (μs)	5.5	1.1	4	1.3	3.5–200
Time to half value (ms)	75	32	35	31	25–2000
Impulse charge(C)	4.5	0.95	0.5	0.6	2–150
$\int I^2 \, dt \ (A^2 \, s)$					-1.5×10^7
Total charge(C)	5.2	1.4	–	–	–
Flash charge(C)	7.5	–	–	–	80
Av. velocity(m/s)	0.7×10^8	0.8×10^8	–	–	–

Note A flash is defined as a sum of individual strokes (i.e. a sum of the first and all subsequent strokes).

Table 5.3 Significant electric field measurements reported for downward negative flashes by different authors

| | Maximum rate of rise of vertical electrical fields at 100 km (kV/m/μs) | | Typical rise times (μs) |
	First strokes	Subsequent strokes	
Tiller et al (1976)	3	2.5	3.0
Lin and Uman (1973), Lin et al. (1979)	2.15, 1.7, 2.8, 1.2	2.3, 1.56, 3.3, 1.08	2.0
Cooray and Lundquist (1982), Cooray (1984)	0.76	–	7.0
Uman (1985), Fisher and Uman (1972)	1.4	1.4	1[a]
Weidman and Krider (1980, 1982), Barry (1980)	45.4	20, 40.6	0.1[a]

[a]10–90% rise times

times (or higher rate of rise of current) when compared to the positive flashes. The energy associated with the flash is the action integral, and the return stroke velocity (which determines the return stroke current) is seen to be about three to four times less than the velocity of light.

Some of the values that characterize lightning-radiated electromagnetic fields or pulses (LEMPs) are given in Table 5.3 with some reservation. References are included to show some of the differences in observed data given by various workers. Some of the differences may be due to the greater accuracy of the observation equipment used in some cases. This is the case for the high value of 45.4 kV/m/μs for the first LRS and 40.6 kV/m/μs (for subsequent LRS) observed for maximum rate of rise of electric field. The difference is due to the sub-microsecond rise times that the measuring equipment was able to capture.

Measurements made in the USA and Sweden may show discrepancies due to the differences in the terrain over which the LEMP travels before being captured by the measuring equipment. Moreover, the values shown in Table 5.3 are for fields measured at close distances to the flash and then normalized to 100 km by a 1/D factor, where D is the distance from the flash. The strokes sampled in Sweden number about five hundred (500 flashes), the other figures are from a sample of around 100 flashes. The standard deviation for the first strokes is higher, and the values thus have a wider spread; this could be a possible reason for the discrepancy in the [(dE/dt) first LRS/(dE/dt) to subsequent LRS] ratio reported by different workers. The higher dE/dt values observed are due to the 10–90% rise times of around 100 ns which were observed, in comparison to the 1 μs observed by others. Another interesting feature is that the first stroke fields have a lower rate of rise than the subsequent strokes.

In Figs. 5.2 and 5.3, the measured electric and magnetic fields at distances of 2 and 15 km from the lightning flash, respectively, are given.

Note that in both cases there are common features in the electric field at 2 and 15 km: an initial peak, then a dip, and finally a closely increasing ramp-like tail. Similarly for the magnetic field: an initial peak, followed by a dip, and then a hump

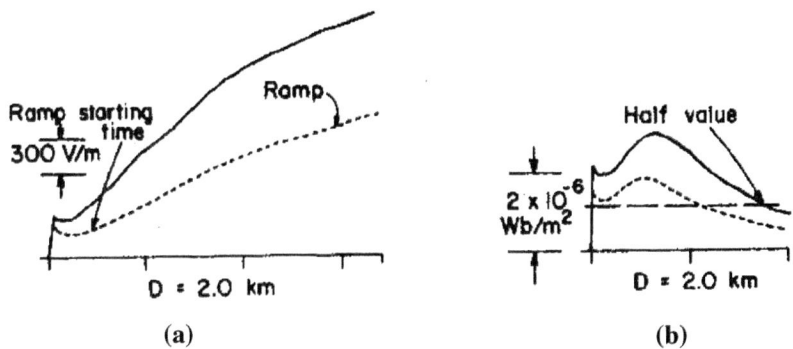

(a)   (b)

Fig. 5.2 Measured **a** electric (in V/m) and **b** magnetic field (in Wbs/m^2) at 2 km from a first (solid line) and subsequent (dotted line) return strokes (Lin et al., J Geophy Res., 1979. Credit: J. Geophys. Res. With permission). The measured fields shown are from 0 to 100 μs

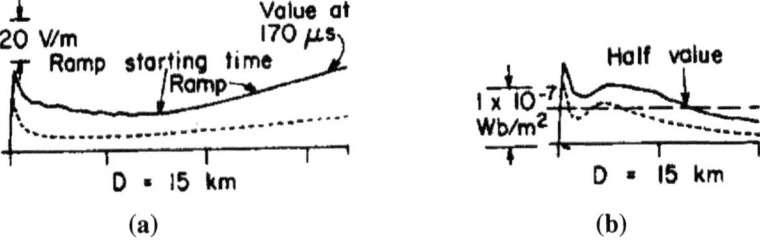

Fig. 5.3 Measured **a** electric field (in V/m) and **b** magnetic field (in Wbs/m^2) at 15 km from a first (solid line) and subsequent (dotted line) return strokes (from Lin et al., 1977; Credit: J Geophys. Res. With permission). The measured fields shown are from 0 to 100 μs

which slowly decays. When we model and implement an LRS model, we will expect the electric and magnetic fields calculated from the LRS currents obtained by the LRS model to resemble the measured electric and magnetic fields.

5.3.3   The Empirical Models: Lumped Circuit Model and the Curve Fitting Model

There are, in general, two different approaches to LRS modeling which continue to be developed and discussed. These are what we may term as the Empirical Models and the Distributed Inductance-Capacitance-Resistance Transmission Line Model (DLCRM). There is a third category, which is the Shock Wave Model (SWM), where gas dynamics theory is applied to the presence of high pressures along the lightning channel axis, associated with the lightning leader tip, the return stroke wavefront, and the subsequent radial shock wave that results in thunder, an acoustic wave. For instance, the LRS is considered a non-linear electron acoustic wave (Fowler 1982). It has received less attention since it has been found inadequate to explain or properly represent the important electrical characteristics observed in LRS, and using the model requires detailed knowledge of thermal and electrical conductivities, and recombination and ionization coefficients, as well as simultaneous solutions of Maxwell's equations, and momentum, energy, and mass equations. Moreover, a convincing case can be made to show that the LRS is a quasi-transverse electromagnetic wave (quasi-TEM) moving along an un-magnetized ionized, plasma channel, which in turn allows the lightning channel to be modeled by a lossy transmission line along which energy flow is sustained by a quasi-TEM wave.

5.3.3.1   The Lumped Circuit Model (LCM)

The origin of the LCM, also called the Bruce-Golde model, may be traced back to the fact that the lightning return stroke currents measured at ground level very much resemble the time-domain waveform of voltages in the long high-voltage sparks produced in the high-voltage laboratories used for testing power system equipment for lightning surges. We may best illustrate this model by using simple circuit theory. Very crude, intuitive models for the leader and the return stroke, as outlined in Chap. 1, consist of a capacitor (C)-resistor (R) circuit to which a stepped voltage V(t) is applied, in the case of the leader stroke. In the case of the return stroke, we have a charged capacitor (C)-series inductor (L)-Resistor (R) circuit. In the case of the leader, as the stepped leader progresses downwards, for each leader step, a step voltage V(t) from the thundercloud is applied to the series CR elements at the cloud end. The leader channel is represented by a resistor R, and the electrostatic energy stored at the tip of the leader is represented by a capacitor C between the leader tip and the ground.

Thus, the nature of the leader is here represented by an RC circuit triggered by a constant voltage source, which produces a leader current I(t), for which we get

$$RI(t) + \frac{1}{C} \int I(t)dt = V(t). \tag{5.19}$$

On differentiating (5.19) and solving the resulting differential equation

$$I(t) = \exp(-t/RC) \int (1/R) \exp(t/RC) \, dV/dt + Io \tag{5.20}$$

where I_o is the continuing current flowing along the lightning leader channel.

After the initial rise of V, the voltage that drives the leader current, if we should consider dV/dt = 0, the leader current is simply Io exp (−t/RC). At each step, the leader will be visible as a bright light pulse, which rapidly decays in intensity, following the current which produces the visible light radiation. If we assume significant magnetic energy in the leader, then the leader current characteristics will also look like the LRS lumped circuit current characteristics.

In the case of the return stroke, assume that all the cloud charge is transferred to the leader as the leader is connected to ground. Thus representing the charge stored in the leader by a capacitor C, and the lightning leader channel being considered as a resistor (R) and inductor (L) in series, the return stroke current flows when the leader channel and becomes attached to the ground. For the lumped LCR circuit, Kirchhoff's law gives

$$LdI/dt + RI + 1/C \int Idt = 0. \tag{5.21}$$

Differentiating (5.21) gives

$$L \, d^2I/dt^2 + R \, dI/dt + I/C = 0 \tag{5.22}$$

This has the solution form

$$I(t) = A \exp\left[\frac{-R - K}{2L}\right]t + B \exp\left[\frac{-R + K}{2L}\right]t, \qquad (5.23)$$

where

$$K = \left(R^2 - \left(4L/C\right)\right)^{1/2} \qquad (5.24)$$

Using $I(t) = 0$ when $t = 0$, we obtain

$$I(t) = I_m\left[\exp\left[-\frac{R - Kt}{2L}\right] - \exp\left[-\frac{R + Kt}{2L}\right]\right], \qquad (5.25)$$

where I_m is the peak LRS current. The form of (5.25) resembles the Bruce-Golde model. Differentiating (5.25) and setting $dI/dt = 0$, we have the rise time given by

$$t_T = \frac{L}{K}\log_n\left[\frac{R + K}{R - K}\right] \text{seconds}, \qquad (5.26)$$

which is a strong function of L/R, when $R^2 \gg 4L/C$ and $K \sim R$, thus $L/K = L/R$. We note that the important parameter t_T depends on careful estimation of L/R. We shall note later that for the current at the earth end, if L/R at the earth end has been set to zero or close to zero then the wavefront will be distorted and could yield singularity solutions. Where these factors have not been attended to, the distribution LCR models become unreliable for rise time estimation. In computation it is important to keep the time step $\Delta t \ll L/R$, and the accuracy is easily checked by ensuring there are computed points on the wave front.

5.3.3.2 The Curve Fitting Model (CFM)

The Lumped Circuit Model (LCM) does not simulate the traveling wave scenario of the LRS. In order to overcome this fundamental weakness in LCM, several papers have been published to specify the current-time and current-height characteristics of the LRS as it travels along the leader. The curve fitting nature of these empirical models goes one important step beyond the LCM model. Whereas the LCM was concerned solely about a waveform that matches the lightning currents measured at ground level, the CFM models searched for a model that will also yield radiated electromagnetic fields that have been measured. The LRS current waveforms are made up of three different current components, of which one is a direct current component, and another is the Bruce-Golde model like double exponential current. Parameters such as the peak current, the time constants, and the velocity of the return stroke may be obtained from ground measurements, including those of currents and/or the lightning-radiated electromagnetic pulse LEMP. The empirically specified parameters of current-time characteristics are adjusted to get the measured fields radiated by the LRS currents along the channel.

Some of the more recent LCM and related models that spill over into modifications of the DLCR models (further complicating, and blurring the issues at times) sought to specify conductivity-time characteristics, or LRS radius-time characteristics in order to get time-varying electric field and magnetic field signatures that closely resemble measured electric and magnetic fields. It is a curve fitting method, without a well-reasoned out or self-consistent LRS model: keep changing the current waveforms, calculate the radiated electromagnetic fields from it, and then get back to adjust the current waveforms and numbers such as peak current, rise time, attenuation along the channel, and direct current, until the calculated radiation (electric and magnetic) fields match the measured radiated fields. However, it is open to question whether such models are true to the LRS physical processes and whether they may rightly be called engineering models of the LRS. We hope that in the future, all those that are at the cutting edge of lightning research will come to a consensus on terms and definitions. The CFM models have been further extended by exploring the effects of an assumed corona layer surrounding the lightning channel, or by assuming a two-component electric charge density flow, with different time constants to get radiated electric fields closer to the measured electric fields.

5.4   The Distributed Circuit, Transmission Line Model (DLCRM)

5.4.1   Background to the DLCRM

The second approach is to model the return stroke by the LCR transmission line, DLCRM. For the model to have self-consistency, it has been shown that it is proper to represent the LRS by a quasi-TEM wave traveling along a lossy transmission line. The L, C, and R elements of the line may be determined from basic electromagnetic principles. In the original work done on DLCRM, the case was made that the DLCRM is attractive for the determination of currents even above ground level, which could also include the presence of an aircraft, lightning conductor, or transmission tower.

The shortcomings of the earlier work, sometimes unknowingly carried on in more recent work, may be worth pointing out so that extra care is taken when developing and coding the DLCRM:

1. The distributed transmission line solution should allow the length of each segment explicitly to play a role in the numerical calculations. The L, C, and R values must remain per unit length values, and the length of the segment must not be multiplied into the L, C, and R values to become lumped elements. During numerical computation, the length of each segment and the time step used are correlated to ensure that during each iteration, the current wave must not travel into the next element. If lumped element segments are used, the lumped

elements add to the numerical error and limit the values of the elements and their layout because of stability problems.

2.   In computing the current in the first segment (stemming from the attachment point to the ground) by considering it as a CR element, or allowing the resistance to be very large so as to allow it to suppress the effect of inductance L, a singularity point is created and the current will be expected to go to infinity. When, for instance, the L/R ratio is about 0.1×10^{-6} s, if the 16 Ω/m resistance value is used, it leads to the same situation as found for lumped circuit models at the attachment segment. This is one reason that a proper wavefront cannot be obtained for the ground-level LRS current. The wavefront may be artificially drawn by the computer as a straight-line jump from zero to the peak current. It is always a good practice to ensure that points along the wavefront are calculated.

3.   The connecting leader must not be assumed to be only made up of a resistance element. The upward leader, with the increase in current flowing in it just before connection with the downward leader from the cloud, would be carrying a significant amount of current. Hence, the energy in the connecting leader magnetic field cannot be ignored, and an L element must be assigned to it as well as the downward leader. Just before the return stroke is initiated, the downward leader transforms the connecting leader into an arc channel which is able to carry the large return stroke current.

4.   It is not necessary to resort to the complex finite difference method (FDM) of computation of electrostatic fields to obtain the distributed capacitance of the lightning channel. The problem involved herein is the arrangement of the electrode system with the cloud charge and the leader charge. An unrealistic cloud structure such as a 100 m sphere as well as a 15 km long plane electrode was used to obtain reasonable values for the capacitance. In order to do an FDM computation, the potential at a height of 15 km is set at 15 MV for a 100 m cloud charge at 100 MV. Instead of such unreasonable assumptions being made, it may be shown that a simple charge simulation computation for the leader channel vertically above a perfectly conducting earth gives reasonable values for the channel. It was also observed that the capacitance close to the earth end will be larger since there is more stored energy expected between the sharp edge of the leader and the ground. This produces the large LRS current at the earth end, and a lower return stroke velocity at the lower end of the lightning channel.

5.   Following the earlier DLRCM using time-varying resistance, some models have also resorted to DLCRM. Braginskii's model for a spark channel is used to obtain a time-varying radius r (proportional to $t^{1/2}$). A curve fit to the work is used to obtain a time-varying conductivity. A clear discussion of the use of theories other than Maxwell's equations needs more extensive discussion. But what is important to point out is that the claim that time-varying resistances and time-varying spatially varying inductances are necessary to obtain calculated electric and magnetic fields resembling measured fields is incorrect. Using Maxwell's equations for static electric fields and static magnetic fields, capacitance and inductance values may be obtained to yield LRS currents that yield convex

wavefront current at ground level, as well as radiated electric and magnetic fields that match the measured fields.

6. The trend to put more and more details into a model must be carefully justified. When such details are put in, it is also important to keep a close check that the computer is giving results that may be checked out, and compared with analytical solutions, such as for the diffusion equation and for different numbers of transmission line segments (e.g. 10 and 30) for the same simulation problem. Time steps must be carefully chosen (see Sect. 5.4.3.1). What is sometimes called the electromagnetic models of LRS are mere variations of the DLCRM, and may be tested using the same kind of verification simulations. When the finite difference method is used with time stepping, it is important to ensure that the time steps are properly coordinated with the size of the grid, and that the different velocities of the wave along the lightning channel (e.g. c/3, where c is the velocity of light) and the velocity of the wave being radiated out into space (which is equal to c) are properly accounted for.

5.4.2 The Transmission Line Dispersion Relation

5.4.2.1 Lightning Channel Resistance, Capacitance, and Inductance

The return stroke channel is assumed to be a perfect cylindrical conductor with a constant radius due to equal and counteracting magnetic and kinetic forces. The resistance of the channel is calculated by $R = 1/(\sigma \pi r^2)$ Ωm^{-1}, where σ represents the conductivity of ionized gas calculated by $\sigma = j/E = eN\mu^-$ where μ^- represents the electron mobility calculated by $\mu^- = e\lambda_e/(mC_T)$, where $\lambda_e = 1/(Nq)$ with $q = e^2/(3kT)^2/(16\pi \ \varepsilon_0^2)$, (the scattered electrons due to the collision), and C_T represents the root-mean-square velocity calculated by $C_T = (3kT/m)^{0.5}$. These substitutions result in $\sigma = 1.5 \times 10^{-5} \times \tau^{3/2}$ $(\Omega cm)^{-1}$. With $T = 20{,}000°$ K and a radius of the channel $r = 1$ cm, an approximate channel resistance of 1 Ωm^{-1} is represented in the transmission line model.

A charge simulation method is used to determine the channel capacitance by calculating the charge-to-potential ratio (q/V) along the channel, where V is the potential of the wire with respective to earth.

Figure 5.4 illustrates the electrode system where the potential at point P due to line charge q Cm^{-1} placed at the center of the conductor of length l and radius a is given by

$$V_p = \frac{q}{4\pi \varepsilon_0} \int_0^1 \left(\frac{1}{\sqrt{(z'-z)^2 + a^2}} \right) dz \qquad (5.27)$$

$$= \frac{q}{4\pi \varepsilon_0} \left(\log_e z_1 - \log_e z_2 \right), \qquad (5.28)$$

Fig. 5.4 Application of the charge simulation method

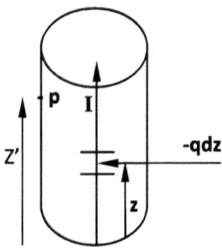

$$\text{where} \quad z_1 = z_A + \sqrt{z_A^2 + 1}, \tag{5.29}$$

$$z_A = \frac{l - z'}{a}, \tag{5.30}$$

$$z_2 = z_B + \sqrt{z_B^2 + 1}, \tag{5.31}$$

$$z_B = -\frac{z'}{a}. \tag{5.32}$$

From (5.28), the ratio q/V will yield the capacitance of the channel. Since the potential will be reduced on the conductor closer to the earth due to the earth's effect, the capacitance C at the lower segment along the channel will be higher than at the other segments. It is found that $C = 24.5$ pFm^{-1}, $L = 3.3$ μHm^{-1} in the vicinity of earth, whereas at a height of 1 km above ground, when the $\frac{1}{\sqrt{LC}} \approx c$ relation applies, $C = 4.3$ pFm^{-1}, $L = 3.3$ μHm^{-1}, where L represents the channel inductance and c represents the speed of light. The charge structure used to determine the capacitance does not portray any significant discrepancy when plugged into the lightning channel.

5.4.2.2 The Lightning Return Stroke Dispersion

Using electromagnetic theory, that the values for L and C determined for the lightning channel are not very different to a crude coaxial system may be seen by considering a 1 cm lightning channel to be surrounded by an outer cylinder of a reasonable cloud radius (e.g. 1 km). The inductance and capacitance are 2 μH/m and 5.5 pF/m, respectively. The capacitance very near the earth is large since the energy stored there will be large. The velocity of the return stroke measured well above the earth (e.g. 200 m up along the channel) shows a wave velocity of about 0.3c where c is the velocity of light. However, at such heights well above ground the computed LRS channel inductance and capacitance values yield $1/\sqrt{(LC)} = c$, the velocity of light. This is the case for overhead power and telecommunications lines. Since measured electromagnetic fields appear to show that the bulk of the energy is transmitted by a group of waves of frequency centered around 5 kHz, we examine here the dispersion curve for a linear transmission line with L,C, and R values close to what has been calculated.

For the approximate equivalent circuit of a short length of line with

R = series resistance per unit lengthof line,
L = series inductance per unit lengthof line,
C = shunt capacitance per unit length, and
G = shunt inductance per unit length,

it may be shown that the attenuation constant of the quasi-TEM wave along the transmission line is

$$\alpha = \frac{1}{\sqrt{2}}\left[\left(RG - \omega^2 LC\right) + \left\{\left(R^2 + \omega^2 L^2\right)\left(G^2 + \omega^2 C^2\right)\right\}^{1/2}\right]^{1/2} \qquad \text{nepers/m.}$$

(5.33)

The phase constant is given by

$$\beta = \frac{1}{\sqrt{2}}\left[-\left(RG - \omega^2 LC\right) + \left\{\left(R^2 + \omega^2 L^2\right)\left(G^2 + \omega^2 C^2\right)\right\}^{1/2}\right]^{1/2} \qquad \text{radians/m.}$$

(5.34)

The wave propagates along the line at a velocity $V_p = \omega/\beta$, which is known as the phase velocity, with the wave amplitude decaying with distance as $\exp(-\alpha z)$. If a group of waves whose frequencies lie between ω and $\omega + d\omega$ is considered, the resultant amplitude envelope of the group, which carries the energy contained in the signals, travels down the line at a group velocity $V_g = d\omega/d\beta$, assuming the $\beta - \omega$ curve to be a straight line between ω and $\omega + d\omega$. For $G = 0$, the $\beta - \omega$ and $\alpha - \omega$ plots are given in Fig. 5.5.

As expected, for $\omega L \gg R$, i.e. at very high frequencies, $\omega/\beta = 1/\sqrt{(LC)}$, the wave is traveling at the velocity of light. This is the region where the $\beta - \omega$ plot becomes a straight line. At low frequencies, with $\omega L \ll R$, we have $\beta = \sqrt{\omega}\,\sqrt{RC}/\sqrt{2}$, i.e. α and β vary as $\sqrt{\omega}$, which is parabolic in shape. The condition where the resistance is negligible is only reached in the frequencies above 1 MHz when the channel resistance $R = 0.8\ \Omega/\text{m}$, and at about 10 MHz for $R = 5\ \Omega/\text{m}$. Therefore, in the frequency ranges of interest in LRS, the phase and group velocities will be less than the velocity of light. Although we do not deal here with very high-frequency signals, it is interesting to note that α from (5.33) goes through a peak and, at very high frequencies, approaches $1/\sqrt{2}(R^2 C/L)$.

Fig. 5.5 The dispersion characteristics of a distributed LCR line. The attenuation constant α is in nepers/m and the phase constant β in radians/m. Adapted from Hoole and Hoole (1993)

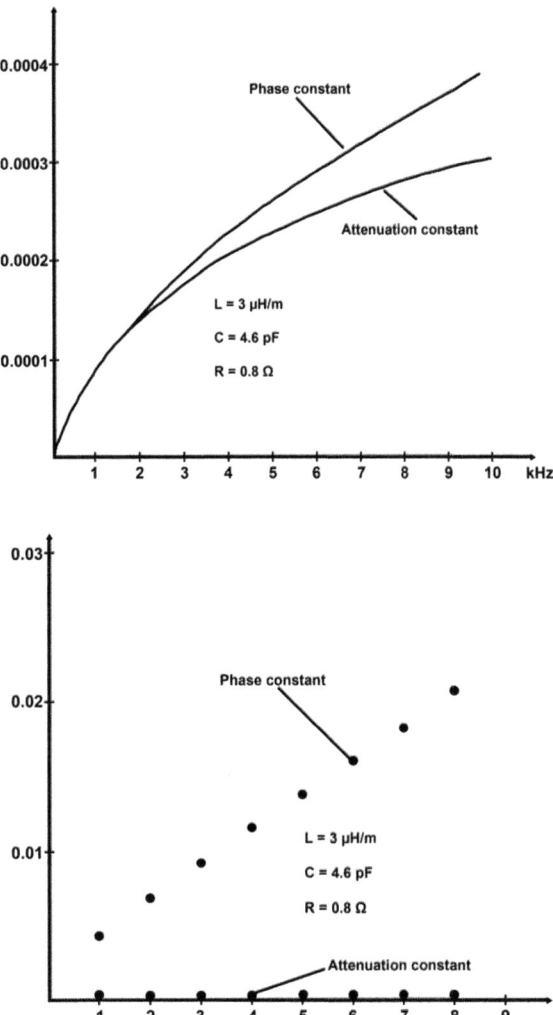

5.4.3 Numerical Solution of the Transmission Line Wave Equation

5.4.3.1 The Finite Difference Solution of the Wave Equation

For conductance G = 0 and a wave traveling along the z-axis, the basic equation we seek to solve numerically is

$$\frac{\partial^2 V}{\partial z^2} - RC\frac{\partial V}{\partial t} - LC\frac{\partial^2 V}{\partial t^2} = 0 \qquad (5.35)$$

Having solved for V, the current may be obtained by integrating the equation

$$I = -C\frac{\partial V}{\partial t}dt \tag{5.36}$$

We retain the partial differential equations since we are interested in the distributed-parameter field phenomena. We may now recast (5.35) using the finite difference approximation as

$$\left(\frac{Rvt}{2} + L\right)V_n(z, t + vt)$$

$$= \left(\frac{Rvt}{2} - L\right)V_n(z, t - vt) + 2\left(L - \frac{vt^2}{Cvz^2}\right)V_n(z, t)$$

$$+ \frac{vt^2}{Cvz^2}(V_{n+1}(z + vz, t) + V_{n-1}(z - vz, t)) \tag{5.37}$$

and (5.36) by

$$I_n(z, t) = I_0 + \sum_{i=1}^{n}\left(\frac{V_i(z, t + \Delta t) - V_i(z, t - \Delta t)}{2\Delta t}\right)C_i, \tag{5.38}$$

where V_n and I_n are, respectively, the voltage and current at the nth segment of the transmission line.

The potential V along the leader is set equal to the cloud potential. This is a valid initial value, since the column electric field drop in an atmospheric air arc carrying 10 A is about 5 V/cm. For a 3 km channel, the total column potential drop will be 1.5 MV, ignored here compared to the 60 MV or more cloud potential. It is useful to note that for a leader current of 300 A and a channel resistance of 2 Ω/m, the column field is 6 V/cm which gives rise to the potential drop of about 1.8 MV over a 3-km-long channel, which is in good agreement with the laboratory arc value. For the perfectly conducting earth, the potential behind the earth resistance is set at zero.

Equations (5.37) and (5.38) could be readily solved by a time stepping process, where the time step is kept small compared to $\Delta z \sqrt{(LC)}$ and $L/(2R)$, in order to obtain a stable solution with sufficient number of calculated points appearing on the wavefront. The distance step Δz is chosen so as to keep it longer than $2\Delta t$. Whence to ensure a stable solution the following two steps are adopted: (i) choose Δt such that it is small compared to $L/(2R)$ and (ii) choose Δz such that it is greater than both $\Delta t/\sqrt{(LC)}$ and $2\Delta t) /(RC)$. These conditions ensure the stability of solution, whatever the magnitudes R, L, and C used in (5.37) and (5.38) are. This is roughly verified by considering the ratios $\Delta z: \Delta t/(RC): \Delta t /(LC)$. Unless the user specifies a time step less than the minimum value of $L/2R$ for each segment of the distributed LCR network, the routine automatically sets it to $L/(10R)$. It is therefore important to ensure that Δz is sufficiently large compared to $L\sqrt{(C/R)}$ and $2L/(RC)$.

5.4.3.2   Testing the DLCRM Computer Code

(a)   Test 1. The accuracy of the LCR transmission line finite difference code was tested by comparing the calculations with the CR routine for the complementary error function. The current along the lightning channel for a diffusion wave is given by

$$\hat{I}(z, t) = (V/R)\sqrt{(CR/\pi t)} \ \exp\left[1 - \left(\sqrt{(CR/\pi t)}z^2\right]\right] \qquad (5.39)$$

Setting the L, C, and R values to obtain the diffusion wave, the numerical solution for currents using (5.37) and (5.38) was compared with that obtained from (5.39). A very good match was found. This test was a double check on the reliability of the finite-difference, computer-based solutions obtained from (5.37) and (5.38).

(b)   Test 2. A further test was performed by using the coded DLCRM Equations (5.37) and (5.38). For the same initial conditions, the LRS currents were computed for a 10-segment lightning channel and a 30-segment lightning channel of the same length as for the 10-segment line. Again, very good agreement between the currents calculated at the same discrete points along the channel was observed.

(c)   Test 3. The influence of time steps chosen on the current wavefront was also studied. For example, the wavefront was calculated with time steps 0.1 and 0.05 μs for a line with an L/R ratio of 0.55 μs. This test also revealed that a time step of 0.1 μs gives good convergence for the values of circuit parameters used in this chapter to simulate the lightning return stroke.

5.4.4   Return Stroke Velocity and the Transmission Line Model

5.4.4.1   Background

A fourth test of the DLCRM and the computer code developed is to observe the velocity of the LRS current pulse (a quasi-TEM wave) along the channel and compare it to the measured LRS velocity. Obviously, the return stroke currents determined from DLRCM are made of a wave train which is influenced in a complex manner by the return stroke channel, including reflections due to the finite length of the channel when currents are computed for a few tens of microseconds. Although we discuss phase velocities for signals of different wavelengths, the precise significance of the phase velocity does not apply to the wave train of finite length generated at the earth end. For LRS currents, it is the group velocity and not the phase velocity that must be calculated from the current pulses, which must be compared with the measured lightning velocity.

Since an electromagnetic field cannot completely be localized in either space or time, there must be an essential arbitrariness about every definition of velocity. For convenience, it is common to talk about the group velocity, phase velocity, and the signal velocity. The group velocity ($d\omega/dk$) is less than the phase velocity (ω/k) for normal dispersion, where k ($= 2\pi/\lambda$) is the wave number and λ is the wavelength.

5.4.4.2 Return Stroke Velocity: from Photography and the DLCRM

Although the lossy transmission line model for quasi-transverse electromagnetic waves is an established tool, the question of the velocity of the current wave train, both that computed from the DLCRM (Sect. 5.2) and that measured, needs some discussion. We take the concentration of electromagnetic fields in space to indicate the energy to be localized in that region. Taking this to be the case, we plotted the times at which current peaks at different points on the line against the height for 1, 2 and 5 Ω/m resistances. The value of $1/\sqrt{(LC)}$ was set equal to 300 m/μs in order to be able to observe the influence of resistance; this setting is not unreasonable since the values for L and C calculated satisfy the relationship $1/\sqrt{(LC)} = c$ away from the immediate vicinity of the ground.

The plots from DLCRM are shown in Fig. 5.6a with measured values for three different LRS (lightning return stroke) in Fig. 5.6b. The photographic measurements give times along the lightning channel when the return stroke luminosity is brightest, and for the transmission line solutions, the arrival times are the times at which DLCRM calculated return stroke current reaches peak values, as shown in Fig. 5.7. Both plots given in Fig. 5.6a and b agree very well, showing a largely constant

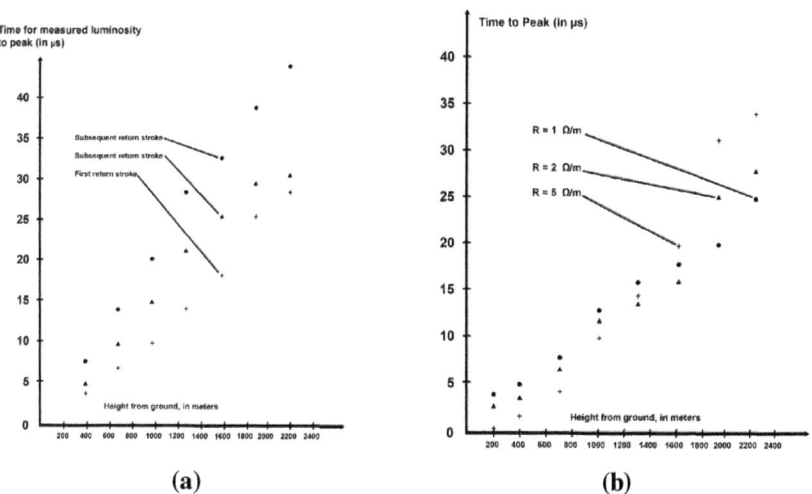

(a)   (b)

Fig. 5.6 Times at which return stroke reaches peak value at different heights along the lightning channel. **a** Calculated from DLCRM Equations (5. 37) and (5. 38) for L = 2μH/m, C = 5.5pF/m and three values of R (5, 2, 1 Ω/m). **b** Measured LRS peak using photography of the luminous pulse moving up the lightning channel. Adapted from (Hoole and Hoole 1988)

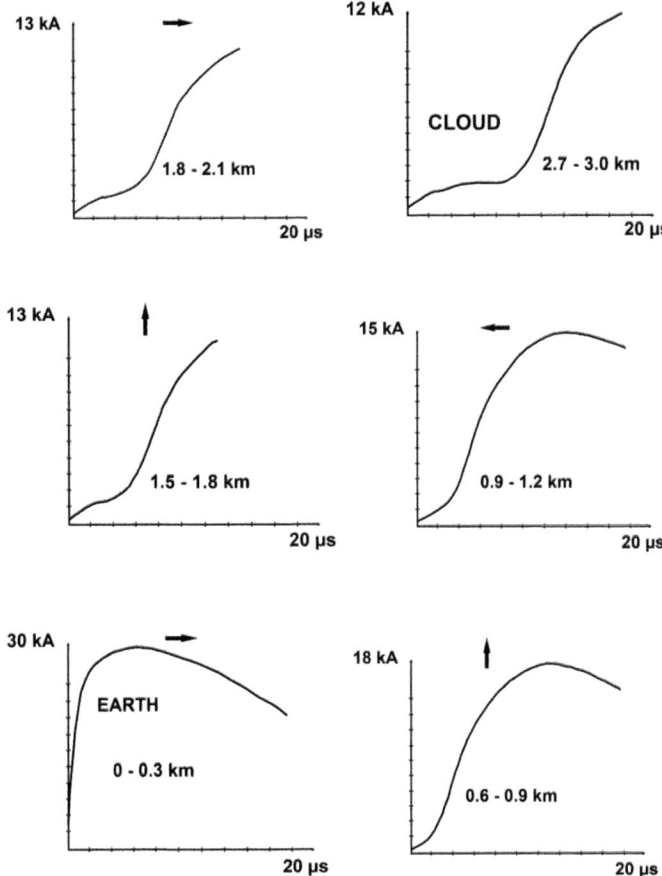

Fig. 5.7 Current-height and current–time characteristics for negative earth flash return stroke. Adapted from (Hoole and Hoole 1993)

velocity along the channel, except for the LRS velocity close to ground. From the DLCRM calculated current waves, it was observed that close to the ground, the return stroke velocity is higher than the velocities calculated or measured above the ground. The average velocity in the case of the transmission line model, with all three cases of different resistance values taken together, is 73 m/μs (roughly c/4, where c is the velocity of light) agreeing well with 100 m/μs (c/3) average measured return stroke velocity. It should be noted that the DLCRM simulation results are not for exact lightning parameters for the LRS which Schonland photographed in 1956, and hence the differences (c/4 and c/3) are understandable and acceptable. A change of lightning channel resistance from 1 to 5 ohms/m results in a 26% increase in the group velocity. This change is within the 5% group velocity change for an LRS current wave packet with a center of gravity at 100 kHz and the 50% change for a wave LRS current packet centered at 5 kHz.

The measured velocities are substantially less than the velocity of light c (= 3 × 10^8 m/s). Although the measured values for the LRS have been questioned, the basis for the case against an LRS velocity less than c is questionable. In some LRS models, the L, C, and R parameters were calculated using assumptions that are not well supported. When the computer-simulated LRS waves were seen to travel at an almost constant velocity close to the velocity of light, it was stated that the error is due to the problems in luminosity measurements. On the contrary, in the DLCR model developed using electromagnetic field principles to determine the circuit parameters, as reported herein, especially L and C, it is seen that the DLCR model LRS current wave yields a velocity which is close to the measured LRS velocity. Moreover, for the case of 1 Ω/m channel, for instance, the LRS velocity changes were calculated from Fig. 5.5a: 10^8 m/s (c/3) close to the ground, which then drops to 0.5×10^8 m/s (c/6) before settling down to a constant velocity of about 0.4×10^8 m/s all the way to the cloud. For an increased value of the resistance, i.e. 5 Ω/m, the LRS velocity, from Fig. 5.6, is reduced to about c/10 and is more constant and diffusion-like over the length of the channel.

5.5 Negative Cloud to Ground Earth Flash Return Stroke: Simulated by the DLCRM

5.5.1 Background

This is the most common type of flash observed. It appears over mountainous regions, as well as over sea. We shall consider the subsequent return strokes for the present. Two types of contact points are considered. In the first case, the flash is to an open ground, with a ground resistivity of 100 Ω m, as in Florida. It is known that from fulgurites in sand, the radius of the contact point is about 0.03–0.52 cm, and that the flash does not progress into the ground for more than a meter. We ignore any movement of the stroke into the ground, since any melting into the earth will take place when the bulk of the charge will be lowered by the continuing current over a few tenths of milliseconds. Since the return stroke exists only for a few tens of microseconds, we take the contact point to be stationary and as a sphere with the radius of the channel. The earth resistance in this case might be in the range of 1–8 kΩ. In the second case, where an earthed electrode provides the return stroke path to the earth, the earth resistance is in the range of 100–250 Ω for a conductor radiusof 0.2–1 cm, buried 1 m in a soil of resistively 100 Ω m.

The DLCRM simulation studies are carried out for the prescribed settings of the following parameters: radius of the cloud spherical electric charge center (500 m), channel resistance R (0.8 Ω/m), inductance L (3 micro-Henry/m), capacitance of the first segment at the earth end (25 picoF/m), capacitance along the segments other than the earth send segment (4.6 picoF/m), earth resistance R_E (1500 Ω), the length of the channel from ground to the base of the thundercloud charge center (3000 m), the number of segments that the channel is divided into (10 segments, 300 m/segment), potential of the thundercloud electric charge center (50 MV), and the initial leader current along the channel (100 A). This data is as obtained from measurements reported by Berger in the 1970s.

5.5.2 LRS Currents from DLCRM Simulation

The calculated currents and potentials of the LRS are given in Figs. 5.7 and 5.8.

The electric field and magnetic field calculated using the currents yielded by the DLCRM simulation are given in Figs. 5.9 and 5.10, respectively. The fields are calculated using the integral method described in Sect. 5.5.3.3, as reported in (Hoole and Hoole 1987a, b).

From an engineering perspective, we are primarily interested in the details of the waveforms over the first few tens of microseconds, when rapid, high current changes occur. Most of the calculations are carried out for the first 20 μs of the LRS. The current wavefront has two distinctive regions at the earth end. The DLCRM predicts the overall concave wavefront of the LRS current as seen in Fig. 5.7. In the current waveforms, we observe an initial slow rise of current, followed by a sudden rise to peak. Thus, the overall current wave has a concave-shaped wavefront. At heights above the earth, there are three regions in the current wavefront; a gradual variation in current, increasing to about 2 kA, before the main return stroke pulse arrives at a point along the lightning channel. Second, the LRS arrives at that point and a sharp rise of current is observed in the wavefront. And third, there is a slower increase towards peak current.

From an engineering perspective, the initial, slow ramp-like increase of current is not the significant part of the LRS. The portion after the ramp current, having a rapid rate of rise, is that which is severe with regard to the induced effects of lightning. This portion of the LRS current wavefront that follows the ramp-shaped current is convex in shape. In a wonderful way, this DLCR model to which no additional currents or curve fitting techniques using time-changing radius or conductivity are added gives an exact representation of the LRS current waveform. All these essential features of the LRS currents are carefully captured by the DLCRM. As expected, the wavefront degenerates with height, and the current crest decays with height.

The overall potential along the lightning channel drops as the LRS current pulse discharges the lightning channel segment over which it has traversed. Within the first 20 μs of the LRS, close to the ground the channel potential may drop from about 50 MV to about 15 MV, and close to the cloud to about 40 MV (Fig. 5.8). This shows rapid discharge of the electric charges deposited on the leader and in the thundercloud charge center. As the electric charges in the thundercloud are emptied into the ground through multiple leader-return stroke occurrences, the potential will drop as the thundercloud becomes discharged. If needed, the electric field inside the channel E_c may be determined from the potential profile in Fig. 5.8, or from the current density $J = \sigma E_c$.

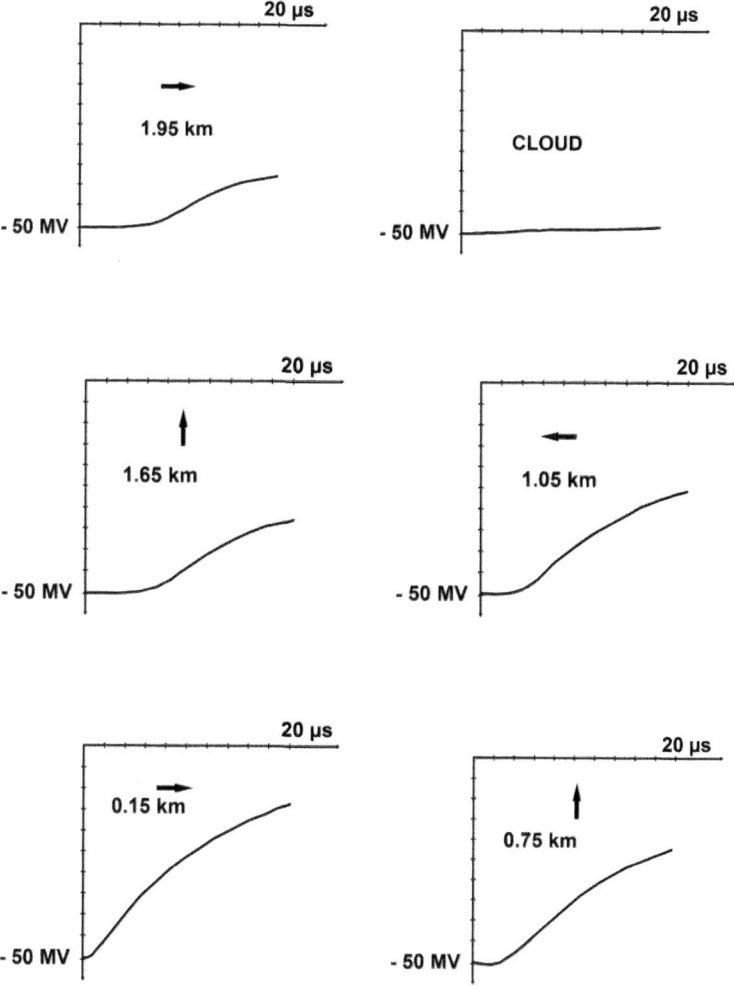

Fig. 5.8 Voltage-height and voltage-time characteristics for negative earth flash return stroke—adapted from (Hoole and Hoole 1993)

5.5.3 Calculation of the Electric and Magnetic Fields Radiated from the Lightning Currents

5.5.3.1 Determining the LEMP at any Distance from the Lightning Flash

We shall first derive the mathematical solutions from Maxwell's equations of the electric fields radiated from a straight wire antenna. The magnetic fields solution is included in Hoole and Hoole (1987a, b). The wire antenna will be used to represent

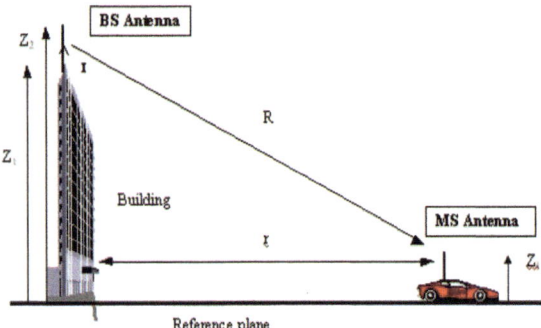

Fig. 5.9 Each segment of the lightning/arc channel modeled by a current-carrying line element antenna

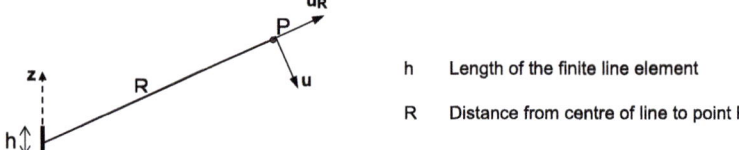

h   Length of the finite line element

R   Distance from centre of line to point P

Fig. 5.10 Orientation of an infinitesimal element carrying current

the segments of the lightning return stroke channel. Once the return stroke currents have been determined, the wire antenna segments, used to represent each segment of the return stroke channel, are used to calculate the total electric and magnetic fields radiated by the lightning flash at any point in space. The fields radiated by each segment are calculated for each point in space, and then summed up to get the total electromagnetic fields at the point in space. The ground, assumed to be a good conductor, may be replaced by the image of each segment.

5.5.3.2 The Transient Electromagnetic Field Pulses (LEMP) Radiated from Lightning

For the finite length of wire antenna element shown in Fig. 5.9, the far field of the electric field radiated by each segment of the line is given by Hoole (2000)

$$E(r, z, t) = \frac{\mu_0}{4\pi} \left[\frac{z_j - z_1}{\sqrt{r^2 + (z_j - z_1)^2}} - \frac{z_j - z_2}{\sqrt{r^2 + (z_j - z_2)^2}} \right] \frac{d[I]}{dt} \qquad (5.40)$$

In the far field, the magnetic field is simply the electric field divided by the free space impedance 377 Ω.

5.5.3.3 Finite Length Antenna: A Basic Building Block for Antenna Simulation

Complex electromagnetic field-radiating sources such as lightning channels, array antennas, and aperture antennas may be thought of as being made up of discrete line elements. In aperture antennas, for instance, the radiating line elements are the electric field lines that appear at the aperture carrying displacement currents. In electromagnetic image reconstruction too, the line element is expected to yield image reconstruction in shorter time and it inherently contains more information about the region it covers. This includes information about the size of the region and its angle of inclination. The equation of the electric field of a finite line element forms the basis of the imaging model. This model is developed and implemented for lightning return stroke, where the long (e.g. 1000 m in length) return stroke channel is divided into finite legths of elements (say 100 m per segment). In this section, we will first set out to derive the equations for the electric field. Studies of the variation of the electric field with the various parameters in the field equation help to understand both the magnitude and waveform of the electric and magnetic fields.

Once the radial and tangential fields for a finite length radiator are derived, the structures of the radiation patterns may be studied. The derived electric field at any point in space has a radial component, $\mathbf{E}_r$, tangential components, $\mathbf{E}_\theta$, and magnitude of the electric field, $\sqrt{(\mathbf{E}_r^2 + \mathbf{E}_\theta^2)}$, each of which is an important component of the LEMP.

5.5.3.4 Derivation of Electric Field Radiated by a Finite Line Element

To derive the electric field for the line element, let us first consider an infinitesimal element of length h carrying current as shown in Fig. 5.10. Let $[I]$ be the retarded current carried by the element and $[Q]$ be the retarded electric charge. Both quantities are a function of $(t - R/c)$, where t is time, R is the distance from the point P to the center of line element, and c is the speed of light. $[Q]$ is related to $[I]$ by the following equation:

$$\frac{d[Q]}{dt} = [I]. \tag{5.41}$$

The three equations governing the electromagnetic field are

$$\mathbf{E}(R, t) = -\nabla \mathbf{V}(R, t) - \frac{\partial \mathbf{A}(R, t)}{\partial t}, \tag{5.42}$$

$$\mathbf{B}(R, t) = \nabla \times \mathbf{A}(R, t), \tag{5.43}$$

$$\nabla \cdot \mathbf{A}(R, t) = -\frac{1}{c^2} \frac{\partial \mathbf{V}(R, t)}{\partial t}. \tag{5.44}$$

The magnetic potential, $\mathbf{A}$, is given as

$$\mathbf{A} = \frac{\mu_0}{4\pi} \int_v \frac{[\mathbf{J}]}{r} dv. \tag{5.45}$$

From the geometry of the problem in Fig. 5.10, (5.45) can be expressed as

$$\mathbf{A} = \mathbf{u}_z \frac{\mu_0}{4\pi} \int_{-h/2}^{h/2} \frac{[I]}{R} dz. \tag{5.46}$$

Assuming that $R \gg h$, (5.46) is approximated to

$$\mathbf{A} = \mathbf{u}_z \frac{\mu_0 h}{4\pi R}[I]. \tag{5.47}$$

Expressing this in spherical coordinates

$$\mathbf{A} = \mathbf{u}_R \frac{\mu_0[I]}{4\pi R} h \cos\theta - \mathbf{u}_\theta \frac{\mu_0[I]}{4\pi R} h \sin\theta. \tag{5.48}$$

Divergence of $\mathbf{A}$ in spherical coordinates is given by

$$\nabla \cdot \mathbf{A} = \frac{1}{R}\frac{\partial}{\partial R}\left(R^2 A_R\right) + \frac{1}{R\sin\theta}\frac{\partial}{\partial}(A_\theta \sin\theta). \tag{5.49}$$

The first terms in the divergence $\mathbf{A}$ can be expressed as

$$\frac{\partial}{\partial R}\left(R^2 A_R\right) = \frac{\mu_0 h \cos\theta}{4\pi}\left(\frac{\partial[I]}{\partial R}R + [I]\right).$$

As $[I] = I(\theta)$, where $\theta = t - R/c$,

$$\frac{\partial[I]}{\partial R} = \frac{\partial[I]}{\partial\theta}\frac{d\theta}{dR} = \frac{\partial[I]}{\partial t}\frac{dt}{d\theta}\frac{d\theta}{dt}$$
$$= -\frac{1}{c}\frac{\partial[I]}{\partial t}.$$

Hence,

$$\frac{\partial}{\partial R}\left(R^2 A_R\right) = \frac{\mu_0 h \cos\theta}{4\pi}\left(-\frac{R}{c}\frac{\partial[I]}{\partial t} + [I]\right).$$

Following the same procedure, the second term is found to be

$$\frac{\partial}{\partial \theta}(A_\theta \; \sin \; \theta) = -\frac{\partial}{\partial \theta}\left(\frac{\mu_0[I]}{4\pi R}h \; \sin^2 \; \theta\right)$$

$$= -\frac{\mu_0[I]}{4\pi R}(2h \; \sin \; \theta \; \cos \; \theta).$$

Hence (5.49) becomes

$$\nabla \cdot \mathbf{A} = \frac{\mu_0 h \cos \theta}{4\pi}\left(-\frac{1}{Rc}\frac{d[I]}{dt} - \frac{[I]}{R^2}\right). \tag{5.50}$$

From (5.44),

$$\frac{\partial V(R,t)}{\partial t} = -c^2 \nabla \cdot \mathbf{A}(R,t).$$

Integrating both sides with respect to time t,

$$V(R,t) = -c^2 \int \nabla \cdot \mathbf{A}(R,t) \; dt.$$

Assuming that when $t < 0$, the integral is zero, i.e. no potential before $t = 0$,

$$V(R,t) = -c^2 \int_0^t \frac{\mu_0 h \; \cos \theta}{4\pi}\left(-\frac{1}{Rc}\frac{d[I]}{dt} - \frac{[I]}{R^2}\right) dt$$

$$= \frac{c^2 \mu_0 \; h \; \cos \theta}{4\pi}\left(\frac{[I]}{Rc} + \frac{1}{R^2}\int_0^t [I] \; dt\right).$$

Now, from the relationship of $[I]$ and $[Q]$ in (5.41),

$$V(R,t) = \frac{c^2 \mu_0 h \; \cos \theta}{4\pi}\left(\frac{[I]}{Rc} + \frac{[Q]}{R^2}\right). \tag{5.51}$$

After finding $V(R,t)$ and $A(R,t)$, the electric field of the line element can be found by substituting these terms in Eq. (5.42). Using

$$\nabla V(R,t) = u_R \frac{\partial V}{\partial R} + u_\theta \frac{1}{R}\frac{\partial V}{\partial \theta} + u_\phi \frac{1}{R \sin \theta}\frac{\partial V}{\partial \phi},$$

and letting

$$k = -\frac{c^2 \mu_0 h}{4\pi},$$

we have

$$\frac{\partial V}{\partial R} = \frac{\partial}{\partial R} \left[k \ \cos \ \theta \left(\frac{[I]}{Rc} + \frac{[Q]}{R^2} \right) \right]$$

$$= k \cos \theta \left(\frac{d[I]}{dR} \frac{1}{Rc} - \frac{[I]}{R^2 c} + \frac{d[Q]}{dR} \frac{1}{R^2} - \frac{2[Q]}{R^3} \right)$$

$$= k \ \cos \ \theta \left(-\frac{1}{Rc^2} \frac{d[I]}{dt} - \frac{[I]}{R^2 c} + \frac{d[Q]}{dt} \frac{1}{R^2 c} - \frac{2[Q]}{R^3} \right)$$

$$= k \ \cos \ \theta \left(-\frac{1}{Rc^2} \frac{d[I]}{dt} - \frac{2[I]}{R^2 c} - \frac{2[Q]}{R^3} \right),$$

$$\frac{\partial V}{\partial \theta} = \frac{\partial}{\partial \theta} \left[k \ \cos \ \theta \left(\frac{[I]}{Rc} + \frac{[Q]}{R^2} \right) \right]$$

$$= -k \ \sin \ \theta \left(\frac{[I]}{Rc} + \frac{[Q]}{R^2} \right),$$

$$\frac{\partial V}{\partial \phi} = 0.$$

Hence,

$$\nabla V(R, t) = u_R \left[k \ \cos \ \theta \left(-\frac{1}{Rc^2} \frac{d[I]}{dt} - \frac{2[I]}{R^2 c} - \frac{2[Q]}{R^3} \right) \right] + u_\theta \left[-k \ \sin \ \theta \left(\frac{[I]}{Rc} + \frac{[Q]}{R^2} \right) \right],$$

$$\frac{\partial \mathbf{A}(R, t)}{\partial t} = u_R \frac{\mu_0 h \ \cos \ \theta}{4 \pi R} \frac{d[I]}{dt} - u_\theta \frac{\mu_0 h \ \sin \ \theta}{4 \pi R} \frac{d[I]}{dt}.$$

Since

$$\mathbf{E}(R, t) = -\nabla V(R, t) - \frac{\partial \mathbf{A}(R, t)}{\partial t},$$

we have

$$\mathbf{E}(R, t) = \mathbf{u}_R \left[k \ \cos \ \theta \left(\frac{1}{Rc^2} \frac{d[I]}{dt} + \frac{2[I]}{R^2 c} + \frac{2[Q]}{R^3} \right) \right] + \mathbf{u}_\theta \left[k \ \sin \ \theta \left(\frac{[I]}{Rc} + \frac{[Q]}{R^2} \right) \right]$$

$$- \mathbf{u}_R \frac{\mu_0 h \ \cos \ \theta}{4 \pi R} \frac{d[I]}{dt} + \mathbf{u}_\theta \frac{\mu_0 h \ \sin \ \theta}{4 \pi R} \frac{d[I]}{dt}$$

$$= \mathbf{u}_R \left[\frac{c^2 \mu_0 h \ \cos \ \theta}{4\pi} \left(\frac{1}{Rc^2} \frac{d[I]}{dt} + \frac{2[I]}{R^2 c} + \frac{2[Q]}{R^3} \right) - \frac{\mu_0 h \ \cos \ \theta}{4 \pi R} \frac{d[I]}{dt} \right]$$

$$+ \mathbf{u}_\theta \left[\frac{c^2 \mu_0 h \ \sin \ \theta}{4\pi} \left(\frac{[I]}{Rc} + \frac{[Q]}{R^2} \right) + \frac{\mu_0 h \ \sin \ \theta}{4 \pi R} \frac{d[I]}{dt} \right]$$

$$= \mathbf{u}_R \frac{h\ \cos\ \theta}{4\pi}\left[\left(\frac{\mu_0}{\varepsilon_0}\right)^{1/2}\frac{2[I]}{R^2} + \frac{2[Q]}{\varepsilon_0 R^3}\right]$$

$$+\ \mathbf{u}_\theta \frac{h\ \sin\ \theta}{4\pi}\left[\left(\frac{\mu_0}{\varepsilon_0}\right)^{1/2}\frac{[I]}{R^2} + \frac{[Q]}{\varepsilon_0 R^3} + \frac{\mu_0}{R}\frac{d[I]}{dt}\right]. \qquad (5.52)$$

Transforming (5.52) into cylindrical coordinates, taking into account

$$\cos\ \theta = \frac{z}{\left(r^2 + z^2\right)^{1/2}}, \qquad \sin\ \theta = \frac{r}{\left(r^2 + z^2\right)^{1/2}}, \qquad R^2 = r^2 + z^2,$$

$$\mathbf{E}(r, z, t) = (E_R\ \cos\ \theta - E_\theta\ \sin\ \theta)\mathbf{u}_z + (E_R\ \sin\ \theta + E_\theta\ \cos\ \theta)\mathbf{u}_r,$$

$$E_z = \mathbf{u}_z \frac{h}{4\pi}\left[\left(\frac{\mu_0}{\varepsilon_0}\right)^{1/2}\frac{2z^2 - r^2}{\left(r^2 + z^2\right)^2}[I] + \frac{1}{\varepsilon_0}\frac{2z^2 - r^2}{\left(r^2 + z^2\right)^{2.5}}[Q] - \mu_0\frac{r^2}{\left(r^2 + z^2\right)^{1.5}}\frac{d[I]}{dt}\right],$$

$$E_r = \mathbf{u}_r \frac{h}{4\pi}\left[\left(\frac{\mu_0}{\varepsilon_0}\right)^{1/2}\frac{3rz}{\left(r^2 + z^2\right)^2}[I] + \frac{1}{\varepsilon_0}\frac{3rz}{\left(r^2 + z^2\right)^{2.5}}[Q] + \mu_0\frac{rz}{\left(r^2 + z^2\right)^{1.5}}\frac{d[I]}{dt}\right].$$

We shall assume that the finite line element is made up of an infinitesimal line element with electric field dE and length dz (see Fig. 5.11).

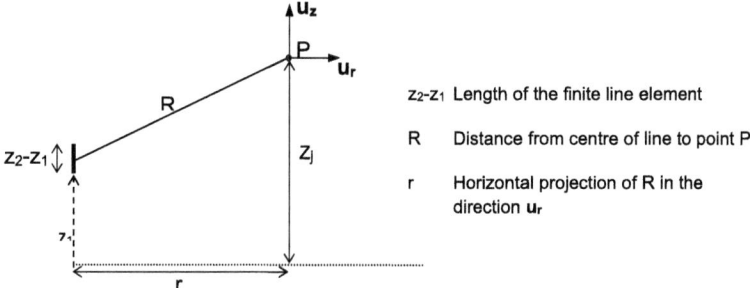

Fig. 5.11 Orientation of a finite line element of length $z_2 - z_1$

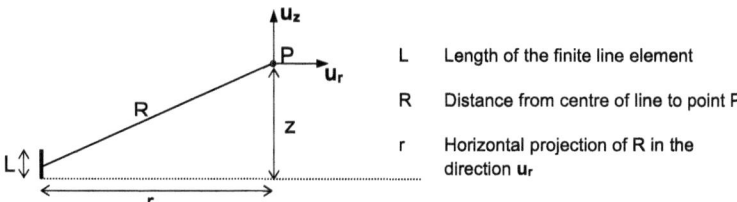

Fig. 5.12 Orientation of a finite line element setup for investigation

Letting $h = dz$, $z = z_j - z$, and $dz = -dz$, the resultant electric field due to a line of length $(z_2 - z_1)$ carrying a current I is given by

$$\overline{\mathbf{E}} = \int_{z_1}^{z_2} \mathbf{E}(r, z, t)\, dz$$

$$= -\mathbf{u}_z \int_{z_1}^{z_2} \frac{1}{4\pi} \left[\begin{array}{c} \left(\frac{\mu_0}{\varepsilon_0}\right)^{1/2} \frac{2(z_j-z)^2 - r^2}{\left(r^2+(z_j-z)^2\right)^2}[I] \\[4mm] + \frac{1}{\varepsilon_0}\frac{2(z_j-z)^2 - r^2}{\left(r^2+(z_j-z)^2\right)^{2.5}}[Q] - \mu_0 \frac{r^2}{\left(r^2+(z_j-z)^2\right)^{1.5}}\frac{d[I]}{dt} \end{array} \right] dz$$

$$- \mathbf{u}_r \int_{z_1}^{z_2} \frac{1}{4\pi} \left[\begin{array}{c} \left(\frac{\mu_0}{\varepsilon_0}\right)^{1/2} \frac{3r(z_j-z)}{\left(r^2+(z_j-z)^2\right)^2}[I] \\[4mm] + \frac{1}{\varepsilon_0}\frac{3r(z_j-z)}{\left(r^2+(z_j-z)^2\right)^{2.5}}[Q] + \mu_0 \frac{r(z_j-z)}{\left(r^2+(z_j-z)^2\right)^{1.5}}\frac{d[I]}{dt} \end{array} \right] dz.$$

Using standard integrals (Hoole and Hoole 1987a, b), the integration of the terms for the electric field can be evaluated. The final form of the electric field is given by the following equations: the r-component of the electric field is given as

$$\overline{\mathbf{E}}_r = -\mathbf{u}_r \left[\begin{array}{c} \frac{3r}{8\pi}\left(\frac{\mu_0}{\varepsilon_0}\right)^{1/2}\left\{ \frac{1}{\left(r^2+(z_j-z_2)^2\right)} - \frac{1}{\left(r^2+(z_j-z_1)^2\right)} \right\}[I] \\[4mm] + \frac{r}{4\pi\varepsilon_0}\left\{ \frac{1}{\left(r^2+(z_j-z_2)^2\right)^{1.5}} - \frac{1}{\left(r^2+(z_j-z_1)^2\right)^{1.5}} \right\}[Q] \\[4mm] + \frac{\mu_0 r}{4\pi}\left\{ \frac{1}{\sqrt{r^2+(z_j-z_2)^2}} - \frac{1}{\sqrt{r^2+(z_j-z_1)^2}} \right\}\frac{d[I]}{dt} \end{array} \right]. \qquad (5.53)$$

and the z-component of the electric field is given as

$$\overline{\mathbf{E}}_z = -\mathbf{u}_z \left[\begin{array}{c} \frac{1}{8\pi}\left(\frac{\mu_0}{\varepsilon_0}\right)^{1/2}\left\{ \begin{array}{c} \frac{3(z_j-z_2)}{r^2+(z_j-z_2)^2} - \frac{3(z_j-z_1)}{r^2+(z_j-z_1)^2} \\[3mm] -\frac{1}{r}\left[\tan^{-1}\left(\frac{z_j-z_2}{r}\right) - \tan^{-1}\left(\frac{z_j-z_1}{r}\right)\right] \end{array} \right\}[I] \\[7mm] + \frac{1}{4\pi\varepsilon_0}\left[\frac{z_j-z_2}{\left(r^2+(z_j-z_2)^2\right)^{1.5}} - \frac{z_j-z_1}{\left(r^2+(z_j-z_1)^2\right)^{1.5}} \right][Q] \\[4mm] + \frac{\mu_0}{4\pi}\left\{ \frac{z_j-z_2}{\sqrt{r^2+(z_j-z_2)^2}} - \frac{z_j-z_1}{\sqrt{r^2+(z_j-z_1)^2}} \right\}\frac{d[I]}{dt} \end{array} \right].$$

$$(5.54)$$

5.5.3.5 Electric Field Radiated by a Line Element

The generalized equations

Based on the line orientation geometry as shown in Fig. 5.11, $z_j = z$, $z_1 = 0$, and $z_2 = L$. The retarded charge $[Q]$ is assumed to be zero. Hence, (5.53) and (5.54) simplify to

$$
\mathbf{E}(r, z, t)
$$

$$
= -\mathbf{u}_r \left\{ \frac{3r}{8\pi} \sqrt{\frac{\mu_0}{\varepsilon_0}} \left[\frac{1}{r^2 + (z - L)^2} - \frac{1}{r^2 + z^2} \right] [I] \right.
$$

$$
\left. + \frac{\mu_0 r}{4\pi} \left[\frac{1}{\sqrt{r^2 + (z - L)^2}} - \frac{1}{\sqrt{r^2 + z^2}} \right] \frac{dI}{dt} \right\}
$$

$$
- \mathbf{u}_z \left\{ \frac{1}{8\pi} \sqrt{\frac{\mu_0}{\varepsilon_0}} \left[\frac{3(z - L)}{r^2 + (z - L)^2} - \frac{3z}{r^2 + z^2} - \frac{1}{r} \left(\tan^{-1}\left(\frac{z - L}{r} \right) - \tan^{-1}\left(\frac{z}{r} \right) \right) \right] [I] + \right.
$$

$$
\left. \frac{\mu_0}{4\pi} \left[\frac{z - L}{\sqrt{r^2 + (z - L)^2}} + \frac{z}{\sqrt{r^2 + z^2}} \right] \frac{dI}{dt} \right\}.
$$

$$
(5.55)
$$

We can write (5.55) in polar coordinates, a form which is used in the studies presented herein. From Fig. 5.13, the equation of the electric field becomes

$$
\mathbf{E}(R, \theta, t)
$$

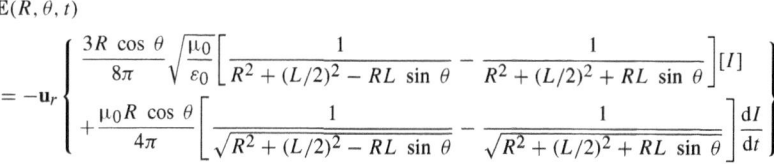

$$
= -\mathbf{u}_r \left\{ \frac{3R \cos \theta}{8\pi} \sqrt{\frac{\mu_0}{\varepsilon_0}} \left[\frac{1}{R^2 + (L/2)^2 - RL \sin \theta} - \frac{1}{R^2 + (L/2)^2 + RL \sin \theta} \right] [I] \right.
$$

$$
\left. + \frac{\mu_0 R \cos \theta}{4\pi} \left[\frac{1}{\sqrt{R^2 + (L/2)^2 - RL \sin \theta}} - \frac{1}{\sqrt{R^2 + (L/2)^2 + RL \sin \theta}} \right] \frac{dI}{dt} \right\}
$$

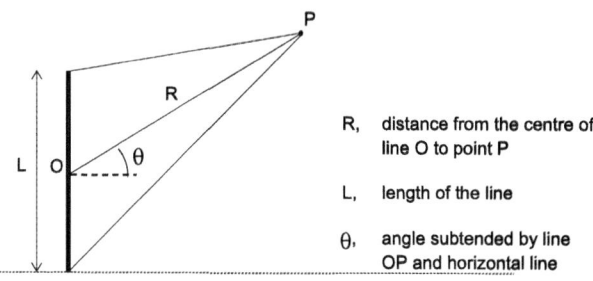

R, distance from the centre of line O to point P

L, length of the line

θ, angle subtended by line OP and horizontal line

Fig. 5.13 Finite line element in polar coordinates

$$
-\mathbf{u}_z \left\{
\begin{array}{l}
\dfrac{1}{8\pi}\sqrt{\dfrac{\mu_0}{\varepsilon_0}}\left[
\begin{array}{c}
\dfrac{3(R\,\sin\,\theta - L/2)}{R^2 + (L/2)^2 - RL\,\sin\,\theta} - \dfrac{3(R\,\sin\,\theta + L/2)}{R^2 + (L/2)^2 + RL\,\sin\,\theta} \\[2mm]
-\dfrac{1}{R\,\cos\,\theta}\left(\tan^{-1}\left(\dfrac{R\,\sin\,\theta - L/2}{R\,\cos\,\theta}\right) - \tan^{-1}\left(\dfrac{R\,\sin\,\theta + L/2}{R\,\cos\,\theta}\right)\right)
\end{array}
\right][I] \\[8mm]
+\dfrac{\mu_0}{4\pi}\left[\dfrac{R\sin\theta - L/2}{\sqrt{R^2 + (L/2)^2 - RL\,\sin\,\theta}} + \dfrac{R\,\sin\,\theta + L/2}{\sqrt{R^2 + (L/2)^2 + RL\,\sin\,\theta}}\right]\dfrac{\mathrm{d}I}{\mathrm{d}t}
\end{array}
\right\}.
$$

$$(5.56)$$

In far fields, (5.55) and (5.56) reduce to (5.40). For further details, see Hoole and
Hoole (1987a, b, 1996).

5.5.4 Computed Electromagnetic Field Pulses LEMPs

Applying (5.56) to each segment of the return stroke, we may determine the electric
field radiated by each segment to a particular point in space. Then by adding these
together, we get the total electromagnetic field from the entire length of the return
stroke at that particular point in space.

5.5.5 LRS Electric and Magnetic Fields Calculated
from Currents Obtained from DLCRM Simulation

The electric fields and magnetic fields radiated by the current pulses are given in
Fig. 5.14 and Fig. 5.15, respectively. The fields are generally in good agreement with
the ground measurements. The fields were calculated using the integral technique
(Hoole and Hoole 1987a, b). It is important to note that the DLCRMgives the correct
picture of the radiated electromagnetic pulse (LEMP) without any artificial, forced
features such as added ramps or time-varying radius or time-varying conductivity
being added on to the model.

Comparing the electric and magnetic fields calculated from the DLCRM LRS
currents (Figs. 5.14 and 5.15) to the measured fields (Figs. 5.2 and 5.3), we note that
there is an overall agreement in the LEMP shape. Consider, for instance, the electric
field measured at a 2000 m distance away from the flash (Fig. 5.2a) and the electric
field computed from the currents yielded by the DLCRM simulation. We note that
both the initial sharp rise to peak of the electric field (because of the convex-shaped
portion of the current we observed) followed by a ramp-like portion to it are observed
in both measured and calculated portions.

Consider now the magnetic field at ground level, and 2000 m away from the
lightning flash. Compare it to the measured magnetic field (Fig. 5.2b). We note that
in both cases a sub-microsecond rise to peak, followed by a hump-shaped decay of
the magnetic field is there. It is important to notice that the initial sharp peak observed

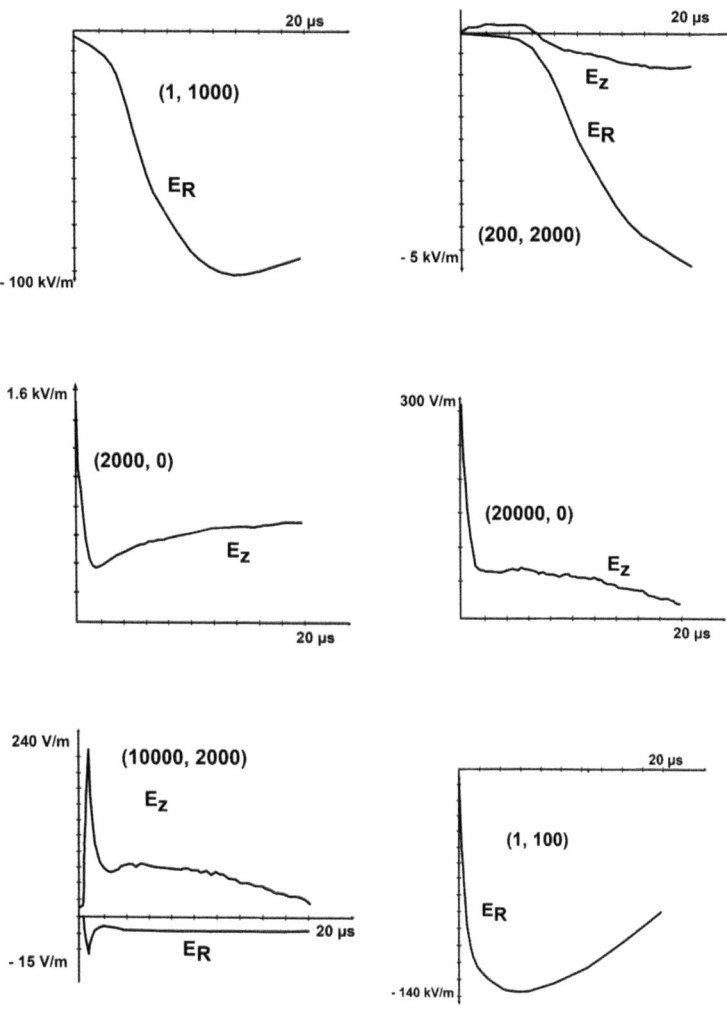

Fig. 5.14 The vertical (E_z) and horizontal (E_R) electric fields the downward negative earth flash return stroke. The bracketed numbers in the form of coordinates (x, y) indicate the spatial point at which the fields were calculated—adapted from (Hoole and Hoole 1993)

for magnetic fields has a sub-microsecond crest time, which in turn is much less than the rise time of the current pulse. This initial peak arises due to the fact that on the sub-microsecond scale, the rate of rise of current on the wavefront was observed to drop sharply as time progressed (Fig. 5.7), i.e. on a microsecond scale the sharp rise to peak may appear as a straight jump after a slow ramp-like increase. The significant part of the LRS wavefront is in reality convex in shape (Fig. 5.7). The rise time to peak current is 4 μs, which is within the 0.22–4.5 μs rise time measured for subsequent strokes striking towers. Moreover, although on a microsecond timescale the DLCRM

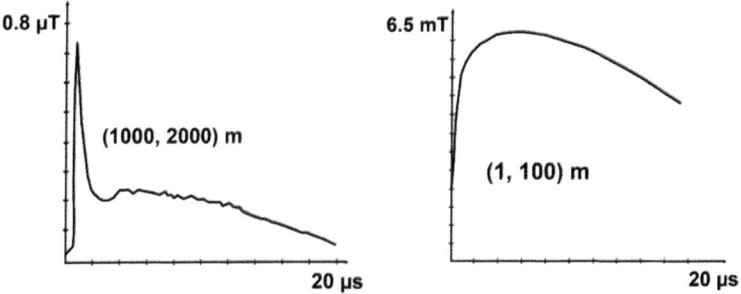

Fig. 5.15 Magnetic fields radiated by the downward negative earth flash return stroke—adapted from (Hoole and Hoole 1993)

calculated LRS current pulse shows that the rate of rise was about 8kA/μs, on a sub-microsecond scale the maximum rate of rise determined is about 98 kA/μs. This high rate of rise of current is due to the convex shape of one part of the LRS wavefront, and is a very important parameter in all engineering considerations, whether they are for electromagnetic compatibility considerations, induced voltage spikes in electronic circuits and power system networks or the threat to fly-by-wire aircraft.

Our identifying the rapid rise time of electric and magnetic fields with a small section of the current wavefront indicates the importance of the convex LRS current wavefront obtained in correct estimates of the sharp, initial peak electric and magnetic fields. The return stroke model indicates that lightning strikes to open ground may be characterized by a sharply convex wavefront on small timescales.

We noted that up to 200 m or so away from the lightning channel, the electric fields are controlled by the negative electric charges along the channel. These have sharp, rapidly changing negative going electric field. Moreover, electric fields very near (e.g. 50 m) to the lightning channel are bipolar. For positive strokes, such bipolar fields are observed at greater distances of the order of a few kilometers. The reason for the bipolar field is that near the channel the electrostatic component of the radiated field is significant. However, as the lightning channel is discharged, the intermediate ($1/r^2$) and radiation ($1/r$) electric field terms dominate. At 2 km above the ground, and 200 m from the flash, the general trend is for the electric field to go negative, since the 0.25 mCoulomb/m or so negative charge along the lightning channel dominates as the current magnitude drops with height. At the earth end, although the charge is about 1.2 mC/m, the current and rate of rise of current are very large. Now this gives us the clue as to why in positive discharges, one observes bipolar fields even at far distances. In positive flashes, the electric charge deposited on the leader is about 10 times higher than the electric charge deposited along the channel of negative flashes. Furthermore, the rate of rise of current for positive lightning discharges is small, an average of about 2.4 kA/μs. Thus, with a smaller rate of rise of currents, which results in a smaller value for the radiation ($1/r$) part of the electric field, the electrostatic portion ($1/r^3$) of the LRS electric field dominates close to the lightning channel. It is indeed encouraging that the DLCRM appears to give very close and exact representation of the downward cloud-to-ground lightning return stroke. This enables us to calculate currents and potentials normally not accessible to measurements, as well as LEMP close to the lightning flashes, and above the ground at heights of interest to aircraft and rocket systems.

To complete our discussions of the radiated lightning electromagnetic fields, more detailed LEMP radiations at different distances from the lightning are shown in Fig. 5.16, for electric and magnetic fields measured close to the lightning flash, at distances of 1–5 km. In Fig. 5.17, the LEMP radiation far away from the lightning flash at distances of 10–100 km is shown.

Referring to (5.54) we note that the electric field has three components, dependent on electric charge Q on the lightning channel (the electrostatic, near-field component,

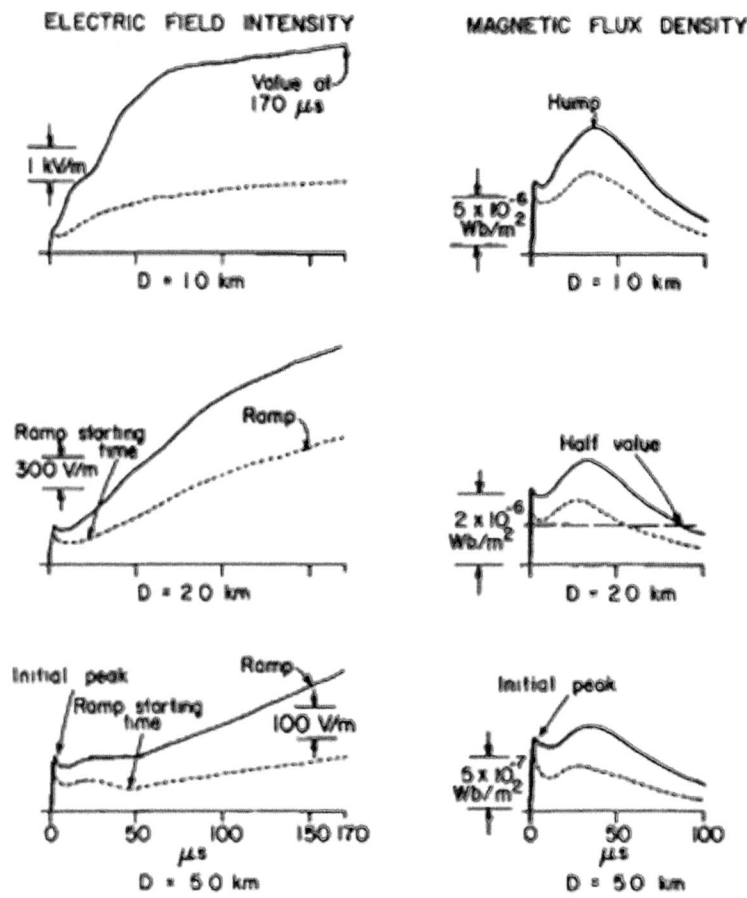

Fig. 5.16 Lightning-radiated electric fields E V/m) and magnetic flux densities (B Wb/m²) at distances 1–5 km. (From Lin et al., J Geophys Res., 1979. Used with permission)

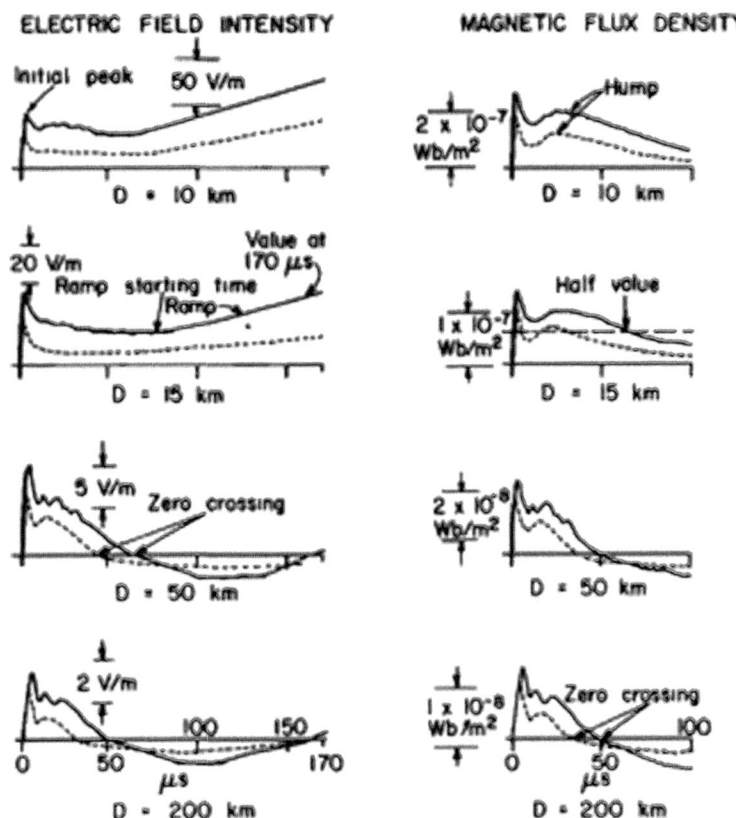

Fig. 5.17 Lightning-radiated electric (E V/m) and Magnetic flux densities (B Wb/m^2) far from the lightning flash (From Lin et al. 1979, J Geophys Res., Used with permission)

roughly decaying as 1/r^3, where r is the distance from the lightning flash), lightning return stroke current I (the intermediate field component, roughly decaying as 1/r^2), and the rate of rise of return stroke current dI/dt (the far-field component, roughly decaying as 1/r). The magnetic field does not have the near-field component due to the electric charge Q. It only has the intermediate field component due to current I and the far-field component due to the rate of rise of current dI/dt. As seen in Fig. 5.16, close to the lightning flash (say 1 km from it) the electrostatic component dominates the electric field (E V/m). At 2 km distance, the electrostatic field still dominates, and the intermediate field component begins to appear, but not yet dominantly so. The humped shape lightning return stroke current I begins to appear at 5 km, though the electrostatic field still influences the magnitude and shape of the electric field E. In case of the magnetic field B Wb/m^2 (magnetic field intensity H = B/μ_o A/m), without the near-field component, at 1 and 2 km, the intermediate field component due to the lightning current I dominates. Hence, the hump-shaped return current of Fig. 5.7 is prominent.

In Fig. 5.17, we can see that at distances of 10–200 km from the lightning flash, the initial shape of the radiated E and B shows sharp peaks to begin with, since this part is dominated by the radiation-filled component due to the rate of rise of current dI/dt. The rate of rise dI/dt of the return stroke current is positive and very large in the initial, wavefront part of the return stroke current. Hence, it dominates the radiation fields, yielding a sharp peak during the initial 1 μs or so of both E and B. At 10 and 15 km distances, the intermediate field components of the magnetic field are still prominent, yielding the hump shape of the return stroke current. The rate of rise dI/dt is negative once the peak of the return stroke current is passed and the current begins to decay. Hence, the negative rate of rise (−dI/dt) yields the zero crossing and negative E and B at 50 and 200 km distances from the lightning flash, where the radiation fields dominate.

5.5.6 Summary

We have presented a DLCRM for subsequent LRS and unbranched first LRS. Its development from a lumped LCR model to distributed LCR model (DLCRM) was considered to obtain the limiting conditions that must be applied when numerical computations are used to solve for the return stroke currents using the DLCRM. Verification tests that could be used to test the accuracy of the numerical solutions were presented. Considering the return stroke currents measured at ground, their measured velocity and measured radiated electric and magnetic fields, it was shown that the DLCRM, simple though in its concepts and parameter specifications, gives a very accurate representation of the subsequent and unbranched first LRS. It has also been seen that the LRS current wavefront possesses a convex-shaped wavefront, and that the sub-microsecond current rate of rise may be as high as 100 kA/μs, whereas the microsecond value may be an order less than this. Moreover, it was found that the near-field, electrostatic portion of the radiated electric fields gives rise to negative electric fields close to the lightning flash, gradually yielding to positive transient electric fields that resemble the LRS current waveform at distances further away from the lightning channel. And in the far-field region, the radiated fields are, as expected, determined by the current rise rates. This simple, easily programmable, fast, and reliable model of the LRS, namely, the DLCRM yields a tool to investigate confidently the engineering parameters of LRS at different heights of the channel and its direct and indirect interactions with power systems, aircraft, and wind turbines.

5.6 A Case Study: Lightning Interaction with Aircraft

5.6.1 Aircraft and Lightning Protection

The usefulness of the computer-based simulation technique of the lightning flash presented in Sect. 5.4 is that it may be readily used with any structures or electric systems to which lightning is attached, to determine the lightning currents and voltages that the structure or system may need to withstand and be protected against when struck by lightning. Moreover, the LEMP radiated into the structure or outside the structure may be determined from the results obtained using the LEMP calculation technique presented in Sect. 5.5. This is crucial in any electrical and electronic systems design, as well as in the protection of structures and people. In this section, we shall illustrate the use of the method with a simple scenario of lightning attached to a low-flying aircraft. As shown in Fig. 5.18, an aircraft is struck by lightning and the path of the lightning flash to ground is completed through the aircraft body. Instead of an aircraft, we may also model and consider a power line tower, tall building, or tree. In this section, we give a lumped circuit model solution for lightning-aircraft interaction. It is much simpler than the more complex, accurate, and distributed LCR model described in Sects. 5.4 and 5.5.

Lightning may directly or indirectly impact on the metallic or composite structure of an aircraft, as well as the electronic communication, command, navigation, and control systems of the aircraft, and its electric power system. The lightning strike in Fig. 5.18 directly discharges through the metallic or carbon composite body of the aircraft. If the body is made out of a good electrical conductor the physical damage is less, though there may appear severe as a few centimeter diameter punctures where the lightning current burns the structure at the point of attachment. If the discharge path through the aircraft runs through a poorly bonded path the physical damage is

Fig. 5.18 Aircraft-lightning interaction

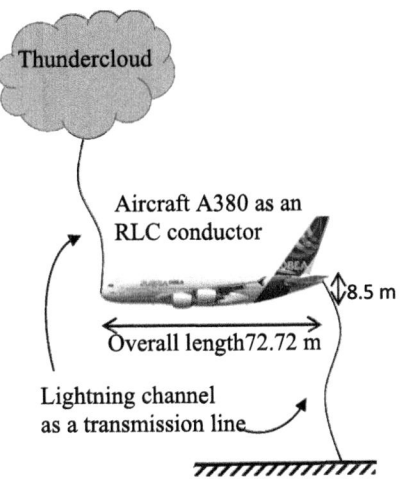

greater. At the poorly bonded point, an arc might form with increased heat dissipation due to a large resistance at the point of bad bonding. Melting could occur, or severe welding, not allowing proper function or movement of metallic parts such as the wing flap or wheel controller. The structure of the aircraft might be deformed due to the force created by the magnetic field of the lightning current. The temperature of the aircraft will increase due to the resistive heat dissipation, and the fuel system or electronic parts may catch fire. Only millijoules of energy is required to ignite and initiate a fire. Antennas and navigation parts are more prone to attracting a lightning strike, causing disruption of the communication, command, and navigation system of the aircraft.

Therefore, it is important to introduce lightning protection of all aircraft. According to the degree of exposure to the lightning effects, sections of the aircraft have been divided into several protection zones. These are

Zone IA: An initial attachment point with a low probability of flash hang-on, such as a leading edge.
Zone IB: An initial attachment point with a high probability of flash hang-on, such as a trailing edge.
Zone 2A: A swept-stroke zone with a low probability of flash hang-on, such as a wing mid-span.
Zone2B: A swept-stroke zone with high probability of flash hang-on, such as a wing inboard trailing edge.

Lightning protection should be designed based on the above-described zone category. Generally lightning protection of aircraft is designed by providing a controlled and safe path for the lightning current to flow in order to reduce the damage from a lightning strike and provide protection to exposed components against arc entry damage. Special attention is given to the protection of the fuel, control, and navigation systems.

5.6.2   Computation of Lightning Currents and Voltage on An Aircraft

As shown in Fig. 5.18, lightning may strike an aircraft's outer extremities, such as nose, wingtip, and tailfin. In this specific case, it strikes the nose and exits through the tail of the aircraft. Most of the strikes occur at take-off or landing. In combat aircraft which have to fly close to the ground, the presence of thunderclouds over the theater of war is a major disadvantage. It may not only be struck by lightning as shown in Fig. 5.18 resulting in damage to the aircraft body, but in addition, the taut electronic flight controller may be driven to instability causing the aircraft to spin off to self-destruction. There is a need to design the aircraft in a way to minimize the damage to it when hit by lightning, and to keep the lightning LEMP caused electronic instability within a manageable, stable margin. The stability of the electronic system as well as

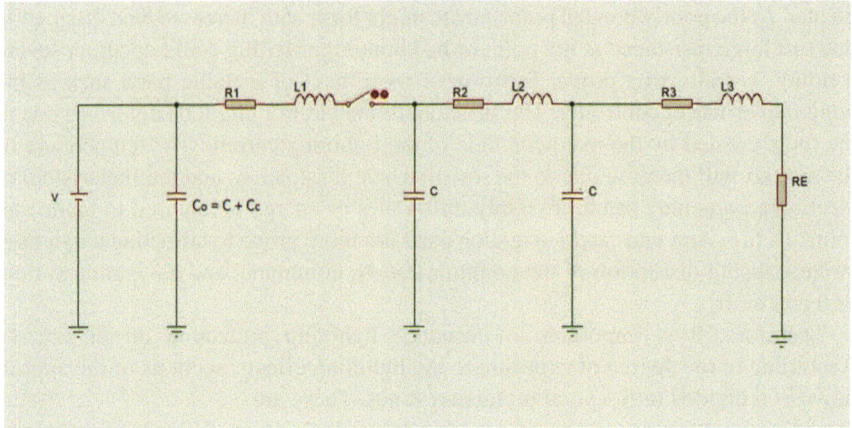

Fig. 5.19 Lumped circuit model of lightning-aircraft dynamics

the stability of the mechanical motion dynamics of the aircraft is interrelated in this scenario. In Fig. 5.19, a simplified, lumped circuit model of the lightning channel attached to the aircraft radome or nose (loop 1), the electric circuit model of the aircraft body (loop 2), and then the lightning channel that is attached to the tail of the aircraft and ground (loop 3), completing the circuit are shown. The cloud voltage V drives the circuit transient once the switch is closed when the lightning makes the final attachment to the aircraft body.

In Fig. 5.19, R1 and L1, of loop 1, are the circuit parameters for the cloud-to-aircraft lightning channel. The aircraft circuit parameters are R_2 and L_2, of loop 2. Finally, loop 3, containing R_3 and L_3, are the circuit parameters for the lightning channel from aircraft to ground. Applying the Kirchhoff's Voltage Law to each loop once the switch is closed:

Mesh # 0:

$$V = \frac{1}{C_0} \int (i_0 - i_1) dt$$

Mesh # 1:

$$\frac{1}{C_0} \int (i_1 - i_0)dt + i_1 R_1 + L_1 \frac{di_1}{dt} + \frac{1}{C_1} \int (i_1 - i_2)dt = 0$$

Mesh # 2:

$$\frac{1}{C_1} \int (i_2 - i_1)dt + i_2 R_2 + L_2 \frac{di_2}{dt} + \frac{1}{C} \int (i_2 - i_3)dt = 0$$

Mesh # 3:

$$\frac{1}{C} \int (i_3 - i_2)dt + i_3 R_3 + L_3 \frac{di_3}{dt} + i_3 R_E = 0$$

Using the above equations, we get

$$\frac{dV}{dt} = \frac{1}{C_0}(i_0 - i_1)$$

$$\frac{1}{C_0}(i_1 - i_0) + R_1 \frac{di_1}{dt} + L_1 \frac{d^2 i_1}{dt^2} + \frac{1}{C_1}(i_1 - i_2) = 0$$

$$\frac{1}{C_1}(i_2 - i_1) + R_2 \frac{di_2}{dt} + L_2 \frac{d^2 i_2}{dt^2} + \frac{1}{C}(i_2 - i_3) = 0$$

$$\frac{1}{C}(i_3 - i_2) + (R_3 + R_E) \frac{di_3}{dt} + L_3 \frac{d^2 i_3}{dt^2} = 0.$$

The assumptions we make are as follows:

- The cloud voltage V is constant.
- Laplace transform initial conditions are

$$L\left\{ \frac{di(t)}{dt} \right\} = SI(s) - i(0)$$

$$L\left\{ \frac{d^2 i(t)}{dt^2} \right\} = S^2 I(s) - Si(0) - i'(0)$$

The MATLAB™ code to solve these coupled equations is given below:

%solving the differential equations obtained in Laplace domain

 syms L1 L2 L3 R1 R2 R3 Re C Co C1 i0 i1 i2 i3 s I1 t   %defining the symbols

V=50*10^6; %50MV constant Voltage in the thunder cloud

 %assuming initial voltage

V0=50*10^6;

 Equation1=V*s-V0==(i0-i1)/Co; %initial V(0-)=50MV

 % assuming initial currents are zero

%defferential equations in the laplace domain

Equation2=(i1-i0)/Co+R1*s*i1+L1*(s^2*i1)+(i1-i2)/C1==0;

Equation3=(i2-i1)/C1+R2*s*i2+L2*(s^2*i2)+(i2-i3)/C==0;

Equation4=(i3-i2)/C+(R3+Re)*s*i3+L3*(s^2*i3)==0;

 %solving equations in the laplace domain for currents

Equations = [Equation1 Equation2 Equation3 Equation4];

Variables = [i0,i1 i2 i3];

[i0,i1,i2,i3] = solve(Equations,Variables);

 Vars = [L1 L2 L3 R1 R2 R3 Re C Co C1];

%pre calculated parameter values are included

%substituting parameter values to equations

i1=subs(i1,Vars,values);

i0=subs(i0,Vars,values);

i2=subs(i2,Vars,values);

```
i3=subs(i3,Vars,values);

%obtaining inverse laplace

I=ilaplace(i1);

I2=ilaplace(i2);

I3=ilaplace(i3);

%ploting graphs(currents Vs time)

tn=0:0.000001:0.003;

In = double( subs(I, symvar(I), tn)) ;

In2 = double( subs(I2, symvar(I2), tn)) ;

In3 = double( subs(I3, symvar(I3), tn)) ;

figure;

plot(tn,In,'r');

hold on;

plot(tn,In2,'b')

plot(tn,In3,'k')

legend('lightning current','current through the aircraft','lightning current form
aircraft to earth')
```

There are several cases of lightning flashes where both return stroke modeling and the calculation of the LEMP-radiated fields may require more complex models. Such complex cases include heavily branched return stroke, bent lightning channel, and multiple, simultaneous lightning flashes. However, the model reported in this chapter may be used as the basic building block to represent these complex situations. In Fig. 5.20a, for instance, a spidery lightning flash event inside the cloud and at a distant a simultaneous cloud to ground flash is shown. It is obvious that such a thundercloud-initiated lightning flash needs a more complex model, where the fundamental model presented in Sects. 5.4 and 5.5 may be used as the basic building blocks. In Fig. 5.20b, a situation where the ground flash is either a delayed picture of the same flash striking the ground multiple times or it may be two different electric charge centers initiating separate lightning flashes to ground is shown. This too would require modification and extension of the lightning model reported in this chapter for a single, straight lightning flash to ground.

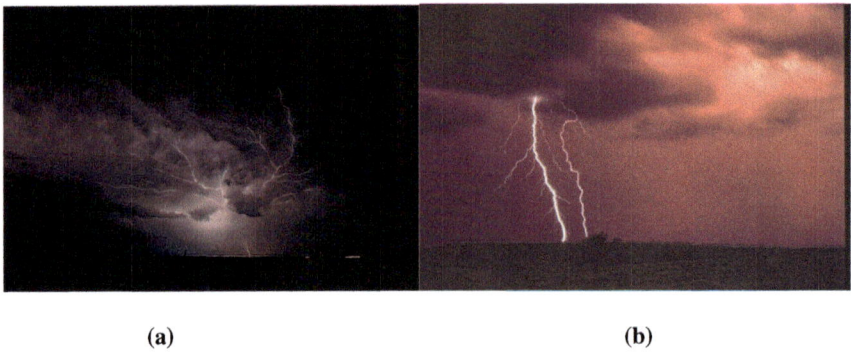

(a)   (b)

Fig. 5.20 Lumped circuit model of lightning-aircraft dynamics

References

Ajayi, N.O.: Acoustic observation of thunder and cloud-to-ground flashes. J. Geophys. Res. **77**, 4586–4587 (1972)

Baba, Y., Rakov, V.A.: On the mechanism of attenuation of current waves propagating along a vertical perfectly conducting wire above ground: application to lightning. IEEE Trans. Electromagn. Compat. **47**(3), 521–532 (2005)

Baba, Y., Rakov V.A.: Electromagnetic models of the lightning return stroke. J. Geophys. Res. **112**, D04102 (2007)

Balachandran, N.K.: Acoustic and electrical signals from lightning. J. Geophys. Res. **88**, 3879–3884 (1983)

Barry, J.D.: Ball Lightning and Bead Lightning, Plenum (1980)

Berger, K.: Novel observations on lightning discharges: results of research on Mount San Salvatore. J. Franklin. Inst. **283**, 478–525 (1967)

Berger, K., Anderson, R.B., Kroninger, H.: Parameters of lightning flashes. Electra. **41**, 23–37 (1975)

Berger, K.: Earth flash, in lightning: physics of lightning. In: Golde, R.H. (ed.), pp. 119–190. Academic Press (1997)

Betz, H.D., Schumann, U., Laroche, P. (eds.): Lightning: Principles, Instruments and Applications, Springer (2009)

Braginskii, S.I.: Theory of the development of a spark channel. Soc. Phys. JEPT (English translation) 34:1068–1074 (1958)

Bruce, C.E.R., Golde, R.H.: The lightning discharge. J. Inst. Electr. Eng. **88**, 487–520 (1941)

Cooray, V.: Further Characteristics of Positive radiation Fields from Lightning in Sweden. J. Geophys. Res. **89**, 11807–11815 (1984)

Cooray, V.: A model for subsequent return strokes. J. Electrostst. **30**, 343–354 (1993)

Cooray, V.: On the concepts used in return stroke models applied to engineering practice. IEEE Trans. Electromagn. Compat. **45**(1), 101–108 (2003)

Cooray, V., Lundquist, P.: On characteristics of some radiation fields from lightning and their further origin in positive ground flashes. J. Geophys. Res. **87**, 11203–11214 (1982)

Cooray, V., Theethayi, N.: Pulse propagation along transmission lines in the presence of corona and their implications to lightning return strokes. IEEE Trans. Antenna Propag. **56**(7), 1948–1959 (2008)

Cooray, V. (ed.): The Lightning Flash. IET (2003)

Cooray, V. (ed.): Lightning Protection. IET (2010)

Cooray, V. (ed.): Lightning Electromagnetics. IET (2012)

da Frota Mattos M.A., Christopoulos, C.: A nonlinear transmission line model of the lightning return stroke. IEEE Trans. Electromagn. Compat. **30**, 401–406 (1988)

da Frota Mattos M.A., Christopoulos C.: A model of the lightning channel, including corona, and prediction. J. Phys., D Appl. Phys. **23**, 40–46 (1990)

Deindorfer, D., Uman, M.A.: An improved return stroke model with specified channel base current. J. Geophys. Res. **95**, 13621–13664 (1990)

von Engel, A.: Electric Plasmas. Taylor and Francis, London (1981)

Feiux, R.P., Gary, C.H., Hutzler, B.P., Eyebert-Berrard, A.R., Hubert, R.A., Meesters, A.C., Pettroud, P.H., Hamelin, J.H., Person, J.M.: Research on artificially triggered lightning in France. IEEE Trans. Power Apparat. Systs. **94**, 725–733 (1978)

Few A.A.: Acoustic radiations from lightning. In: Volland, H. (ed.): Handbook of Atmospherics, vol. 2, CRC Press (1981)

Fisher, R.J., Uman, M.A.: Measured electric field rise times for first and subsequent return strokes. J. Geophys. Res. **77**, 399–407 (1972)

Fowler, R.G.: Lightning. Appl. Collision Phys. **5**, 31–67 (1982)

Golde, R.H. (Ed.): Lightning, Vol 1: Physics of Lightning. Academic Press (1977)

Golde, R.H.: Lightning, Vol. 2: Engineering Applications. Academic Press (1977)

Guo, C., Krider, E.P.: The optical and radiation electric field signatures produced by lightning return strokes. J. Geophys. **87**, 8913–8922 (1982)

Hoole, P.R.P.: Doctor of Philosophy Thesis. Oxford University, Department of Engineering Science (1987)

Hoole, P.R.P.: Simulation of lightning attachment to open ground, tall towers and aircraft. IEEE Trans. Power Delivery **8**(2), 732–738 (1993)

Hoole, P.R.P.: Modeling the lightning earth flash return stroke for studying its effects on engineering systems. IEEE Trans. Magnet. **29**, 1839–1844 (1993)

Hoole, P.R.P.: Electromagnetic Imaging in Science and Medicine. WIT Press, UK (2000)

Hoole, P.R.P., Hoole, S.R.H.: Guided waves along an un-magnetized lightning plasma channel. IEEE Trans. Magn. **24**(6), 3165–3167 (1988)

Hoole, P.R.P., Hoole, S.R.H.: A distributed transmission line model of cloud-to-ground lightning return stroke: model verification, return stroke velocity, unmeasured currents and radiated fields. Int. J. Phys. Sci., UK **6**, 3851–3866 (2011)

Hoole, P.R.P., Hoole, S.R.H.: Charge simulation method for the calculation of electromagnetic fields radiated from lightning. In: Conner, J.J., Brebbia, C.A. (eds.) Boundary Element Technology, pp. 153–169. Computational Mechanics Publications, Southampton (1986)

Hoole, P.R.P., Pirapaharan, K., Hoole, S.R.H.: An electromagnetic field based signal processor for mobile communication position-velocity estimation and digital beam-forming: an overview. J. Jpn. Soc. Appl. Electromagn. Mech., Japan **19**, S33–S36 (2011)

Hoole, P.R.P., Pirapaharan, K., Hoole, S.R.H.: Waveguide and circuit EM models of lightning return stroke currents. J. Jpn. Soc. Appl. Electromagn. Mech., Japan **19**, S167–S170 (2011)

Hoole, P.R.P., Pirapaharan, K., Hoole, S.R.H.: Waveguide and circuit EM models of lightning return stroke currents. J. Jpn. Soc. Appl. Electromagn. Mech., Japan **19**, S167–S170 (2011)

Hoole, P.R.P., Pirapaharan, K., Hoole, S.R.H.: Electromagnetics Engineering Handbook. WIT Press, UK (2013)

Hoole, P.R.P., Pirapaharan, K., Kavi, M., Fisher, J., Aziz, N.F., Hoole, S.R.H.: Intelligent localisation of signals using the signal wavefronts: a review. In: Lightning Protection (ICLP), International Conference, pp. 474–479. IEEE Xplore Library (2014)

Hoole, P.R.P., Hoole, S.R.H.: Simulation of lightning attachment to open ground tall towers and aircraft. IEEE Trans. Power Delivery **8**(22), 732–740. Institute of Electrical and Electronics Engineers (1993)

Hoole, P.R.P., Hoole, S.R.H.: Finite element computation of magnetic fields from lightning return strokes. In: Cendes, Z.J. (ed.) Computational Electromagnetics, North Holland, pp. 229–237, July 1986

Hoole, P.R.P., Pearmain, A.J.: A review of the finite-difference method for multidielectric electric field calculations. J. Electr. Power Syst. Res. **24**(11), 19–30. Elsevier (1992)

Hoole, P.R.P., Thirukumaran, S., Hoole, S.R.H., Harikrishnan, R., Jievan, K.: Ground to cloud lightning flash currents and electric fields: interaction with aircraft and production of ionospheric sprites. In: Proceedings of the 28th International Review of Progress in Applied Computational Electromagnetics, 6 p., Michigan, USA (2012)

Hoole, P.R.P., K. Pirapaharan, K., Hoole, S.R.H.: Waveguide and circuit em models of lightning return stroke currents. J. Jpn. Soc. Appl. Electromagn. Mech. Japan **19**, S167–S170 (2011)

Hoole, P.R.P., Thirikumaran, S., Ramiah, H., Kanesan, J., Hoole, S.R.H.: Ground to cloud lightning flash and electric fields: interaction with aircraft and production of ionospheric sprites. J. Comput. Eng., Article ID 869452 (2014). https://doi.org/10.1155/2014/869452

Hoole, P.R.P., Thirukumaran S., Hoole, S.R.H.: A software testbed for electrodynamics of direct cloud to ground and ground to cloud lightning flashes to aircraft. Int. J. Appl. Electromagn. Mech. **47**(4), 911–925 (2015)

Hoole, P.R.P., Fisher, J., Pirapaharan, K., Al K. H. Othman, Julai, N., Aravind, C.V., Senthilkumar, K.S., Hoole, S.R.H.: Determining safe electrical zones for placing aircraft navigation. Measurement and microelectronic systems in static thunderstorm environment. Int. J. Control Theory Appl. 10(16) (2017)

Hoole, P.R.P., Balasuriya, B.A.A.P.: Lightning radiated electromagnetic fields and high voltage test specifications. IEEE Trans. Magnet. **29**(2), 1845–1848 (1993)

Hoole, P.R.P., Hoole, S.R.H.: Computer aided identification and location of discharge sources. J. App. Phys. **61** (1987b)

Hoole, P.R.P., Hoole, S.R.H.: Computing transient electromagnetic fields from lightning. J. Appl. Phys. **61**, 3473 ff (1988)

Hoole, P.R.P., Hoole, S.R.H: Stability and accuracy of the finite difference time domain (FDTD) method to determine transmission line traveling wave voltages and currents. J. Eng. Technol. Res. (2011)

Hoole, P.R.P.: Smart Antennas and Electromagnetic Signal processing for Advanced Wireless Technology: with Artificial Intelligence and Codes. River Publisher (2020)

Hoole, P.R.P.: Smart Antennas and Signal Processing for Communication, Medical and Radar Systems. WIT Press, UK (2001) (See IEE review of this book close to the end of this document)

Hoole, S.R.H., Hoole, P.R.P.: A Modern Short Course in Engineering Electromagnetics. Oxford University Press, USA (1996)

Hoole P.R.P., Hoole, S.R.H.: Computing transient electromagnetic fields radiated from lightning. J. Appl. Phys. **61**(8), 3473–3475 (1987a)

Idone I.P., Orville R,E.: Correlated peak intensity light intensity and peak current in triggered lightning subsequent strokes. J. Geophys. Res. **90**(D4), 6159–6164 (1985)

Jordon, D.M., Uman, M.A.: Variations in light intensity with height and time from subsequent return strokes. J. Geophys. Res. **88**, 6555–6562 (1983)

Lin, Y.T., Uman, M.A.: Electric radiation fields of lightning return strokes in three isolated florida thunderstorms. J. Geophys. Res. **78**, 7911–7914 (1973)

Lin, Y.T., Uman, M.A., Tiller, J.A., Brantley, R.D., Beasley, W.H., Krider, E.P., Weidman, C.D.: Characterization of lightning return stroke electric and magnetic fields from simultaneous two-station measurements. J. Geophys. Res **84**, 6307–6314 (1979)

Little, P.F.: Transmission line representation of a lightning return stroke. J. Phys. D: Appl. Phys. **11**, 1893–1910 (1978)

Master, M.J., Uman, M.A., Lin, Y.T., Standler, K.B.: Calculations of lightning return stroke electric and magnetic fields above ground. J. Geophys. Res. **86**, 12127–12132 (1981)

Moosavi S.S., Moini, R., Sagdeghi, R.: Representation of a lightning return stroke as a nonlinearly loaded thin-wire antenna. IEEE Trans. Electromagnet. Compat. **51**(3), 488–498 (2009)

Mosaddeghi, A., Pavanello, D., Rachidi, F., Rubenstein, A.: On the inversion of the electric field at very close range from a tower struck by lightning. J. Geophys. Res **112**, D19113 (2007)

Nayak, S.K., Meledash, T: Lightning induced current and voltage on a rocket in the presence of its trailing plume. IEEE Trans. Electromagn. Compat. **52**(1), 117–127 (2010)

Orville, R.E., Idone, V.P.: Lightning Leader characteristics in Thunderstrom research International program (TRIP). J. Geophys. Res. **87**, 11177–11192 (1982)

Plooster, M.N.: Numerical model of the return stroke of the lightning discharge. Phys. Fluids. **14**, 2124–2133 (1971)

Price, G.H., Pierce, E.T.: The modeling of channel current in the lightning return stroke. Radio Sci. **12**, 381–388 (1977)

Rachidi, F., Janischewsky, W.A., Hussein, A.M., Nucci, C.A., Guerrieri, S., Kordi, S.B., Chang, J.S.: Current and electromagnetic field associated with lightning return strokes to tall towers. IEEE Trans. Electromag. Compat. **43**(3), 356–367 (2001)

Rakov, V.A., Uman, M.A.: Review and evaluation of lightning return stroke models including some aspects of their applications. IEEE Trans. Electromagn. Compat. **40**(4), 403–426 (1998)

Rakov, V.A., Uman, M.A.: Lightning Physics and Effects. Cambridge University Press, USA (2003)

Rakov, V.A., Uman, M.A., Rambo, K.J.: A review of ten years of triggered lightning experiments at Camp Blanding, Florida. Atmos. Res. **76**, 503–517 (2005)

Rakov, V.A.: Fundamentals of Lightning. CUP (2016)

Schonland, B.F.: The lightning discharge. Handb. Phys. **22**, 576–628 (1956)

Schonland, B.F., Malan, D.J., Collens, H.: Progressive lightning II. Soc. London Ser. A **152**, 595–625 (1935)

Shumpert, T.H., Honnell, M.A., Lott, G.K.: Measured spectrum amplitude of lightning Sferics in the HF, VHF and UHF bands. IEEE Trans. Electromagn. Compat. **24**, 368–372 (1982)

Spitzer, L.: Physics of Fully Ionized Gases. Interscience, New York (1961)

Strawe D.F.: Non-linear modelling of lightning return stroke. In: Proceedings of the Federal Aviation Administration/Florida Institute of Technology Workshop on Grounding and Lightning Technology. Report FAA-RD-79.6: 9–15 (1979)

Theethayi, N., Cooray, V.: On representation of the lightning return stroke process as a current pulse along a transmission line. IEEE Trans. Power Deliv. **20**(2), 823–837 (2005)

Thotappillil, A., Uman, M.A.: A lightning return stroke model with height-variable discharge content. J. Geophys. Res. **99**, 22773–22780 (1994)

Tiller J.A., Uman M.A., Lin Y.T., Brantley R.D., and E.P. Krider E.P.: Electric field statistics for close lightning return strokes near Gainesville, Florida, J. Geophys. Res. **81**, 4430–4434 (1976)

Uman, M.A.: Lightning return stroke electric and magnetic fields. J. Geophys. Res. **90**, 6121–6130 (1985)

Uman, M.A., Krider, K.P.: A review of natural lightning: experimental data and modelling. IEEE Trans. Electromagn. Compat. EMC **24**, 79–112 (1982)

Uman, M.A., Standler, R.B.: Lightning return stroke models. J. Geophys. Res. **85**, 1571–1583 (1980)

Uman, M.A.: The Art and Science of Lightning Protection. CUP (2008)

Uman M.A.: Lightning. McGraw Hill (1969)

Uman M.A.: The Lightning Discharge. Academic (1987)

Weidman, C.D., Krider, E.P.: Sub microsecond structure of the return stroke waveforms. Geophys. Res. Lett. **7**, 955–958 (1980)

Weidman C.D., Krider K.P.: The fine structure of lightning return stroke waveforms. J. Geophys. Res. **87**, 6239–6247 (1982). Correction, J. Geophys. Res. **87**, 7351

Chapter 6
Localization and Identification of Acoustic and Radio Wave Signals Using Signal Wavefronts with Artificial Intelligence: Applications in Lightning

K. Pirapaharan, P. R. P. Hoole, and S. R. H. Hoole

Abstract Localization of lightning flash occurring at an unknown location may be done by measuring the electromagnetic pulse (LEMP) radiated by the lightning flash, or the sound waves associated with thunder generated by the high current lightning return stroke. Instruments, placed at three locations, measure the sound wave emitted by lightning at an unknown location and localize it. Alternatively, lightning localization may be done by measuring at three or more locations the LEMP radiated by a lightning flash. Such localizers are used, for instance, at airports. We may develop a system to locate a source of sound by studying the possible localization technique used by bats. Bats typically are capable of laryngeal echolocation that enables them to identify their position and move in complete darkness. The bat sound contains a signal with multiple frequency components, as is the case with thunder. Also, the acoustic signal propagation in the atmosphere deviates from a spherical wave propagation due to a number of different factors including absorption of sound in air, non-uniformity of the propagation medium due to meteorological conditions, and interaction with absorbing solid obstacles with acoustic properties that are influenced by the frequency of the acoustic signal. Hence, a test acoustic signal with multiple frequency components is modeled and tested for the application of acoustic signal localization using signal wavefronts. A computer simulation is made to compare the received signal patterns at different distances using the empirical atmospheric attenuation model for acoustic signal attenuation provided in the ISO 9613-2:1996 standard. Further, this chapter also describes a technique for lightning localization using the LEMP measured at three different locations. An appropriate empirical model, similar to the Bruce-Golde model, is proposed for the lightning return stroke current represented by a dipole antenna. The return stroke current proposed, unlike the Bruce-Golde model, is both time and frequency dependent. The return stroke current thus emits multiple frequency components for LEMP radio wave signals. In addition, closed form equations for the signal propagation for the proposed LEMP radio wave field components, in terms of the distance from the origin, are given here. The wavefront of the overall LEMP changes with distance traveled from the lightning flash. The frequency- and distance-dependent LEMP wave shape is used to detect and localize the lightning flash. A computer-coded

© The Author(s), under exclusive license to Springer Nature Switzerland AG 2022   209
P. Hoole and S. Hoole, *Lightning Engineering: Physics, Computer-based Test-bed, Protection of Ground and Airborne Systems*,
https://doi.org/10.1007/978-3-030-94728-6_6

lightning localizing technique is illustrated to detect the lightning location from the wavefronts of LEMP at different distances from the lightning flash.

6.1 Introduction

An acoustic wave is defined as a phenomenon whereby a transient elastic wave is generated by the rapid release of energy from a localized source. Such is the case when the lightning return stroke releases a large amount of heat energy along the ionized channel of the lightning leader stroke. The process of locating the source of these acoustic waves, by recording the propagating acoustic signals from various sensors and properly analyzing them, is commonly known as the acoustic source localization technique. Kundu (2014) has reviewed the research status of acoustic source localization research technology. Tobias (1976) is a pioneer in the study of acoustic source localization in isotropic materials. The triangulation method proposed by Tobias (1976) is the most commonly used method for isotropic materials. It determines the location of the acoustic source based on the time difference of arrival (TDOA) between the acoustic waves reaching different (a minimum of three) sensors. But only when the precise wave velocity of the elastic wave propagation in the material is obtained, an accurate location result can be obtained. In response to this limitation, the triangulation method has been improved, and an acoustic source localization method suitable for isotropic materials without wave velocity has been proposed by Kundu et al. (2008). However, the localization accuracy of the TDOA method is affected by noise, dispersion effect, energy attenuation, and other factors affecting the wave during the propagation process. The localization accuracy of these methods depends on the measured wave velocity, and the relevant properties of the material need to be obtained in advance. Although the TDOA method is mature and easy to use, its localization accuracy is highly dependent on the accuracy of the measured TDOA.

In this chapter, we propose a new method of localization of acoustic sources for acoustic signals, which contain a range of frequency components. Acoustic energy is dissipated in air by the following two major mechanisms: (a) viscous losses due to friction between air molecules, which results in heat generation and (b) the relaxation processes by which the acoustic energy is momentarily absorbed by the air molecules and causes the molecules to vibrate and rotate. These molecules can then re-radiate sound at a later instant, which can partially interfere with the incoming acoustic signal. Hence, the atmospheric absorption that takes place is a function of frequency, atmospheric properties, and the distance propagated. Thus, the atmospheric absorption coefficient is very dependent on the frequency component of the signal. Moreover, the atmospheric absorption varies for different frequency components at different distances from the acoustic source as per the empirical atmospheric attenuation model for acoustic signal attenuation standard provided in ISO 9613-2:1996. Ultimately, the shape of the received signal variation depends on the distance traveled. This phenomenon is used to localize the source distance from the receiver location.

A similar technique is used to localize the radio wave signal source since the radio wave propagation amplitude also depends on the frequency components of the current to the dipole antenna model of the return stroke. A current model with different frequency components may be used and the closed form equation for the propagating radio wave field components in terms of the frequency and the distance from the origin is presented. Once again the shape of the radio field components varies, depending on the distance from the origin of the source.

6.2 Methodology: Test Signals and Wavefronts

6.2.1 Methodology for Acoustic Signals

For the novel method of wavefront estimation, the test signal model and the respective wavefront models vary for acoustic and radio wave signals. The wavefront functions for the respective test signal functions are used.

In order to have multiple frequency components, the test signal is selected as follows.

$$S(t) = \sum_{i=1}^{N} A_i e^{-f_i t} \tag{6.1}$$

where A_i, f_i, and N are defined to have $S(t)$ as an energy signal with multiple frequency components. Also, the proposed signal satisfies the condition

$$S(t)|_{t=0} = 0. \tag{6.2}$$

The atmospheric attenuation factor is given as stated in ISO 9613-2:1996. For a standard pressure of one atmosphere, the absorption coefficient α (in dB/m) can be calculated as a function of frequency f (Hz), temperature T (degrees Kelvin), and molar concentration of water vapor h by:

$$\alpha = 8.69 f^2 \left\{ \begin{array}{l} 1.84 \times 10^{-11} (T/T_0)^{1/2} + \\ (T/T_0)^{-5/2} \left[\dfrac{0.01275 e^{-2239.1T}}{F_{r,O} + f^2/F_{r,O}} + \dfrac{0.1068 e^{-3352T}}{F_{r,N} + f^2/F_{r,N}} \right] \end{array} \right\} \tag{6.3}$$

where $T_0 (293.15^0 K)$ is the room temperature in Kelvin while $F_{r,O}$ and $F_{r,N}$ are respective oxygen and nitrogen relaxation frequencies as given below:

$$F_{r,O} = 24 + 4.04 \times 10^4 h \left(\frac{0.02 + h}{0.391 + h} \right) \tag{6.4}$$

$$F_{r,N} = \left(\frac{T}{T_0} \right)^{1/2} \left(9 + 280h\, e^{\left\{ -4.17 \left[\left(\frac{T}{T_0} \right)^{-1/3} -1 \right] \right\}} \right). \tag{6.5}$$

Thus, the atmospheric attenuation factor A_{atmos} is obtained as:

$$A_{atmos} = e^{-\alpha r} \tag{6.6}$$

where r is the distance over which the wave has been propagated.

Taking the fast Fourier transform (FFT) of $S(t)$ and applying the atmospheric attenuation factor for the respective frequencies and distance traveled, it will yield the FFT of the wavefront at different distances. Finally, by taking inverse FFT, the wavefront is determined at different ranges for the given test signal $S(t)$.

6.2.2   Methodology for Radio Wave Signals

The current to the dipole antenna could be modeled as a sum of exponential terms given below to obtain the current with multiple frequency components:

$$I(t) = \sum_{n=1}^{N} I_n e^{-f_n t} \tag{6.7}$$

where I_n, f_n, and N are defined to give the current signal $I(t)$ as an energy signal with multiple frequency components. Also, the current modeled as such satisfies the following conditions:

$$\left. \begin{array}{l} I(t)|_{t=0} = 0 \\[2mm] \dfrac{dI(t)}{dt}\bigg|_{t=0} = 0 \end{array} \right\}. \tag{6.8}$$

Selecting the dipole as a vertical dipole of length L, the electrical field components at a distance r could be derived in the spherical coordinate system as:

$$E_r = 2\eta \frac{L \cos\theta}{4\pi} \sum_{n=1}^{N} k_n^2 \left(\frac{1}{(k_n r)^2} + \frac{1}{(k_n r)^3} \right) I_n e^{-f_n t} \tag{6.9}$$

$$E_\theta = \eta \frac{L \sin\theta}{4\pi} \sum_{n=1}^{N} k_n^2 \left(\frac{1}{k_n r} + \frac{1}{(k_n r)^2} + \frac{1}{(k_n r)^3} \right) I_n e^{-f_n t} \tag{6.10}$$

where $k_n = j\sqrt{\mu_0 \varepsilon_0} f_n$ and $\eta = \sqrt{\frac{\mu_0}{\varepsilon_0}}$.

Hence, the resultant electrical field could be expressed as

$$\boldsymbol{E} = E_r \hat{\boldsymbol{r}} + E_\theta \hat{\boldsymbol{\theta}}. \tag{6.11}$$

Thus, the amplitude of the resultant field is

$$|\boldsymbol{E}| = \sqrt{|E_r|^2 + |E_\theta|^2}. \tag{6.12}$$

The resultant electric field is the result of the static charges of the dipole (electrostatic), the DC current in the dipole (magneto-static) and time-varying current in the dipole (radiation). Thus, different contributions predominate the resultant field component at different distances. Consequently, we attempt to use the amplitudes of the resultant field pattern to identify the distance of the source point from the observation point. Thus, the shape and size of the wavefront of the electric (or magnetic) field component could be used to identify the distance from the LEMP source for the empirically modeled lightning current.

We are grateful to Bamunusinghe B.A.A.R. and Dushmantha W.S for the MATLAB™ codes listed below. Moreover, credit for the development of the code in Sect. 5.6.2 is due to Abeywardhana S.A.Y., Senarathne L.R., Subhashini H.A.A. These codes may be used to develop, implement and experiment with the techniques presented in Sects. 5.6.2 and 6.2.

1. Ultrasonic signals: Stationary transmitter

```
Main function:

clear all
clc
%input data
h = 70;%humidity
Fs=400000;%sampling frequncy
%input signal
t=0:1/(2*Fs):(0.0001(1/(Fs)));
y=(23*exp(-2*pi*20000*t)-29*exp(-2*pi*40000*t)+5*exp(-2*pi*60000*t)+1*exp(-
2*pi*80000*t)+23*exp(-2*pi*100000*t)-29*exp(-2*pi*120000*t)+5*exp(-
2*pi*140000*t)+1*exp(-2*pi*160000*t));
B = max(y);
%time domain graph of orginal signal
figure(1)
plot(t,abs(y/B))
xlabel('time(s)');
ylabel('Nomalized values');
title('time domain signal of original')

y1=fft(y);% fast fourier transform of input signal
L2=length(y1);
y2=y1(1:L2/2);
f = Fs*(0:((L2/2)-1))/L2;
A = max(y2);
%frequncy domain graph of orginal signal
figure(2)
plot(f,abs(y2/A))
title('frequency domain singal of original')
xlabel('frequency');
ylabel('Nomalized values');

d=1;%distance between transmiter and object/2
[K] = attenuation_effect(f,h,d,y2,Fs,L2,3);% attenuation fact with graphs
```

Attenuation effect function:

```
function [K] = attenuation_effect(f,h,d,y2,Fs,L2,num)
F_ro = 24 + (40400*h*(0.02+h)/(0.391+h));
F_rn = 9 + (280*h);
A = (6.1425*(10^-6))./(F_ro + (f.^2)/F_ro);
B = (1.0817*(10^-5))./(F_rn + (f.^2)/F_rn);
C = A + B + (1.84*(10^-11));
alpha = C .*(f.^2);
k = exp(-alpha * d);
K = k.*y2;
A = max(K);
%frequency domain graph of attenuated signal
figure(num)
plot(f,abs(K/A));
title(['frequency domain signal when d =',num2str(d)]);
xlabel('frequency');
ylabel('Nomalized values');

t0=(0:2:(L2-2))/Fs;
x = ifft(K);%inverse fourier transform
B = max(x);
%time domain graph of attenuated signal
figure(num+1)
plot(t0,x/B);
title(['time domain signal when d=',num2str(d)]);
xlabel('time(s)');
ylabel('Nomalized values');
end
```

2. Ultrasonic signals: Moving transmitter

```
Main function:

clear all
clc
%input data
Vair=330;    %ultrasonic signal velocity in Air

h = 70;   %humidity
Fs=400000;%sampling frequncy
%input signal
t=0:1/(2*Fs):(0.0001-(1/(2*Fs)));
y=(23*exp(-2*pi*20000*t)-29*exp(-2*pi*40000*t)+5*exp(-2*pi*60000*t)+1*exp(-
2*pi*80000*t)+23*exp(-2*pi*100000*t)-29*exp(-2*pi*120000*t)+5*exp(-
2*pi*140000*t)+1*exp(-2*pi*160000*t));
L=length(y);%length of signal
%time domain graph of orginal signal
figure(1)
plot(t,abs(y))
xlabel('time(s)');
ylabel('Nomalized values');
title('time domain signal of original')

y1=fft(y);% fast fourier transform of input signal
L2=length(y1);
y2=y1(1:L2/2);
f = Fs*(0:((L2/2)-1))/L2;
%frequncy domain graph of orginal signal
figure(2)
plot(f,abs(y2))
title('frequency domain singal of original')
xlabel('frequency');
ylabel('Nomalized values');

Vobject=100;%object velocity
d=1;
%doppler shift
t0 = doppler_shift(Vair,Vobject,1/Fs);

%doppler shifted signals
Fs_new = 1/t0;
f = Fs_new*(0:((L2/2)-1))/L2;
t=0:1/(2*Fs_new):((L-1)/(2*Fs_new));
num=3;

[K] = attenuation_effect(f,h,d,Vobject,y2,Fs_new,L2,num);
```

Attenuation effect function:

```
function [K] = attenuation_effect(f,h,d,Vobject,y2,Fs,L2,num)
F_ro = 24 + (40400*h*(0.02+h)/(0.391+h));
F_rn = 9 + (280*h);
A = (6.1425*(10^-6))./(F_ro + (f.^2)/F_ro);
B = (1.0817*(10^-5))./(F_rn + (f.^2)/F_rn);
C = A + B + (1.84*(10^-11));
alpha = C .*(f.^2);
k = exp(-alpha * d);
K = k.*y2;
A = max(K);
%frequency domain graph of attenuated signal
figure(num)
plot(f,abs(K/A));
title(['frequency domain signal when d,Velocity =',num2str(d),',',num2str(Vobject)]);
xlabel('frequency');
ylabel('Nomalized values');

t0=(0:2:(L2-2))/Fs;
x = ifft(K);%inverse fourier transform
B = max(x);
%time domain graph of attenuated signal
figure(num+1)
plot(t0,x/B);
title(['time domain signal when d,Velocity =',num2str(d),',',num2str(Vobject)])
xlabel('time(s)');
ylabel('Nomalized values');
end
```

Doppler function:

```
function [t0] = doppler_shift(Vair,Vobject,Ts)
if  Vobject>0
   t0=Ts*(Vair-Vobject)/(Vair+Vobject);
else
   t0=Ts*(Vair+abs(Vobject))/(Vair-abs(Vobject));
end
end
```

3. Microwave signals

```
Main function:

clear all
clc
Fs=8000000000;%sampling frequncy
%input signal
angle1 = 0.00000002;
angle = pi/6;
t=0:1/(10*Fs):(0.00000001-(1/(Fs)));
y=(23*ex p(-2*pi*250000000*t)-29*exp(-2*pi*300000000*t)+5*exp(-
2*pi*400000000*t)+1*exp(-2*pi*950000000*t));

A1=23000;
A2=-29000;
A3=5000;
A4=1000;
f1=250000000;
f2=300000000;
f3=400000000;
f4=950000000;
k1=250000000j*(2*pi*3.4*(10^-9));
k2=300000000j*(2*pi*3.4*(10^-9));
k3=400000000j*(2*pi*3.4*(10^-9));
k4=950000000j*(2*pi*3.4*(10^-9));
r=2;
E_r = 2*cos(angle)*(((k1^2)*A1*exp(-
2*pi*f1.*t)*((1/(k1*r)^2)+(1/(k1*r)^3)))+((k2^2)*A2*exp(-
2*pi*f2.*t)*((1/(k2*r)^2)+(1/(k2*r)^3)))+((k3^2)*A3*exp(-
2*pi*f3.*t)*((1/(k3*r)^2)+(1/(k3*r)^3)))+((k4^2)*A4*exp(-
2*pi*f4.*t)*((1/(k4*r)^2)+(1/(k4*r)^3))));
E_t = sin(angle)*(((k1^2)*A1*exp(-
2*pi*f1.*t)*((1/(k1*r)^2)+(1/(k1*r)^3)+(1/(k1*r))))+((k2^2)*A2*exp(-
2*pi*f2.*t)*((1/(k2*r)^2)+(1/(k2*r)^3)+(1/(k2*r))))+((k3^2)*A3*exp(-
2*pi*f3.*t)*((1/(k3*r)^2)+(1/(k3*r)^3)+(1/(k3*r))))+((k4^2)*A4*exp(-
2*pi*f4.*t)*((1/(k4*r)^2)+(1/(k4*r)^3)+(1/(k4*r)))));

E = ((abs(E_r).^2)+(abs(E_t).^2)).^(0.5);

figure(1)
plot(t,abs(E))
xlabel('time(s)');
ylabel('Electrical field(E)');
title("Time domain signal when r=2");
```

6.3   Test Results

6.3.1   Test Results of Acoustic Signal Model

The test results are simulated for the acoustic model and the radio wave model of LEMP. The results have proven that the two model-based signal processing approaches could be utilized for the localization of the signal source point.

Table 6.1 Parameters of the acoustic signal in the basic model Eq. (6.1).

i	A_i	f_i
1	23	4000 π
2	−29	8000 π
3	5	12000 π
4	1	16000 π
5	23	20000 π
6	−29	24000 π
7	5	28000 π
8	1	32000 π

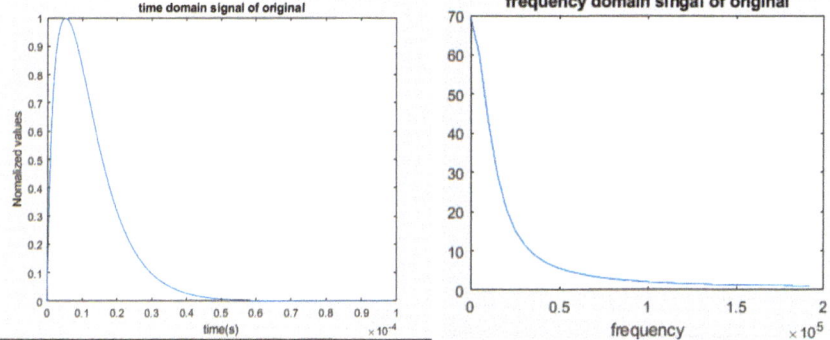

Fig. 6.1 Time and frequency domain pattern at the origin

The test signal is modeled as given in (6.1) by satisfying the conditions given in (6.2). A_i, f_i, and N values are selected as in Table 6.1.

The basic acoustic signal parameters are selected as given in Table 6.1, and the FFT of the signal is obtained. The atmospheric attenuation factor obtained in (6.6) is used in the FFT terms of the signal, and the multiple frequency component signal output of the FFT at different distances is combined to get the resultant signal in the frequency domain. Finally, the inverse FFT is obtained to get the time domain wavefront of the signal at different distances. The wavefront patterns at different distances are shown in Figs. 6.1, 6.2, 6.3, 6.4 and 6.5.

The results in Figs. 6.1, 6.2, 6.3, 6.4 and 6.5 show distinct changes of field patterns with respect to the distance from the source. Therefore, this change of field pattern characteristics could be used to localize the source location from the observation points.

6.3.2 Test Results of Radio Wave Model

The empirical model of the lightning current is as given in (6.7), with the conditions given in (6.8) satisfied. I_n, f_n, and N values are selected as in Table 6.2.

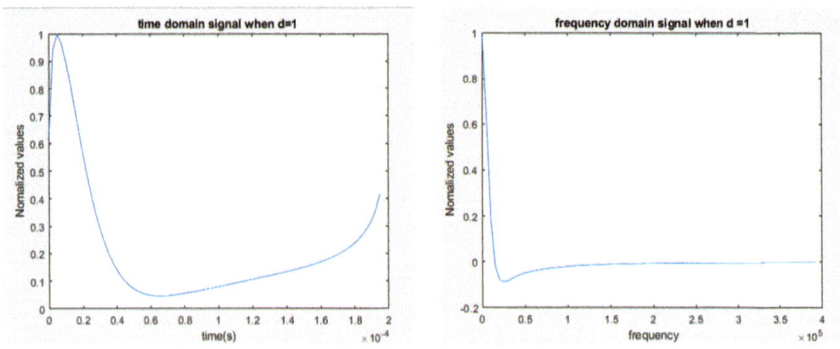

Fig. 6.2 Time and frequency domain pattern at 1 m from the source

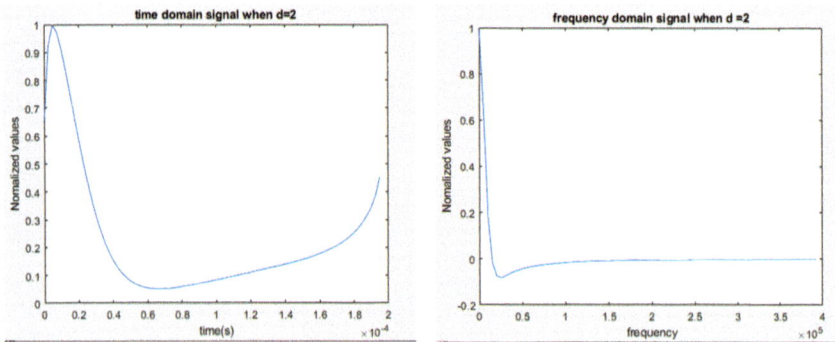

Fig. 6.3 Time and frequency domain pattern at 2 m from the source

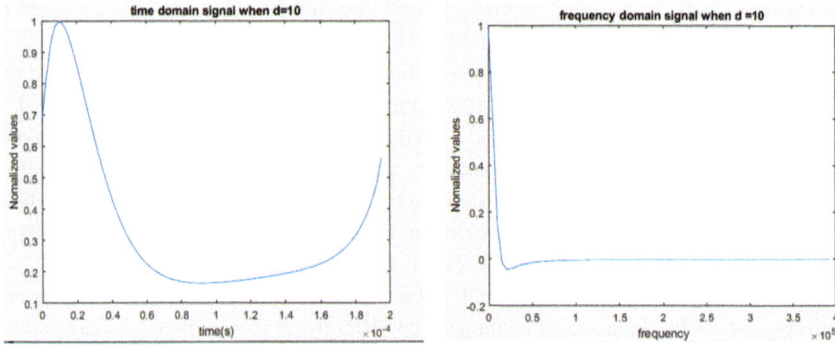

Fig. 6.4 Time and frequency domain pattern at 10 m from the source

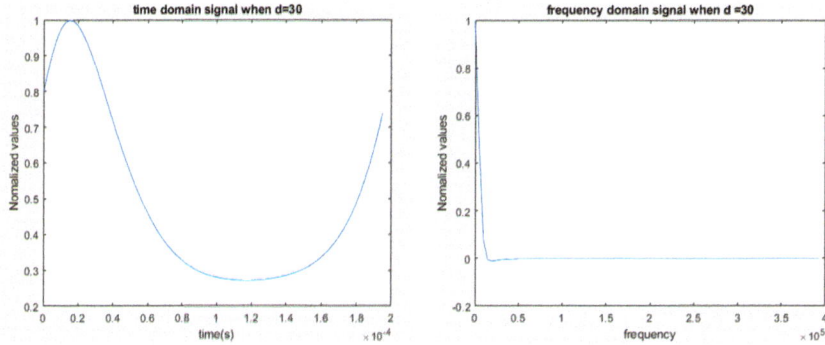

Fig. 6.5 Time and frequency domain pattern at 50 m from the source

Table 6.2 Parameters of the current in the basic model

n $N = 4$	I_n	f_n	Time Constant $\tau_n = \frac{1}{f_n}$ μs
1	-29×103	3×10^6	0.333
2	23×103	2.5×10^6	0.400
3	5×103	6×10^6	0.167
4	1×103	9.5×10^6	0.105

The basic return stroke current signal model parameters are selected as given in Table 6.2. The wavefront of the electrical field signal patterns is calculated using the governing Eqs. (6.9)–(6.12) derived for the current signal defined in (6.7). The return stroke current pattern is shown in Fig. 6.6. The LEMP electrical field wavefront patterns are shown in Figs. 6.7, 6.8, 6.9, 6.10 and 6.11 for a range of distances from 2 m to 10 km from the lightning flash. Only the magnitudes are plotted in Figs. 6.7, 6.8, 6.9, 6.10 and 6.11.

Fig. 6.6 Current Signal Pattern

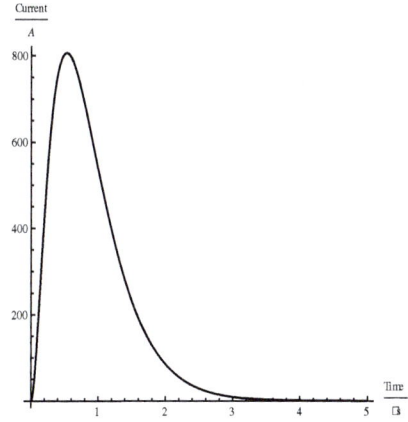

Fig. 6.7 Electrical field at 10 m

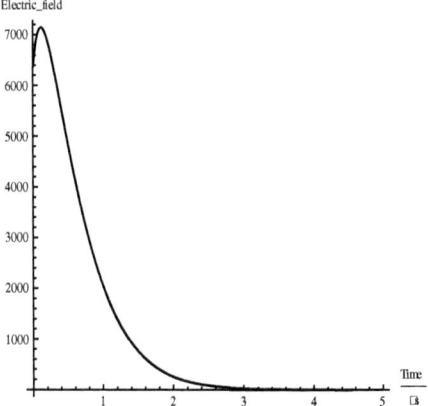

Fig. 6.8 Electrical field at 50 m

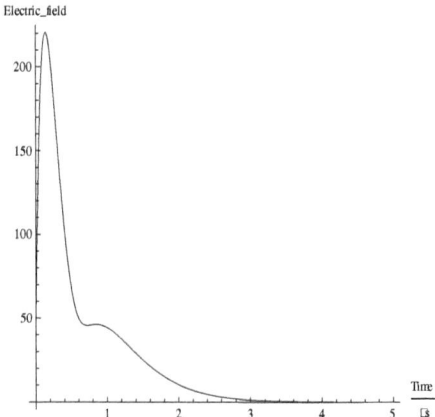

Fig. 6.9 Electrical field at 100 m

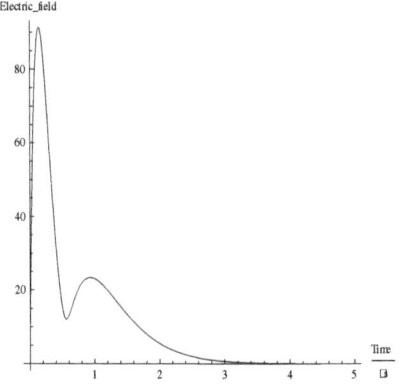

Fig. 6.10 Electrical field at 200 m

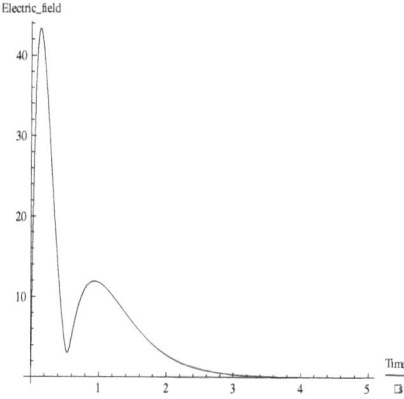

Fig. 6.11 Electrical field at 500 m

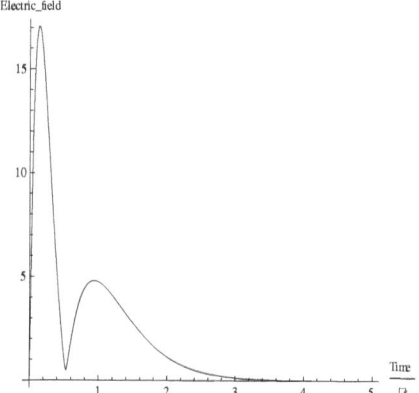

The radiated electric field results of Figs. 6.7, 6.8, 6.9, 6.10, and 6.11 show distinct changes in time-domain electric field patterns depending on the distance from the current source (Fig. 6.6), that is, the point of lightning flash. Therefore, this difference in electric field pattern characteristics could be used to localize the source with respect to the observation points where the time-domain electric fields are measured.

6.4   An Array Antenna for Direction and Identity of Lightning Radiated Signals

We have presented in the above sections how lightning may be localized by using both the LEMP and acoustic signals from a lightning flash. However, two issues have not been addressed in the above sections. These are:

Fig. 6.12 The flowchart of the simultaneous scan Adapted from Singkang et al. (2021) courtesy of The Electromagnetics Academy

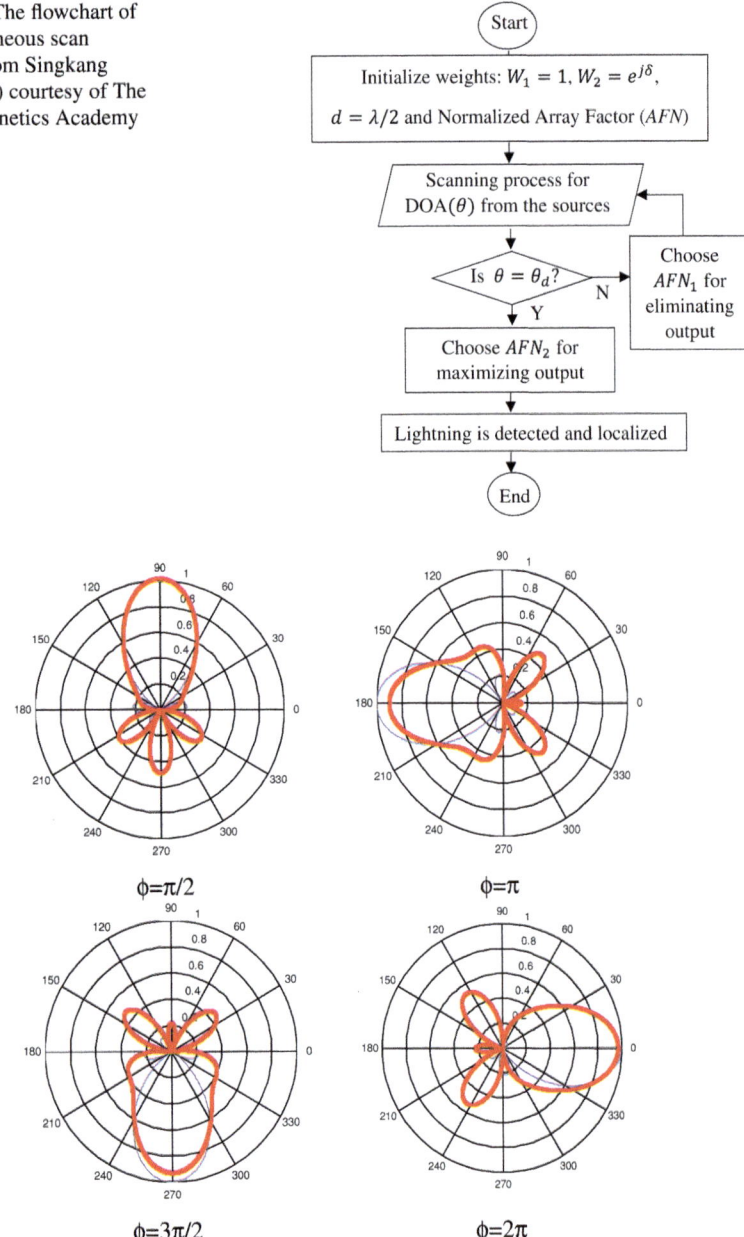

Fig. 6.13 ANN-driven rotating smart antenna beam (from Hoole 2020)

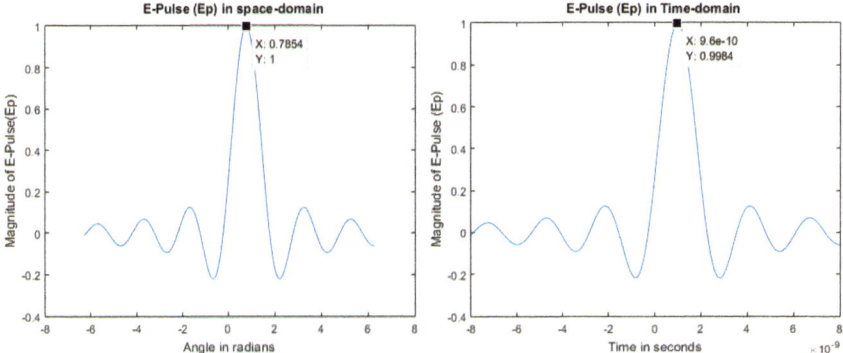

Fig. 6.14   The waveforms of *maximum electric field* generated at 125 MHz (Singkang et al. (2021) reproduced courtesy of The Electromagnetics Academy)

1. Having determined the distance of the lightning flash from the point at which the signal has been measured, how do we know from which specific direction the signal came? In communication systems, this is called the Angle of Arrival (AoA) and is extensively described in Hoole (2020). Here, we shall outline the basic technique to determine the direction from which the lightning impulse has arrived at the observation or measurement point. More specifically, we shall outline the use of an Artificial Neural Network driven smart antenna technique to determine the direction from which the LEMP arrives. The same technique, with acoustic sensors, may be used to detect the direction from which the acoustic signal, associated with thunder, arrives.

2. How do we know that the signal captured is from lightning, and is not from another source, such as a nuclear explosion radiating nuclear electrometric pulse (NEMP) or acoustic sound from a gunshot or bomb explosion? The technique we outline here is based on recognizing the waveform of a signal specifically from lightning. The applications of these techniques are manifold, for instance, identifying gunfire in a mass social gathering where there are firecrackers being lit.

Consider a steerable beam, two-element array antenna being used to measure the LEMP. For a two-element array antenna in receiving mode with one of the two elements containing a digital beam steering (beamforming) weight (w), our task is to scan the region around the lightning area or in a smart city with smart antennas, in the plane that is parallel to the earth surface. The total output from the two-element array antenna is given by

$$E_T = w_1 E_1 + w_2 E_2 \qquad (6.13)$$

where E_1 and E_2 are the electric fields picked up by the two antenna elements, with the received signal of the second element multiplied by the electronic weight, w. For a lightning flash occurring at a distance R from the apparatus, and at an azimuth angle θ, the dipoles pick up electric fields of

$$E_1 = \mu_0 \, j\eta \frac{kIh}{4\pi R} \, e^{-jkR} \, \sin\theta \qquad (6.14)$$

$$E_2 = \mu_{0\theta} \, j\eta \frac{kIh}{4\pi R} \, e^{-jkR} \, \sin\theta \left(e^{jkd\, \sin\theta\cos\varphi} \right); \qquad (6.15)$$

where η is the free space intrinsic impedance; h is the length of the antenna (for half-wavelength dipole, $h = \lambda/2$, and wave number $k = 2\pi/\lambda$) and I is the current flowing along with the lightning flash. Therefore, the total electric field picked up by the array antenna is as defined in (6.13) where the weights w_1 and w_2 are determined by the Perceptron (a single-layer artificial neural network), which being trained based on the specific location of each apparatus, we must cyclically keep under observation. The rotating antenna beam captures the maximum LEMP in the direction of $\theta = \theta_d$ and $\varphi = \varphi_d$ where the angles (θ_d, φ_d) indicate the direction of the lightning flash. The best moment of capturing the lightning flash activity is when the antenna beam is pointed directly towards the location where the lightning flash is generated, when$AF_m = 2$. Since the LEMP has the rise times of the order of a sub-nanosecond (1 GHz) to microsecond (1 MHz), the sampling frequency of the signals received by the array antenna will be in the microwave spectrum for nanosecond radiation from the lightning flash. For the antenna beam to enable the sampling of received signals at every 30° (the segment angle used in fourth-generation wireless communication systems), the microprocessor (depending on the clock frequency) should change the w_1, w_2 values to ensure that the beam rotates, searching for the lightning flash and sample LEMP at 2 GHz. One further step is to use a single beam smart antenna instead of two symmetrical beams generated by the linear array antenna because two symmetrical beams require a reflector to fold one of the beams over. Thus, we can use the non-linear three-element smart antenna, which is capable of generating a single, steerable beam, as described in Hoole (2020).

6.5 Application of the Perceptron ANN for UHF Lightning Flash Detection

An ANN is a numerical structure, which comprises of interconnected artificial neurons, on a substantially littler scale, and works like that of a natural neural system or brain. An ANN can gain from information either in a managed or unsupervised way and can be utilized as a part of assignments, for example, arrangement, relapse, grouping, and many more from there. The human brain gets signals from sensors, for example, the eye, ear, and touch. These signals are then processed by the brain. In ANN, the sensors might be genuine image sensors (camera), sound sensors (microphone), or capacitive touch sensors, which are the inputs to the ANN. Figure 6.16 simplifies the concept of applying the Perceptron ANN training algorithm for UHF lightning flash LEMP angle of arrival (DoA) Detection based on a two-element array antenna to form the smart antenna (Hoole 2020). On account of the Smart Model-Based Testing as proposed in Singang (2021), the inputs are normally transmitted signals from transmitting antennas. The input signals are processed mathematically, for example by multiplying each input signal by a number (weight, w) and phase-shifting the signal (complex weights, b), at that point sum up the input signals and place it at the output as a transfer function that will yield the final output signals. On account of the human brain, the final output signals might be activating signals to the muscles, for example, to move the human body for activity. In Fig. 6.12, the use of an ANN-driven smart antenna where the beam is rotated and the direction of arrival DOA of a lightning LEMP is detected.

For the smart antennas, the final output signals might be to divert the beamforming towards the desired users. An ANN is structured of a substantial number of highly interconnected processing elements called artificial neurons organized in layers. The weights (w) and biases (b) are known as adjustable scalar parameters (also known as hyperparameters) of the neuron. The parameters can be adjusted to meet the desired behavior as part of the network training process. An ANN is preferable in many applications as it was fast convergence and adaptive to any complex changes. The hyperparameters (i.e. the parameters used to control the learning process), the learning rate (step size) and the bias of the perceptron control the values of weights calculated and the convergence rate of the perceptron. Therefore, selecting the initial values for these hyperparameters is crucial. However, this Perceptron ANN algorithm has some limitations. These include some series of sufficient trainings to obtain the optimum perceptron hyperparameters if one needed to apply more than the two-element array antenna to meet the desired behavior. This algorithm is designed to detect, localize, and identify the EMP but not to classify the EMP signal. The Perceptron ANN was incorporated with a linear array antenna to form a smart antenna. The linear structure of the array antenna was used due to its low complexity and it can perform beamforming in a single plane within the angular sector. This smart antenna can simultaneously scan and detect any abnormal electrical discharge activity such as the lightning flashes, using the Direction of Angle (DOA) of the LEMP from the lightning flash. Two conditions were considered in this investigation, i.e. to eliminate the interferer signals (signal from an undesired direction) and to maximize the desired signal (from the desired direction).

The Electromagnetic Pulse (*EP*) is used in representing the LEMP for simulation purposes. The radiation pattern of the desired output LEMP signals of Figs. 6.7, 6.8, 6.9, 6.10, 6.11, 6.12, 6.13 and 6.14 may be detected by a time-domain frame that is represented by a *sinc* function in the time-domain. The beam is steered by a Perceptron ANN trained to different locations by another spatial-domain using a *sinc* function. The spatial radiation pattern of the desired output signals traversing in different directions, defined is by $y_d = sinc(\varphi - \varphi_m)$. For a series of DOA (directions of signal arrival), randomly selected at $\varphi_m = 45°$, $120°$, and $300°$ there will be a unique time delay of DOAs due to the small differences in the distances between the transmitter and the observation points at which the signals are observed (Hoole, 2020). The similarity of the desired output waveforms in space and time-domain makes the technique efficient, demanding minimal computational complexity. This similarity shows the same ANN used in Smart Antenna (SA) spatial-beam optimization and *e*lectric field detection. In Fig. 6.17, the smart antenna single beam is shown being rotated by the Perceptron ANN over any desired direction around the observation point. The antenna beam may be cyclically rotated by pre-trained weights to detect and to identify lightning activities, as also the case with electrostatic discharges inside a power substation. As observed, the beam is being focused in the specified DOA.

Figure 6.13 shows the Perceptron ANN-generated beam, focused towards the spatial location of 45° and the simultaneous time-domain search to match the received signal with the time-domain pulse for LEMP detection. The waveforms showed that there is a similarity of the waveforms in space-domain (the beam pointing towards a power apparatus at 45°) and time-domain (a UHF lightning flash current generated LEMP). The maximum peak of *electric field* in space-domain denoted as θ_m, representing the delay angle of DOA. Details of complex methods for electric signal source localization may be found in Hoole (2020).

Figure 6.14 shows the spatial antenna beam and time-domain pulse picked up in the band encompassing 125 MHz at DOA = 45°. In the space-domain of the antenna beam, the maximum magnitude, y = 1 observed at an angle of 0.7854 rad, while in time-domain pulse, the peak magnitude, y = 0.9984 showed peak time, $t_p = 9.6e^{-10}s$. Thus, at the peak time of 0.96 ns, the frequency allowed into the antenna receiver is 1.04 GHz. As observed, for the 1 ns peak LEMP, the error is 4%. Therefore, the accuracy for LEMP signal detection and localization is about 96%.

Following are the MATLABTM computer codes used for ANN training and beam forming:

(Main)

```
%% Program to predict the antenna output
clear;
clc;

%% Data Generation
angle = 0:0.1:2*pi;
Xin = [exp(1i*pi*cos(angle))' ones(size(angle,2),1) exp(-
1i*pi*cos(angle)')];
Yd = cos(2*angle)';

%% Network Training
learningRate = 0.9;
maxEpoch = 1000;
C = spso_train(Xin,Yd,learningRate,maxEpoch);
disp('Trained weights with bias = ');
disp(C);

%% Network Test
out = spso_classify(C,Xin);
error = abs(sum(sqrt((out-Yd).^2)));        % Total squared error
disp(['Error = ' num2str(error)]);
```

Training

```matlab
function C = spso_train(data,labels,alpha,maxEpoch)
%
% Initialize all the variables
[n,m] = size(data);
O = 1;      % Single output
W = randn(m+1,O);        % Random initial weights with bias
W = max(min(W,0.5),-0.5);        % Limit weights to -0.5 until 0.5
%
% Training phase
for epoch = 1:maxEpoch    % Loop until max epoch is reached
    error = 0;  % Initialize error to 0

    for k = 1:n  % Present each and every data
        x = horzcat(1,data(k,:));    % Load the current data as x and
add bias signal
        y = x * W;  % Claculate the output

        % Activation function  (sigmoidal)
        y = (1/(1+exp(-y)));

        % If error occured, update the weights
        if y ~= labels(k)
            error = error + (labels(k) - y);
            W = W + alpha * labels(k) * transpose(x);
        end
    end

    % If no error occured during this epoch, exit
    if error < 0.1
        break;
    end
end
%
% Return the final weight values
C = W;
```

RMSError

```matlab
%% Simple_Perceptron_Single_Output
%-------------------------------------------------------------
----%
function labels = spso_classify(C,data)
%
% Initialize all the variables
[n,~] = size(data);
y = zeros(n,1);

for k=1:n  % Present each and every data
    x = horzcat(1,data(k,:));    % Load the current data as x and add
bias signal
    y(k) = x * C;    % Calculate the output y

    % Activation function (bipolar sigmoidal)
    y(k) = (1/(1+exp(-y(k))));

end
%
% Return the labeled output
labels = y;
%-------------------------------------------------------------
----
```

For further details on rotating beam antennas for sensing, the reader may consult Hoole (2020).

6.6 Conclusion

Lightning energy signal models covering a range of frequencies are proposed as the source signal for lightning radiated acoustic (thunder) and radio wave signal (LEMP)-based source localization. The governing equations for the propagation of acoustic and radio wave signals for the proposed signal models are based on the respective general equations of acoustic and radio wave propagations. Having simulated the proposed model signal for the governing equations, distinct changes of received signals for different distances from the source were observed. It was shown that the received lightning generated acoustic and radio wave (LEMP) signals can be used to localize the signal source (i.e. the lightning strike point return stroke current) using the the signal models and source localization techniques proposed.

Bibliography

Ahadi, M., Bakhtiar, M.S.: Leak detection in water-filled plastic pipes through the application of tuned wavelet transforms to acoustic emission signals. Appl. Acoust. **2010**(71), 634–639 (2010)
Golde, R.H. (ed.): Lightning, vol 1: Physics of Lightning, Academic Press (1977)

Hoole, P.R.P.: Smart Antennas and Electromagnetic Signal processing for Advanced Wireless Technology: with Artificial Intelligence and Codes. River Publisher (2020)

Hoole, P.R.P; Pirapaharan, K; Kavi, M; Fisher. J; Nur Farah Aziz; Hoole, S.R.H..: Intelligent localisation of signals using the signal wavefronts: A Review. In: International Conference on Lightening Protection(ICLP), Shankai, China (2014)

ISO 9613-2:1996 (en): Attenuation of sound during propagation outdoors-Part 2: General method of calculation (1977)

Kundu, T., Das, S., Martin, S.A., Jata, K.V.: Locating point of impact in anisotropic fibre reinforced composite plates. Ultrasonics 2008(48), 193–201 (2008)

Kundu, T.: Acoustic source localization. Ultrasonics 54, 25–38 (2014)

Kunsei, H., Pirapaharan, K., Hoole, P.R.P.: A new fast, memory efficient wireless electromagnetic beamformer antenna with fast tracking for 5/6G systems. Progress in Electromagnetics Research PIER B (2021)

Kunsei, H., Pirapaharan, K. Hoole, P.R.P. Hoole, S.R.H.: Tracking Everyone and everything in smart cities with an ANN driven smart antenna. In: Jude Hemanth D. (ed.) Mobile Learning techniques for Smart City Applications: Trends and Solutions. Springer Nature (2021)

Pirapaharan, K., Kunsei, H., Senthilkumar, K., Hoole, P. R. P., Hoole, S.R.H.: A single beam smart antenna for wireless communication in a highly reflective and narrow environment. In: International Symposium on Fundamentals Electrical Engineering, pp. 1–5 (2016)

Senthilkumar, K.S., Pirapaharan, K., Hoole, P.R.P., Hoole, S.R.H.: Single perceptron model for smart beam forming in array antennas. Int. J. Electric. Comput. Eng. 6(5), 2300 (2016)

Singkang, L.M.B., Ping, K.A.H., Hoole, P.R.P.: Electric discharges localization for substation fault monitoring using two elements sensor. J. Comput. Theor. Nanosci. 17(2–3), 1009–1013 (2020)

Singkang, L.M., Ping, K.A.H., Kunsei, H., Senthilkumar, K., Pirapaharan, K., Haidar, A.M., Hoole, P.R.P.: Model based-testing of spatial and time domain artificial intelligence smart antenna for ultra-high frequency electric discharge detection in digital power substations, vol. 99, PIER, pp. 91–101 (2021). https://doi.org/10.2528/PIERM20090301. http://www.jpier.org/PIER/, (CHECK LAS AND FIRST NAMES HERE, 2021)

Tobias, A.: Acoustic-emission source location in two dimensions by an array of three sensors. Non-Destr. Test 9, 9–12 (1976)

Chapter 7
Lightning Electrodynamics: Electric Power Systems and Aircraft

Joseph Fisher, Paul R. P. Hoole, Kandasamy Pirapaharan, and Samuel R. H. Hoole

Abstract Lightning impact on electric power systems and telecommunication systems is an increasing concern with the widespread use of microelectronic devices in these systems, such as in digital power substations and in smart city networks. This chapter reviews the important aspects of lightning protection of electric power systems. Whereas much of the past measurements and modeling captured microsecond changes in lightning return stroke currents and radiated electromagnetic fields, the importance of sub-microsecond changes and the detailed capturing of the current wavefronts are also addressed in this chapter. To this end, lightning-attached power system models are revisited and simulated to capture and study the wavefronts. Major findings related to climate change appear to indicate an overall increase in lightning activity or electric intensity as global temperatures are set to increase. Thus, the severity of the electric parameters of electric storms (including the lightning return stroke current I, electric charge induced, the rate of rise of current dI/dt, and thundercloud potential) is set to increase. This poses an increased threat to the electric power transmission and distribution apparatus and systems, and hence demands a more accurate modeling and computation of the interaction between lightning and the electric power grid. This chapter presents a more exact lightning–electric power system interaction model and critical insights into I and dI/dt values that are used in high voltage testing, as well as in protection of highly sensitive Internet of Things (IoT) equipment and systems. Furthermore, it presents the different techniques used when power transmission and distribution systems are designed with reference to lightning threats. In the second part of this chapter, we present the modeling and computation of lightning aircraft electrodynamics. The large commercial Airbus A380 aircraft and the smaller F16 military aircraft are used to illustrate the principles and parameters involved; these are critical when designing the aircraft geometry, lightning zoning, and protection of aircraft. The material presented here forms a complete lightning interaction testbed for both lightning electric current, including sub-microsecond details, and radiated electromagnetic field pulses (LEMP). The numerical, computational results given in this chapter are meant to give a rough guide to how the computer-based testbed for lightning simulation and testing may be used to study realist lightning electrodynamics, but the final values will be dependent on the particular geometry or network being investigated.

© The Author(s), under exclusive license to Springer Nature Switzerland AG 2022
P. Hoole and S. Hoole, *Lightning Engineering: Physics, Computer-based Test-bed, Protection of Ground and Airborne Systems*,
https://doi.org/10.1007/978-3-030-94728-6_7

7.1 Introduction

7.1.1 Lightning and Electric Power Systems

Electric power transmission and distribution lines are routed over several tens to hundreds of kilometers along electricity highways in open environments through different topography of different soil characteristics and resistivity. The exposure of transmission and distribution lines becomes prone to such atmospheric disturbances as lightning strikes, tornadoes, high winds, high humidity and temperature rise, and other disturbances such as geological hazards. With transmission line pylons being the tallest structures along the electricity highways, they become prone to lightning strike. That is, the tall structures become the shortest routes for cloud-to-ground flashes to discharge to earth, generating high transient voltage and current pulses. Thus, lightning-generated impulsive transients remain a potent source of many inadvertent power supply outages on high voltage transmission lines, sub-transmission lines, and the distribution lines, which are able to bring down the entire electric grid.

Power supply outages occur when a lightning strike causes a voltage flashover in a network. A voltage flashover is an electric discharge or a spark-over occurring between two live conductors or through a shield wire through a pylon structure across a string insulator to a live conductor. There are two types of flashovers: the direct stroke flashover and indirect stroke or induced voltage flashover. The direct stroke occurs when lightning hits a shield wire, a pylon/tower, or phase conductor, which leads to back flashover or a shielding failure flashover, respectively. Conversely, in the indirect stroke, lightning does not directly hit any part of the transmission or distribution lines or tower structures. Instead, lightning strikes the ground, or nearby objects, leading to a large impulse current. The large current generates magnetic fields, which, through inductive couplings, induce voltage impulses on the transmission or distribution lines. The voltage impulses induced can cause flashovers leading to power supply outages. The effect of the indirect stroke is more severe for sub-transmission and distribution lines with lower critical flashover (CFO) voltage levels compared to those for transmission lines with higher transmission voltages.

With transmission line pylons being the tallest structures along electricity highways, they become prone to lightning strikes. The structures become the shortest routes for cloud-to-ground flashes to discharge to earth-generating large impulse current and voltage transients. Thus, lightning-generated impulsive transients remain a potent source of many inadvertent power supply outages on high voltage transmission lines, sub-transmission lines, and distribution lines.

As stated above, lightning interaction with structures is considered as either direct effects and indirect effects. The direct effects of lightning stroke comprise high return stroke currents. The current peak magnitudes are of the orders of several tens to hundreds of kilo-amperes. The commonly encountered lightning current waveform was discussed in Chap. 5. The current is usually divided into components A, B, C, and D waveforms, which represent the first return stroke current (A), intermediate current (B), continuing current (C), and the subsequent return stroke current (D), respectively.

Component A is the high-current impulse known as the first return stroke current. The magnitudes of the first return stroke currents recorded have reached up to 200 kA lasting up to 200 microseconds. Other studies have shown that maximum first return stroke currents could be as high as 500 kA. The steepness of lightning current rise ($dI/dt = \frac{\Delta i}{\Delta t}$), which is effective during the interval Δt, defines the intensity of the electromagnetically induced voltages. These voltages are induced in all open or closed conductor loops located in the vicinity of conductors carrying lightning current. Typical rate of current rise with respect to time is 3×10^{10} A/s, but it could reach a higher value close to 2×10^{11} A/s.

The components A and D contribute to electromagnetic forces and the development of high voltages due to the fast rise time of the pulse and the high peak current. The magnetic forces arising from the high currents can cause damages from puncturing, vaporization, and crushing, or drive together/pull apart conductors in the case of striking an aircraft, buildings, and other structures.

Component B comprises the intermediate current, which is a transition phase of the order of several thousand amperes. Component C is a continuing current of approximately 300–500 amperes that lasts up to 0.75 s. The charge transfer during the B and C components of the lightning event is usually larger than that from components A and D, with common effects being melting, hole burning, and hot spots on surface. The last component, D, is a restrike current surge that is typically a half that of component A in a given strike. It has generally the same duration and effects as component A.

On average, the first return stroke currents measured for ground-based structures typically rise to an initial peak of about 30 kA in some microseconds and decay to half-peak value in some tens of microseconds. The four specific effects of lightning current due to direct effects considered to be of high severity in producing damages are:

(1) The peak current that is the high current pulse flowing through a conducting surface. This can induce a large voltage magnitude on the surface of structures, ($V = IR$), where I is the current pulse, and R is the surface resistance of the structures,

(2) The maximum rate of change of current is referred to as the current steepness which gives rise to an electromagnetically induced voltage ($v = M\frac{di}{dt}$), where M is the mutual inductance between conductors,

(3) The integral of the current over time ($Q = \int i\,dt$) which is the electric charge transferred) is responsible for the mechanical force, melting, and the heating effects, and

(4) The integral of the current squared over time ($\frac{W}{R} = \int i^2 dt$), where W is the energy dissipated into a 1 Ω resistor (R), which is referred to as the specific energy or the action integral. R is the temperature-dependent DC resistance of the conductor and R/W is the specific energy, which is responsible for the

melting effects, hole burning, and hot spots. It is this energy, often referred to as action integral, that generates heat in the object struck by lightning.

The indirect effects of lightning threats are due to the radio frequency interferences and lightning electromagnetic pulses (LEMPs). The LEMPs can induce disruptive voltages (v = Ldi/dt) and currents (i = Cdv/dt) that can disrupt or damage electrical and electronic systems through resistive and or electromagnetic couplings.

The impact of lightning indirect effects becomes increasingly important with the advent of digital electronic technologies. The evolution of digital substations, and the convergence of smart technologies utilizing modern communication and information technologies and the application of wireless sensor network in smart grid and micro-grid operations become susceptible to LEMPs. The protection of these technologies will heighten the need for a professional approach to lightning protection. LEMP threats can have serious damaging effects. The electrical and electronic systems are susceptible to LEMPs at frequencies between 1 and 500 MHz and produce internal field strengths of 5–200 V/m or greater. Internal field strengths greater than 200 V/m with pulse widths less than 10 μs can result in induced voltages and currents, ranging from 50 V and 20 A to over 3000 V and 5000 A. Electrical/electronic system susceptibility to LEMPs has been suspected as the cause of "nuisance disconnects," "hardovers," and "upsets" in electronic systems. Generally, such malfunctions in digital electronic systems occur at lower levels of LEMP field strength than that which could cause component failures if no proper shielding or protection system is utilized.

7.1.2   Lightning and Aircraft

The need to analyze lightning strike to horizontal and vertical earth structures, and airborne structures becomes pivotal in order to study accurately the impacts on these structures for protection and shielding coordination. In particular, modeling of the lightning return stroke as described in Chap. 5 is vital in acquiring a better understanding of the nature of lightning and related phenomena, since the return stroke causes the most destructive disturbances to electrical and telecommunication networks. In analyzing lightning-aircraft electrodynamics, the return stroke model is crucial for lightning protection and shielding coordination as it (i) quantifies the electromagnetic fields induced by the lightning stroke in order to analyze the transient voltages and currents induced within the aircraft, (ii) it gives a good indication of the magnitude of direct injection of lightning current, and (iii) it provides a good statistical distribution of peak currents and the current derivatives that can couple into nearby electronics. In order to have a good understanding of the return stroke stage of lightning-aircraft electrodynamics, mathematical modeling is done for implementing on computers and has the following two advantages. First, experimenting in the field is very time consuming, as well as expensive and often takes a number of years or up to a decade to get reliable data (due to the random nature of the lightning phenomenon). Even in-flight data measurements are limited to the cloud charge center sizes and cannot be attested to be true reflection of lightning phenomena. Secondly, laboratory experiments do not reproduce the same effects of lightning return stroke due to limitation of generators reaching very high megavolts and rates of rise of current in laboratory setups. Thus, although the mathematical model is only an approximate model, depending on its accuracy of representing the real aircraft-lightning electrodynamics, it reproduces certain crucial aspects of the interaction normally inaccessible to field or laboratory experiment of the lightning phenomena.

7.2 Circuit Elements Used in Back Flashover and Shielding Failure Performances

7.2.1 Preamble

The circuit components that affect the lightning current response are quantified in order to observe the current waveforms for the lightning channel, the arcing current across the string insulator, the current through the shield wires and tower, and the lightning current along conductors at the substation tower where the lightning strikes and from which point the lightning current flows along the power line to the next substation further from the strike point. The circuit parameters are discussed with reference to the various circuit elements for the tower, conductors, shield wires, effect of corona envelope on the conductors, and the tower footing resistance, and the surge impedances of the substations.

7.2.2 Tower Surge Impedance

A Class 3 type tower of 500 kV overhead transmission line is modeled to analyze lightning stroke parameters. The conductor type is a pheasant with a diameter of 35.103 mm with two conductors in bundle. The two shield wires are of 9.78 mm in diameter. From the data based on the conductors and the 500 kV tower geometry, the tower surge impedance is calculated using:

$$Z_{tower} = 60 \cdot \left(\ln\left(\sqrt{2} \cdot \frac{2 \cdot h_{tower}}{r_{tower}} \right) - 1 \right) \tag{7.1}$$

where h_{tower} is the height of the tower and r_{tower} is the radius of the tower. The tower radius is estimated to be 3.05 m. From the tower geometry, the surge impedance of the tower is calculated to be 153 Ω.

7.2.3 Shield Wire Surge Impedance

The equivalent of the two shield wires that make up the surge impedance is calculated using the self-impedance and the mutual impedance as defined in (7.2). The calculations take into account the effect of the corona envelope that forms around the shield wire when a high voltage is produced by lightning contact to the wire.

$$Z_{shield} = \frac{Z_{shield-self} + Z_{shield\_mutual}}{2} \tag{7.2}$$

where Z_{shield_self} is the self-impedance with corona effects and Z_{shield_mutual} is the mutual impedance of the shield wire as defined in (7.3) and (7.8).

$$Z_{shield\_self} = 60 \cdot \sqrt{\ln\left(\frac{2 \cdot y_g}{r_{corona}}\right) \cdot \ln\left(\frac{2 \cdot y_g}{r_{shield}}\right)} \tag{7.3}$$

where r_{corona} is the radius of the corona sheath (around the conductor), r_{shield} is the radius of the metallic shield wire, and y_g is the height of the shield or ground wire. The radius of the corona sheath is determined from Eqs. (7.4), (7.5), (7.6), and (7.7). Equation (7.4) gives the electric field (E_{max}), which is the limiting corona electric field below in which the corona envelope can no longer grow. This is assigned to be the breakdown electric field in air (above sea level), which is 15 kV/cm. The corona envelope contains a charge Q due to the lightning voltage induced on the shield wire. By Gauss's law

$$Q = 2 \cdot \pi \cdot \varepsilon_0 r_{corona} \cdot E_{max}. \tag{7.4}$$

Equating for E_{max}.

$$E_{max} = \frac{Q}{2 \cdot \pi \cdot \varepsilon_0 \cdot r_{corona}}. \tag{7.5}$$

The tower top voltage at flashover is calculated for the minimum distance to tower from the middle conductor of the V-string insulator and is taken to be the length of the string insulator. This is based on the dimension and geometry of the tower, and the basic impulse level (BIL), where BIL is the lightning-impulse withstand voltage. The equation for the voltage ($V_{top-2\mu}$) in (7.6) gives the flashover voltage of the tower reaching a crest in about 2 μs. The flashover voltage calculated through the gap across the string insulator to the top conductor is 4.352 MV at 2 μs rise time.

$$V_{top-2\mu s} = \frac{Q}{2 \cdot \pi \cdot \varepsilon_0} \cdot \ln\left(\frac{y_c}{r_{corona}}\right) \tag{7.6}$$

where y_c is the height of the phase conductor. Substituting for Q from (7.4) into (7.6) yields (7.7).

$$V_{top-2\mu s} = E_{max} \cdot r_{corona} \ln\left(\frac{y_c}{r_{corona}}\right). \tag{7.7}$$

Equation (7.7) is a non-linear equation given in terms of r_{corona}. The equation can be solved computationally using the iterative Newton–Raphson approach employing an initial guess value. The corona sheath radius is calculated using a mathematical solver and found to be about 0.74 m. Thus, using the corona sheath radius and substituting in (7.3) yields the self-impedance of the shield conductor, which is calculated to be 400.113 Ω.

The mutual impedance of the shield conductor is given by,

$$Z_{shield\_mutual} = 60 \cdot \ln\left(\frac{d_{S1\_S2}}{D_{S1\_S2}}\right) \tag{7.8}$$

where d_{S1_S2} is the separation distance of the two shield wires, and D_{S1_S2} is the distance between the shield wires and their images as given in (7.9).

$$D_{S1\_S2} = \sqrt{(d_{S1\_S2})^2 + (2 \cdot y_g)^2}. \tag{7.9}$$

The mutual surge impedance is calculated from (7.8) to be 124.21 Ω. Thus, substituting the values of the self-impedance and mutual-impedance of the shield wires in (7.2) yields the value of the surge impedance of the shield wires. The equivalent impedance is calculated to be 262.16 Ω.

7.2.4 Tower Ground Resistance

The tower footing resistance is a very significant parameter in determining lightning back flashover rates. The varying geography and topography of the earth and the nonlinearity of the physics of the soil conduction can influence back flashover rates. A high footing resistance can result in reflected voltage impulses traveling back and forth from the tower and the ground. This can raise the voltage at the cross-arms thus stressing the insulator and causing a flashover to the phase conductors resulting in inadvertent power outages and damages to equipment.

The tower has a crowfoot grounding with four vertical rods used. A ground resistivity of 1000 Ω.m is chosen for a desert with a dry-sandy environment. A rod of 3 cm diameter and a length of 3.05 m is used. The tower ground resistance is calculated using):

$$R_f = \left(\frac{1}{4}\right) \cdot \left(\frac{\rho}{4 \cdot \pi \cdot l_{rod}}\right) \cdot \ln\left(\frac{2 \cdot l_{rod}}{r_{rod}}\right) \tag{7.10}$$

where ρ is the soil resistivity (Ω.m), r_{rod} is the radius of the rod, and l_{rod} is the length of the rod. From the parameters given, the tower footing resistance is calculated to be about 34.69 Ω. However, the resistance at 60 Hz frequency has to be reduced to avoid back flashover at high current impulse. Using the values from the graph for the 60 Hz power frequency resistance, a resistance of 23 Ω is selected.

7.2.5 Conductor Circuit Elements

The circuit elements for the conductors are selected based on the conductor type and the number of conductors in bundled conductors and the geometry of the tower, which influences the coefficient of potential taking into account the conductor and its image. The self and mutual impedances of the conductors are also determined based on the tower geometry. The resistance of the conductor is selected at 75 °C for a Pheasant conductor and is 0.0881 Ω/mile. The line inductance is computed from the self and mutual impedances of the line with respect to the geometry of the tower, the conductors on the tower and also taking into account the effect of ground resistance, loop reactance, and the effect of the transposition of the line. The capacitance is calculated from the coefficient of potential with respect to the conductor height, the tower geometry, and the conductor images.

7.3 Lightning Fash Parameters

7.3.1 Ground Flash Density

Good knowledge of the frequency of occurrence of lightning incidence is important in the design of power system protection against lightning strikes. Lightning activities in a given area is referred to as the lightning flash density or the ground flash density (GFD) occurring in a unit area in a unit time. GFD gives a measure of the number of lightning flashes per square kilometer per year (km^2/year) denoted by N_g. It provides a degree of exposure of an object, in this case, transmission and distribution lines, to potential lightning risk. The average value of N_g is determined directly from lightning detection sensors. However, where there are no lightning detection instruments available, the value of N_g can be estimated from the keraunic level data. The keraunic level is defined as the average annual number of thunderstorm days or hours recorded or observed in an area. These are estimated for different countries and regions around the world.

An area through which transmission and distribution lines traverse may have a certain keraunic level or isokeraunic level, which can be used to calculate the GFD, N_g. Equations (7.11) and (7.12) give an estimate of N_g:

$$N_g = 0.04T_d^{1.25} \tag{7.11}$$

$$N_g = 0.04T_h^{1.1} \tag{7.12}$$

where T_d is the number of thunderstorms days/year and T_h is the number of thunderstorm hours/year. The location is taken to be in the southern part of USA where the number of thunder days and hours are 20 and 25, respectively. In regions close to the equator, the values of N_g will be much higher, as highlighted in Chap. 1.

7.3.2 Number of Lightning Strokes to the Line

As transmission lines cast electrical shadows over the land beneath, a lightning flash that would terminate on the land will be intercepted by the transmission line. Any flashes outside the shadow will miss the line. The cone of protection indicates that there will be the electrical shadow of the tower and conductor arrangements with two shield wires, where any lightning leader that approaches within that shadow will be attracted to the shield lines. The width of the shadow of the line is denoted by W. The width is calculated using

$$h = y_g - \left(\frac{2}{3}\right) \cdot \left(y_g - y_{gsag}\right) \tag{7.13}$$

where h is the mean height of the shield wire and y_{gsag} is the shield wire mid span clearance to ground. The electrical shadow width is then calculated using h and is given in.

$$W = b + (4 \cdot h) \tag{7.14}$$

where b is the distance between the shield wires. For a single shield wire, b becomes zero. The relationship for the number of lightning flashes to the line, N_L, per 100 km/year becomes

$$N_L = 0.012 \cdot T_d \cdot \left(b + 4 \cdot h^{1.09}\right). \tag{7.15}$$

From (7.13), (7.14), and (7.15), the number of lightning flashes to the line is calculated to be 95.3 per 100 km per year.

7.4 Simulations of Lightning Flash to a Transmission Line

7.4.1 Back Flashover Analysis for 500 kV Transmission Line

The lightning stroke is represented by a voltage source of -50 MV for a negative flash to ground. The downward leader is intercepted by the substation tower and shield wires as intended, However, due to a critical flashover or gap flashover voltage of 4.532 MV (calculated at 2 μs) which far exceeds the 1800 kV BIL of a 500 kV system, the line insulation cannot withstand the magnitude of the voltage induced, thus causing a flashover from the tower structure to a phase conductor.

Figure 7.1 shows the DLCRM-based (see Chap. 5) computation of the back flashover current waveforms for the lightning channel, the arcing current across the string insulator, the current through the shield wires and tower, and the lightning current along conductor at the substation tower (short line) and the lightning current along the conductor to the next substation, which is 9.5 km away from lightning strike point substation.

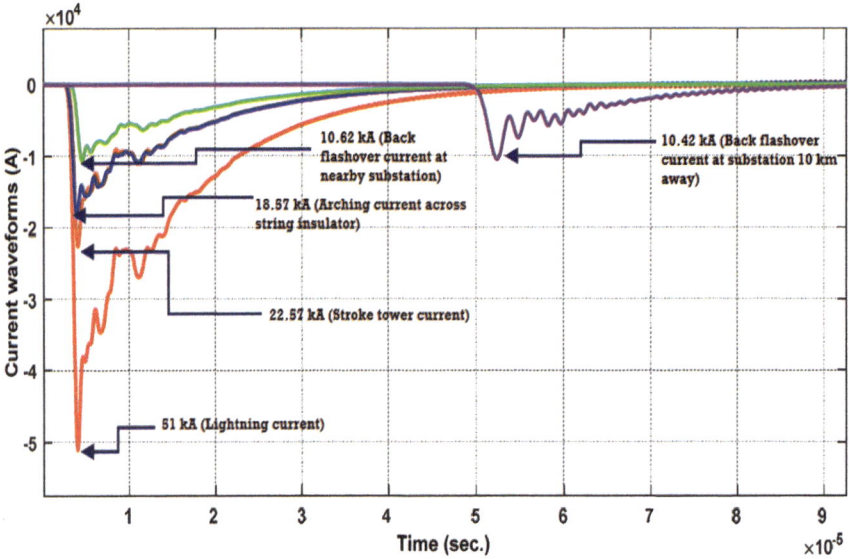

Fig. 7.1 The back flashover first return stroke current waveforms

7.4.2 Sub-microsecond Analysis of Conductor Back Flashover Current at Substation Tower

Figure 7.2 shows the flashover current waveform along the conductor at the substation tower. The substation tower is the tower that is struck by the lightning. The flashover current along the conductor reaches a high peak of 10.62 kA. The peak current falls below 95% of the statistically measured and estimated value of 14 kA. The peak current possesses a front reaching peak within a period of 4.642 µs as shown in Table 7.1. The peak times fall within the 95% and 50% frequencies of the statistically measured and modeled values of 1 and 5 µs, respectively, for tall, grounded towers. These towers are usually the landmark towers located in major cities around the world with height reaching over 100 m.

The rate of rise from 10 to 90% of the peak values is 11.28 kA/µs. Similarly, the rate of rise from 30 to 90% of the peak value, with the sharp front becoming prominent, is 12.983 kA/µs. The rates of rise times from 10 to 90% and of 30 to 90% of the peak values are 0.753×10^{-6} s and 0.491×10^{-6} s, respectively. These rates of rise times are very important in determining the response time of surge protection devices (SPD).

Another important aspect of the current waveforms in Figs. 7.2 and 7.3 is that the greatest rate of rise occurs near the current waveform peak, reaching a value of 14.37 kA/µs in about 4.2 µs. The peak values of the current derivatives fall below the statistical values of 50% of the measured and modeled values of 100 kA/µs. However, the time derivatives can induce high voltage ($v = L\frac{di}{dt}$) in the inductive components

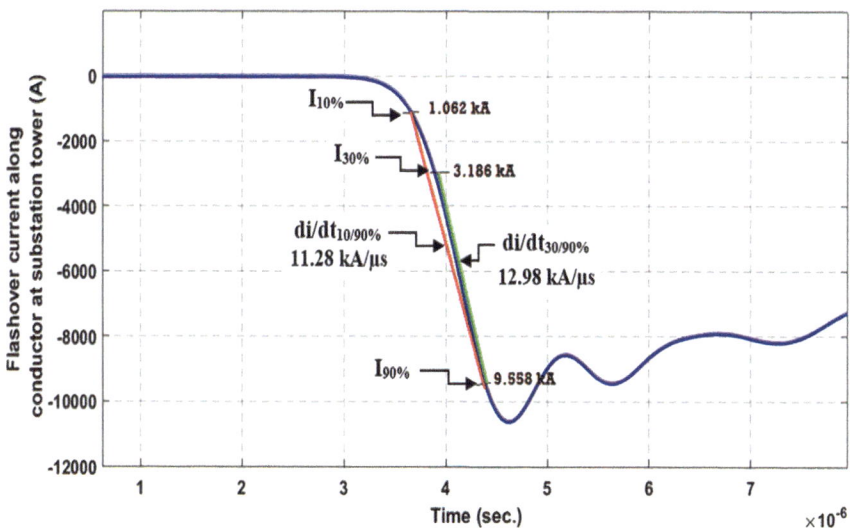

Fig. 7.2 Waveshape of negative first return stroke current at substation tower: sub-microsecond changes of wavefront

Table 7.1 First return stroke current at substation tower

Current parameters	Values
Main (negative) peak current (kA)	10.62
Time to 10% of negative peak (s)	3.653×10^{-6}
Time to 30% of negative peak (s)	3.915×10^{-6}
Time to 90% of negative peak (s)	4.406×10^{-6}
$di/dt_{10/90}$ (kA/μs)	11.28
$di/dt_{30/90}$ (kA/μs)	12.983
Time from 0 to peak (s)	4.642×10^{-6}
Main peak di/dt (kA/μs)	2.29
Action integral A^2s	660
Total charge (C)	−0.13

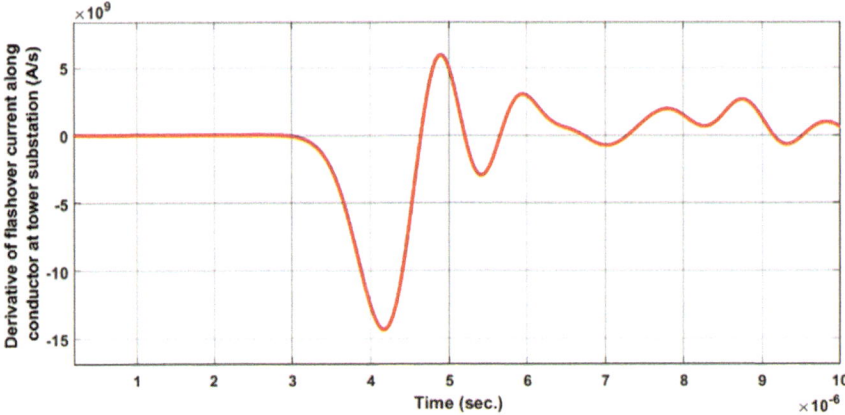

Fig. 7.3 Waveshape of current derivative of current at substation tower: sub-microsecond changes

of the substation circuits, which can have adverse effects on the electrical systems, especially in the protection and control circuitries. Further, the lightning-induced rate of rise $\left(\frac{di}{dt}\right)$ is capable of producing a time-varying magnetic field that can couple into the wiring of the electronics and communication systems. This can pose threats to substation equipment especially for digital equipment and intelligent electronics devices used in measurements and operations of smart technologies in smart grid and microgrid systems.

The lightning parameters in Table 7.1 show a charge transfer $\left(\int i \, dt\right)$ of -0.13 C. This value falls within the statistically measured and modeled values of 5 C and 5.5 C, respectively. The charge transfer is responsible for the effects of severe heating or burn-through and can have adverse effects on substation components of insulation materials. On the other hand, the 'action integral' value of 660 A^2s falls below 95% of the statistically measured and modeled values of 6×10^3 A^2s as listed in Table 7.2. However, this can induce a mechanical force on the surface of the substation

Table 7.2 Shielding failure current at substation tower

Current parameters	Values
Main (negative) peak current (kA)	25.57
Time to 10% of negative peak (s)	3.709×10^{-6}
Time to 30% of negative peak (s)	3.977×10^{-6}
Time to 90% of negative peak (s)	4.482×10^{-6}
di/dt$_{10/90}$ (kA/μs)	26.463
di/dt$_{30/90}$ (kA/μs)	30.38
Time from 0 to peak (s)	4.698×10^{-6}
Main peak di/dt (kA/μs)	5.44
Action integral A^{2}s	3.6×10^3
Total charge (C)	-0.324

components resulting in heating $(I^2 \times R)$ and vaporization and melting effects $(R \times \int i^2 dt)$ of electrical components.

7.4.3 Sub-microsecond Analysis of Shielding Failure

Figure 7.4 shows the shielding failure current waveform along the conductor at the substation tower. The current along the conductor reaches a high peak of 25.57 kA.

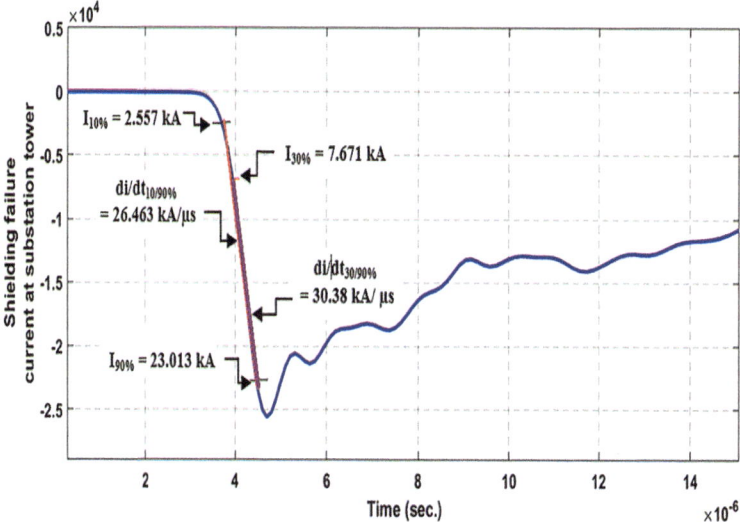

Fig. 7.4 The shielding failure current waveform at substation tower: sub-microsecond changes of wavefront

The relevant values are summarized in Table 7.2. Further, the peak current possesses a front reaching its peak within a period of 4.698 μs.

The rate of rise from 10 to 90% of the peak values is 26.46 kA/μs. Similarly, the rate of rise from 30 to 90% of the peak value gives a sharp front of 30.38 kA/μs. The rise times from 10 to 90% and from 30 to 90% of the peak values are 0.773 μs and 0.505 μs, respectively. These rise times are very important in determining the response time of surge protection devices (SPD).

7.4.4 Back Flashover, Shielding Failure Current Parameters, and Mitigation of Flashovers

Some important observations of the back flashover and shielding failure waveforms (from Figs. 7.4 and 7.5) show the variation in current peaks at the substation tower. The back flashover current peak of 10.62 kA is relatively low due to multiple paths in the branching of the lightning current. The lightning current branches from the point of attachment on the tower top through the shield wires, the tower to ground, and across the string insulator to the phase conductor. Further observation shows both currents possess a front reaching its peaks at almost the same time of 4.643 and 4.698 μs. From Figs. 7.3 and 7.5, we observe that the greatest rate of change occurs near the peak values. Further, the high currents observed due to shielding failure show a high rate of rise, which could have a significant damaging effect on

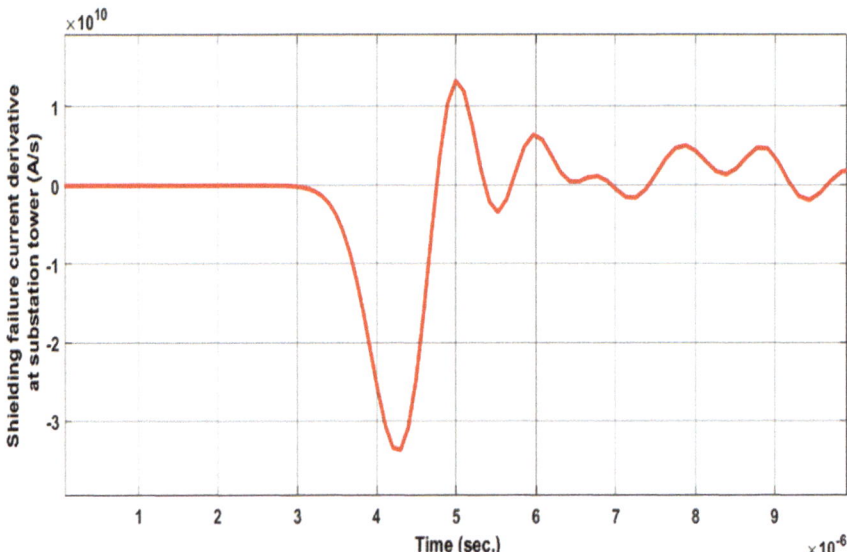

Fig. 7.5 Waveshape of current derivative of shielding failure current at substation tower: sub-microsecond changes

the substation equipment. Thus, the high current due to shielding failures on high voltage transmission lines at 500 kV can be mitigated by installing transmission line arresters and or having additional shield wires for effective shielding. Other methods such as improvement of the grounding system can be effective, especially foundation grounding for towers near substations

7.4.5 Summary

Lightning strikes to transmission lines have been the cause of many inadvertent outages in power systems around the world. Moreover, these cause immediate- or long-term damage to power equipment and microelectronic devices and systems. This section illustrated the importance of sub-microsecond of lightning-induced currents in power lines due to both shielding failure and back flashover. The sub-microsecond changes in both lightning currents and rate of rise need to be accounted for in future protection device and system design. The need to harden modern power systems infrastructures and their components, especially in modern digital systems in a power system environment, against severe electric storms becomes an eminent solution. The threats can be contained through proper protection measures such as the SPDs, lightning arresters, and proper shielding and grounding practices that are specifically designed and chosen to take into account sub-microsecond threat levels. However, with the probabilistic nature of lightning phenomena, protection measures may not provide 100% protection in shielding direct and indirect effects of lightning flashes. Thus, new sub-microsecond measures in protections and standards will have to drive protection to a higher level for mitigation of lightning threats to power systems.

7.5 Lightning Flashover on Transmission and Distribution Lines

7.5.1 Back Flashover

A back flashover originates from the pole or tower shield wire onto the phase conductor. The downward lightning leader is intercepted by the shield wires as intended; however due to a critical flashover (CFO), the line insulation may not withstand the magnitude of the voltage impulse thus causing a flashover from the tower structure to the conductor. Another significant factor to note is that the lightning-induced current or voltage impulses can travel along the transmission lines reaching the nearby substation. This can eventuate in considerable damages if no proper protection system is in place.

7.5.2 Shielding Failure

A shielding failure occurs when a lightning leader misses the shielding wire and hits a phase conductor. A correct design should improve the shielding effect, thus reducing such failures through the use of appropriate shielding angle of protection.

7.5.3 Probability and Intensities of a Flashover

There are two distinct possibilities when a lightning leader comes into contact with a transmission line; these are categorized as a success or a failure. A success indicates no flashover, which means all lines remain in service. However, a failure results in a flashover, which compromises the security of the grid. According to a basic theorem of binomial statistics, if the probability of a success (no flashover) for a single flash is p, then for n number of flashes to the line, the probability that there will be exactly k successes and n-k failures is defined as:

$$P_k = \frac{n!}{k!\,(n-k)!}\,p^k q^{n-k}. \tag{7.16}$$

Equation (7.16) gives the probability of lightning flashover in a year. Lightning flashover level of transmission line insulators, transformers and other power apparatus are defined by the basic impulse level, which is the peak value of the withstand voltage of a standard impulse voltage wave generated by a direct or indirect lightning strike. It is a measure of a electric power apparatus insulation system to withstand high surge voltage. Even when there is no electric break down, the bearings of power apparatus are subject to large mechanical forces by the high lightning currents that flow through them, threatening to rip them apart. In practice, more accurate data on the flashover can be obtained by keeping record of lightning interruptions to the network. This is more site-specific and will require daily monitoring over some years. Further, the issue of global warming and greenhouse effects has contributed to severe weather storms resulting in increase in lightning frequencies and severity of lightning flash intensities. The evidence for the relation between lightning activities and climate change points to the fact that climate change influences the global lightning frequencies and intensities.

The multiple lightning channels observed in lightning strikes to electric power lines indicate that there is not only one flash, but following the first return stroke, there are subsequent strokes to the same point of the power line. When subsequent strokes occur there will be a series of destructive high voltage transient pulses that will travel one after the other, along the line in both directions, that is, towards the power generating station and towards the substations.

7.5.4   Factors Influencing Lightning Strike to Transmission Lines

7.5.4.1   Listing of Factors

Several factors can have an influence on lightning strikes to transmission lines. These factors include ground flash densities, route selection, environmental shielding, tower heights, and soil resistivity. Each of these factors is discussed separately in the following sections.

7.5.4.2   Frequency of Lightning Incidence

A good knowledge of the frequency of occurrence of lightning is important in the design of power system protection against lightning strikes. Lightning activities in a given area are referred to as the lightning flash density or the ground flash density (GFD) occurring in a unit area in a unit time.

7.5.4.3   Transmission Line Route Selection

The transmission line highways are part of power system network planning. The transmission line routes are dictated by whether loads and generations are situated far apart. Transmission of bulk electric power requires selection of appropriate tower structures for either AC or DC transmission system, the voltage levels, circuit requirements, and bundling of conductors. Many transmission lines often make use of multiple circuits to utilize the same right-of-ways. Thus, cost optimization often becomes the deciding factor in routing overhead transmission lines. Environmental factors such as topographic terrains, swamps, rivers, and oceans also play a key part in the route selections. Further, government policies may influence the transmission route selection, especially in isolating highly polluting power plants and nuclear power plants from from cities. Thus, given these factors, lightning ground flash densities are often given least priority in the selection of overhead transmission lines routes. Furthermore, with the non-probabilistic nature of lightning strike rates, transmission line routes are often chosen based on least cost route selections. However, these practices could lead to severe lightning-related problems when the system is put in operation.

7.5.4.4   Environmental Shielding

The surrounding natural environment can provide shielding to transmission lines. Rugged terrains through mountainous and forest regions have the advantage in providing natural shielding against lightning strikes and also serve as windbreaks to transmission lines. The trees can intercept lightning and divert it to ground without hitting the electric power lines.

Conversely, tall trees have a disadvantage in that they may fall on the lines or the branches swaying during high winds touching lines causing a ground fault. This can be overcome by constant pruning of trees and maintaining a safe clearance of tree branches from the power line route. Further, tall structures within the vicinity of right-of-way (ROW) often become a source of induced lightning voltage impulses generated through inductive couplings on the transmission line. However, very high voltage transmission networks tend to possess sufficient impulse strength that makes them immune to the induced voltage impulses. This holds true only if transmission lines are well maintained. A polluted string insulator, for instance, can become a source of leakage current and flashover across the conductor to the tower structure causing a ground fault especially during precipitation. Similarly, a breakdown due to tracking in which dry conducting tracks are formed on the insulator surface can lead to gradual breakdown along the surface of the insulator.

Other artificial environments of tall, grounded structures such as communication towers, wind turbines, and buildings within close vicinity of transmission line corridors can shield the lines from lightning strikes. Further, for multiple transmission lines of different heights sharing the right-of-way, the shorter towers are shielded, thus reducing the strike rates.

Similarly, multiple transmission towers of same heights and design along the same route share the lightning strikes, thus the strike rate is lower than it would be for a single transmission line. Although sharing right-of-way reduces the strike rates, it can lead to a greater risk of multiple transmission line outages since severe electric storms with multiple lightning strikes can be a major concern for any transmission network with multiple lines in the same area of lightning activity.

Other tall structures such as wind turbines located on high terrains and further tall iconic towers and buildings can shield transmission lines from lightning strikes. Many of the iconic towers and buildings stand up to about several hundreds of meters tall. Such tall structures become the attachment point for downward lightning flashes. Conversely, these tall structures often become the frequent source of upward lightning flashes (GC flashes). The reason that tall structures become the source of upward lightning flashes has often been attributed to the fact that extremities of structures tend to generate high electric fields. In Chap. 2, the dipole theory was used to explain how the electric fields at the extremities of a conducting object reach large values when it becomes polarized, even when the ambient electric fields are small. This is due to the small separation distances between the dipoles. That is, the electric fields at the extremities are larger than those of the downward leader, resulting in upward lightning flashes or ground-to-cloud GC) flashes. As presented in Chap. 2, the algorithm based on the dipole theory takes into account the electric fields induced by other electric charges on the surface and their image charges in the calculation of the overall electric field.

7.5.4.5 Effect of Tower Heights on Lightning Strike

Tall structures become the shortest route for a charged cloud to discharge to ground in the case of cloud-to-ground (CG) flash or become the emanating point for the upward leader in the case of a ground-to-cloud (GC) flash. For the overhead transmission line highways routed through a topography of flat plains, the towers become the shortest path for the CG flashes. The highest tips of the tower become the emanating point for the upward positive electric discharge, known as positive streamers. The streamers approach the downward-stepped leaders in the air. The point where the leader and the streamer meet is the attachment point, which paves way for the first return stroke. The first return stroke current measured at ground typically rises to an initial peak of about 30 kA in some microseconds and decays to half-peak value in some tens of microseconds. The first return stroke waveform often shows the slowly increasing current of the upward streamer, before the sudden increase in current after the return stroke is initiated. The return stroke effectively lowers to ground several Coulombs of charge originally deposited on the stepped leader channel including all the branches. It is possible for another leader to travel along the same channel that has been ionized by the stepped leader. This leader is referred to as the dart leader and can result in subsequent return strokes. The time interval between the pre-breakdown and the subsequent strokes could last for about 62.5 ms.

The shield wires and the tower structure are the attachment points or the flash collection points on a tall transmission line tower where the return stoke currents and any other subsequent stroke currents are dissipated. The flash collection rate denoted by N_s is

$$N_s = N_g \cdot \left(\frac{28 \cdot h^{0.6} + b}{10} \right) \qquad (7.17)$$

where h is the tower height (m), b is the overhead ground wire separation distance (m), N_g is the ground flash density (flashes/km^2/year), and N_s is the flashes/100 km/year. The value of b is zero for a single ground wire. Ng can be calculated based on the keraunic level using either (7.11) or (7.12) if no direct value is defined.

7.5.4.6 Effects of Soil Resistivity on Lightning Strikes to Towers

Transmission line towers are grounded for protection against current and voltage transients generated by lightning when it strikes the OHGW (Overhead Ground Wire) or the tower. The current and voltage transients have a steep front. The current or voltage transients as traveling waves will travel at almost a third of the speed of light. The incident traveling waves will travel in different directions from the strike point over the transmission line media of different impedances or even points of discontinuities such as open circuit breakers or substation transformers. The traveling waves will be partly transmitted and reflected depending on the impedances of the terminating media. The traveling waves become additive if the crest of the reflected waves traversing through a medium coincides with the incoming reflected crests. These result in large voltage spikes causing a flashover across the string insulator to the phase conductors. Thus, a phase-to-ground fault is generated. Further, the traveling waves can reach the substations if they are not completely damped. This can cause damages to transformers and other substation equipment.

The damping can be achieved through low footing resistances of the towers. The soil resistivity plays a pivotal role in dissipating these high transient-induced currents or voltages to ground. However, the soil resistance fluctuates depending on the topography and terrain and the nonlinearity of the conductivity of the earth.

7.6 Protection Measures to Reduce Impacts of Lightning on Transmission Lines

Protection against lightning strike on any power system transmission lines can be achieved via the following three methods: Overhead ground wire (OHGW) of Shield Wires, Transmission Line Grounding System and Surge Arresters.

As the charged cloud-stepped leader travels downward to earth, it looks for the shortest path to discharge to ground. Tall tower structures in open fields become the shortest paths for a charged cloud to discharge to ground in the case of a cloud-to-ground (CG) and ground-to-cloud flashes. A well-designed transmission line tower geometry will have the OHGW effectively placed at the highest point and aligned at an angle such that it will intercept any lightning flashes, thus reducing the shielding failure rate to an acceptable level. This makes OHGW most effective in transmission line protection against a direct lightning strike. Specific positions of the OHGW are chosen to provide proper shielding of the tower. A shielding angle can be positive or negative depending on the position of the OHGW. A negative shielding angle has the horizontal placement of the OHGW outside the phase conductors. Conventional tower designs normally have a single shield wire aligned at an angle of about 30°. However, this is not effective for tall tower structures of higher voltage levels exceeding 230 kV with multiple circuits. Thus, a two-OHGW arrangement is proven to be very effective for lightning flash protection for high voltage.

The effectiveness of the angles between the placement of shield wire and the phase conductors can reduce the shielding failure rates. This is illustrated using the well-known electrogeometric model (EGM) theory. EGM theory combines electrical parameter and geometry methods in shielding failure calculations in transmission lines.

As a downward lightning-stepped leader descends seeking the shortest paths to discharge along the electricity highways, the tall transmission line tower structures would normally become the target. A transmission line tower structure directly beneath a charged cloud base would become polarized. That is, for a negative downward stepped leader, the ground structures become positively charged. The electric field at the top of the tower would increase such that it will induce an upward positive stepped leader. The point where the upward stepped leader from the tower meets the downward leader is referred to as the attachment point or the stroke point. The distance from the tower top to the strike point is the distance of interest in determining the angle of protection of the transmission line phase conductors. This distance is often referred to by some authors as the final jump. It can also be referred to as the strike distance, which extends from the tower tip to the stroke point. The electric field at the tower tips increases such that the positive charges migrate upward towards the descending negative stepped leader. The point of interception of the two leaders is where the lightning flash becomes visible to the naked eyes. The distance is denoted as S. Some texts denote this as r_g, the distance from the OHGW to the stroke point.

The return stroke current pulse induced at the stroke point travels up the ionized path to the cloud base and down to the tower. At the tower tip, the current pulse splits at wave peak depending on the surge impedance as per Ohm's Law and travels along the shield wire in both directions and also down the tower structure to the ground. The current pulse as a traveling wave between adjacent towers will be transmitted and reflected along the towers and OHGW on both sides. The traveling wave will gradually decay as it travels further down the transmission line towers. At certain point, the crest of the traveling waves will be in phase with each other, and when summing up reaching a new peak which may cause a flashover if it exceeds the critical flashover of the line voltage. Similarly, the crests and troughs of the traveling waves coinciding will reduce the peak wave. Thus, eventually, the current pulse will decay. However, for lightning striking the transmission towers within close vicinity to substations, the current pulse (i.e. surge current) will reach the substations with destructive energy. In such cases, surge arresters will operate to divert the surge current to ground. If the surge arresters fail to operate, the circuit breakers will operate to isolate the circuit thus preventing damages to transformers and other high voltage equipment. The lightning strike attractive radius expression is defined by the strike distance S.

The strike distance S is a function of the electric charge and consequently the return stroke current I. There are several equations, yielding slightly different values, that are used in determining the distance S:

$$S = 2 \cdot I + 30 \cdot \left(1 - e^{-\left(\frac{I}{6.8}\right)}\right) \quad \text{Darveniza et al. (1975)} \qquad (7.18)$$

$$S = 10 \cdot I^{0.65} \quad \text{Love (1973)/IEEE (1997)} \qquad (7.19)$$

$$S = 9.4 \cdot I^{\frac{2}{3}} \quad \text{Whitehead (1974)} \qquad (7.20)$$

$$S = 8 \cdot I^{0.65} \quad \text{Anderson at al. (1982)/IEEE (1985)} \qquad (7.21)$$

$$S = 3.3 \cdot I^{0.78} \quad \text{Suzuki et al. (1981)} \qquad (7.22)$$

$$S = \left(\frac{0.338}{v_1}\right)^{1.34} \cdot I^{1.42} \text{ Taniguchi et al. (2010)} \qquad (7.23)$$

where S is the strike distance in meters and I is the return stroke current in kA, and v_1 is the ratio of the lightning stroke velocity to the speed of light. The current I is a function of the surge impedance (Z_s) and the basic impulse level (*BIL*), and is given by

$$I = 2.2 \cdot \frac{BIL}{Z_s} \cdot \qquad (7.24)$$

7.7 Lightning-Aircraft Electric Circuit Models

7.7.1 The Basic Equations

Lightning return stroke models as discussed in Chap. 5 are categorized into four classes. The engineering model, using the electric transmission line model of the return stroke electromagnetic wave phenomena, will receive more attention here as it is applied here with respect to airborne vehicles. The transmission line modeling method is widely used in electromagnetic modeling in the microwave and even in Terahertz applications. The DLCRM method is based on the analogy between the propagation of electromagnetic fields and circuit networks and is characterized by its ability to guide propagation of electromagnetic energy. It is a time-domain numerical method that solves differential forms of Maxwell's equations. The four Maxwell's equations that describe classical electromagnetic theories are highlighted:

$$\nabla \times \vec{E} = -\frac{\partial \vec{B}}{\partial t}$$

$$\nabla \times \vec{H} = \vec{J}_C + \frac{\partial \vec{D}}{\partial t} \qquad (7.25a)$$

$$\nabla \cdot \vec{D} = \rho_v$$

$$\nabla \cdot \vec{B} = 0$$

where the symbols $\vec{E}$, $\vec{B}$, $\vec{H}$, $\vec{J}_C$, $\vec{D}$, and ρ are the electric field intensity (V/m), the magnetic field density (T), the magnetic field intensity (A/m), conduction current density (A/m^2), the electric flux density (C/m^2), and the scalar electric charge density (C/m^3). Applying the following constitutive relations

$$\vec{D} = \varepsilon_0 \vec{E}$$
$$\vec{B} = \mu_0 \vec{H} \tag{7.26}$$

where μ_0 and ε_0 are the permeability and the permittivity of the medium in free space. However, in a medium of propagation such as a metal, and other materials, the relative permeability (μ_r) and the relative permittivity (ε_r) are taken into account. In free space, the constitutive parameters are $\rho = 0$, $\mu_r = 1$, and $\varepsilon_r = 1$ so Ampere's and Faraday's laws become:

$$\nabla \times \vec{H} = \varepsilon_0 \frac{\partial \vec{E}}{\partial t}$$
$$\nabla \times \vec{E} = -\mu_0 \frac{\partial \vec{H}}{\partial t} \tag{7.25b}$$

Equations (7.25a) and (7.25b), after differentiation can be expressed as

$$\nabla \times \nabla \times \vec{E} + \frac{1}{c^2}\left(\frac{\partial^2 \overline{E}}{\partial t^2}\right) = -\mu_0\mu_r\left(\frac{\partial \vec{J_c}}{\partial t}\right) \tag{7.27}$$

$$\nabla \times \nabla \times \vec{H} + \frac{1}{c^2}\left(\frac{\partial^2 \vec{H}}{\partial t^2}\right) = \nabla \times \vec{J_c}$$

where c is the speed of light which is given by

$$c = \frac{1}{\sqrt{\varepsilon_0\mu_0}}. \tag{7.28}$$

It can be seen from the classical electromagnetic equations above that an electromagnetic wave generated by lightning or any other sources comprises both electric and magnetic fields components. Whenever an electromagnetic (EM) wave strikes a conductive object, it excites the electrons on the surface and current is generated. This surface current transmits electromagnetic energy. The energy will be absorbed or radiated by the object depending on the characteristics of the medium and its electrical parameters relating to the transmission and reflection coefficients of the medium. The propagation of the electromagnetic signals along the medium can be modeled as a transmission line using the distributed circuit model, or DLCRM. With reference to lightning-induced electromagnetic waves on an aircraft surface, the lightning-induced bidirectional leader that connects the aircraft to cloud and to ground is modeled as shown in Fig. 7.6. Figure 7.7 shows a finite length of the transmission line, Δz, comprising a vertical segment, with the coordinate z specifying the lightning channel or the aircraft with respect to the ground attachment point of the

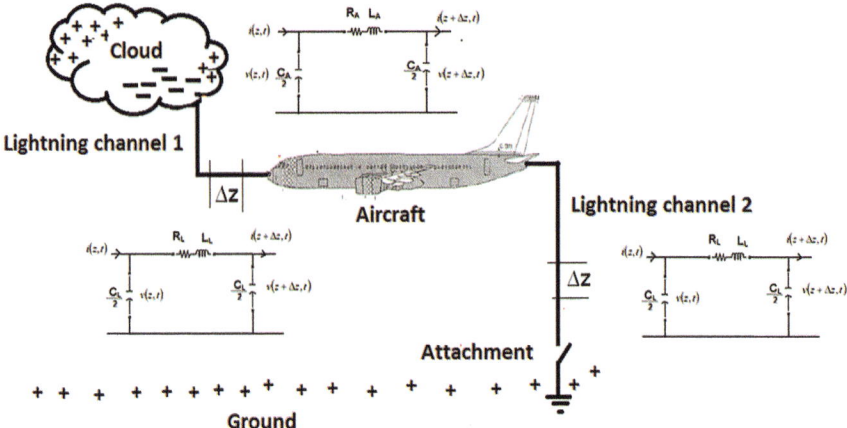

Fig. 7.6 Transmission line representation of lightning and aircraft channels and the distributed circuit parameters

Fig. 7.7 A finite length of
the lightning channel

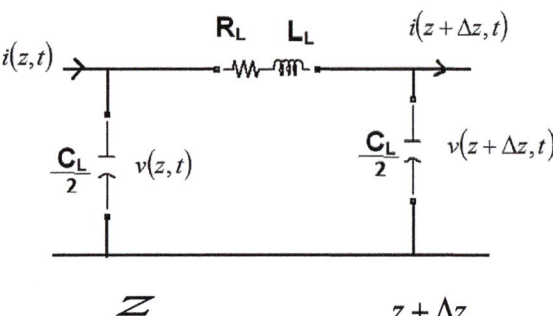

lightning flash. The equations governing the current and voltage distributions along a finite length (Δz) of the line are derived based on Kirchoff's voltage and current laws.

Applying Kirchoff's voltage law, the lightning channel voltage is determined as a function of both position z, and time t as follows:

$$v(z,t) \; - \; v(z+\Delta z, t) \; = \; i(z,t)R\Delta z \; + \; L\Delta z \frac{\partial i(z,t)}{\partial t}. \tag{7.29}$$

Dividing (7.29) by Δz and taking the limit as Δz approaches zero

$$\lim_{\Delta z \to 0}\left(\frac{v(z,t) \, - \, v(z+\Delta z, t)}{\Delta z}\right) = i(z,t)R \; + \; L\frac{\partial i(z,t)}{\partial t} \tag{7.30}$$

remembering that the limit is the definition of first derivative, (7.30) becomes

$$- \frac{\partial v(z,t)}{\partial t} = i(z,t)R + L\frac{\partial i(z,t)}{\partial t}. \tag{7.31}$$

Applying Kirchoff's current law, a similar expression is found for current:

$$i(z,t) - i(z + \Delta z, t) = i(z,t)G\Delta z + C\Delta z\frac{\partial v(z,t)}{\partial t}. \tag{7.32}$$

Since the conductance, G, is very small and is negligible, (7.32) reduces to

$$i(z,t) - i(z + \Delta z, t) = C\Delta z\frac{\partial v(z,t)}{\partial t}. \tag{7.33}$$

Dividing (7.33) by Δz and taking the limit as Δz approaches zero

$$\lim_{\Delta z \to 0} \left(\frac{i(z,t) - i(z + \Delta z, t)}{\Delta z} \right) = C\frac{\partial v(z,t)}{\partial t} \tag{7.34}$$

$$- \frac{\partial i(z,t)}{\partial t} = C\frac{\partial v(z,t)}{\partial t}. \tag{7.35}$$

Equations (7.31) and (7.35) are known as the transmission line equations. The two equations can also be derived from Maxwell's equations if the electromagnetic wave propagating on or guided by the line exhibits a quasi-transverse electromagnetic field structure and line elements R, L, and C are time-invariant.

The following relationships defined relate distributed circuit parameters L and C to Maxwell's equations for the case of a non-linear transmission line, where (7.31) and (7.35) are still valid if L and C were dynamic.

$$L = \frac{\partial \varphi}{\partial I} \tag{7.36}$$

$$C = \frac{\partial \rho}{\partial V} \tag{7.37}$$

where φ is the magnetic flux linking the channel, and ρ is the channel electric charge density. Figure 7.8a and b depicts a simplified block diagram of a long transmission line model (DLCRM) for the lightning channel through an aircraft. The equivalent RLC distributed model of the various segments is shown in Fig. 7.8(c). It comprises several cascading segments of a π-network.

A typical DLCRM model for a lightning-aircraft system is in Fig. 7.8c, showing only three segments for convenience. However, to model accurately the aircraft A380 airbus of 72.72 m length, the lightning channel is uniformly subdivided into several segments of 50 m π sections. From the strike point, the return stroke current propagates along the lightning channel to the aircraft and cloud, and through the earth resistance to the ground. The π-segment parameters for the lightning are determined

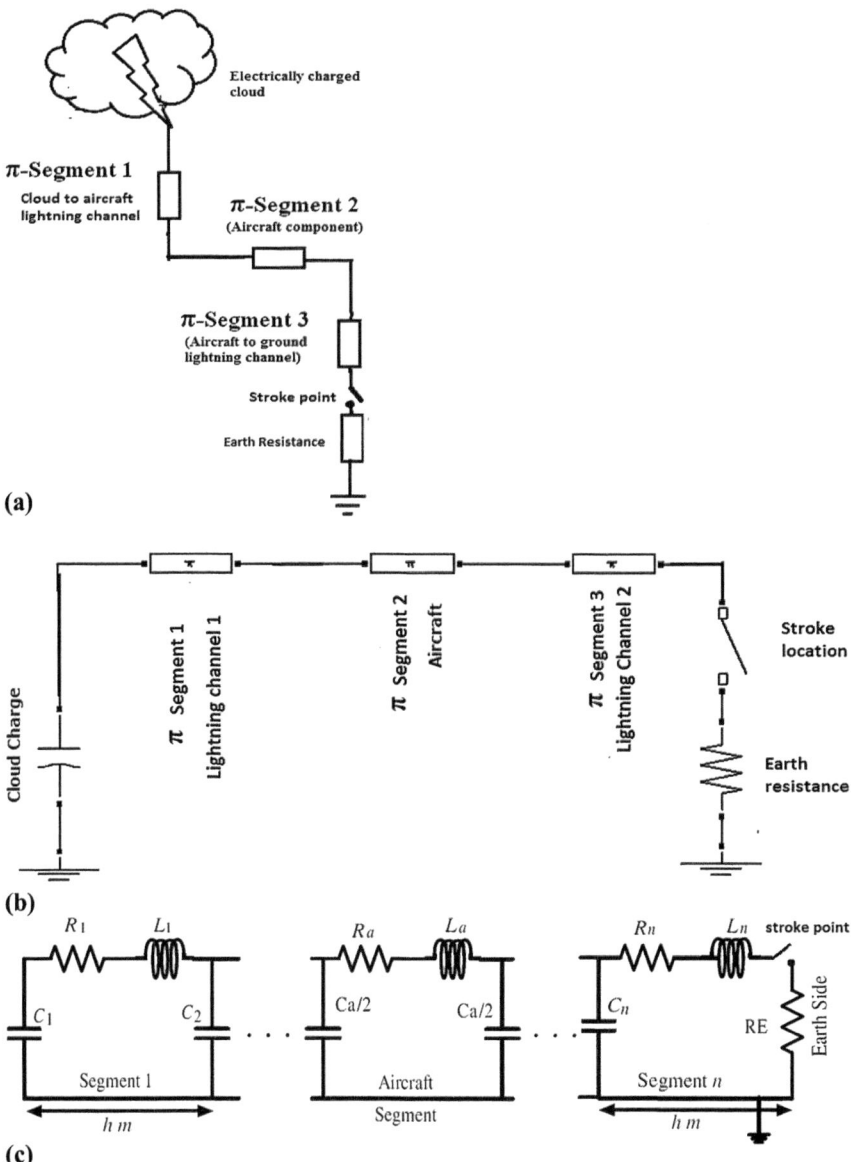

Fig. 7.8 Lightning aircarft transmission line circuit model **a** simplified block diagram **b** the cascaded π-segments, and **c** the distributed RLC network

Table 7.3 RLC values for aircraft and lightning channels

	Resistance (Ω/m)	Inductance (H/m)	Capacitance (F/m)
Lightning channel	1×10^{-6}	3×10^{-6}	4.60×10^{-12}
A380 Aircraft	2×10^{6}	8.27×10^{-7}	1.34×10^{-11}

from the lightning channel parameters. The aircraft segment resistance, inductance, and capacitance are determined from the position of the aircraft using its material properties and the cloud charge and the aircraft charge geometries and their images. Table 7.3 shows the circuit parameters for the lightning channel and the aircraft. The aircraft parameters will vary with aircraft sizes, geometries, and the materials used in the construction of the airframes.

7.7.2 The DLCRM Parameters of the Lightning channel

The RLC values for the lightning return stroke channel obtained for a lightning channel radius of a $= 0.010$ m and a $= 0.004$ mm are given in Table 7.4. The per unit length values obtained for L and C are obtained from (7.38) and (7.39), respectively. The lightning channel lengths between the cloud and aircraft, and aircraft and ground are taken to be 500 m each.

$$C = \frac{2\pi \varepsilon_0}{\ln\left(\frac{2h}{r}\right)} \tag{7.38}$$

$$L = \left(\frac{\mu_0}{2\pi}\right)\left(\ln\left(\frac{2h}{r}\right)\right) \tag{7.39}$$

$$R = \pi r^2 \sigma. \tag{7.40}$$

Equation (7.40) gives an estimated value of the return stroke channel per unit length resistance. In (7.40), the value of conductivity σ is based on the lightning channel temperatures. Temperatures in the range of 15, 000–24,000 K are used. The discrepancies in the lightning channel temperatures account for the variations in the values of resistances used. The value of the resistance adopted herein is based on the conductance value calculated for a temperature of 20,000 K, yielding a conductance

Table 7.4 Lightning RLC parameters based on channel radius and height

RLC calculations		
Channel radius (m)	0.01	0.004
Resistance Ω/m	0.8	4.7
Inductance μH/m	3	3
Capacitance pF/m	4.6	4.6

of 4242 Ω^{-1} m^{-1}. Using (7.40) with the cylindrical lightning channel radius of 1 cm, and R = 1 Ω/m. The lightning channel is assumed to be constant in radius throughout the presence of the return stroke, with equal and counteracting magnetic and kinetic forces keeping it stable. The CG lightning strikes always dissipate much of the return stroke energy into the earth, and not in the lightning channel itself. The earth resistance can be calculated using (7.41), for a typical soil conductivity of 0.01 Ω^{-1} m^{-1} for wet ground. Using a lightning channel radius of 1 cm, the earth resistance calculated from (7.41) is 200 Ω.

$$R_E = \frac{1}{2\pi \sigma_E} \left(\log_e \left(\frac{8l_E}{1.36a} \right) \right)$$
(7.41)

where R$_E$ is the earth resistance, σ_E is the conductivity of the earth, and l$_E$ is the length of the return stroke channel penetrated into the earth.

7.7.3 The DLCRM Parameters of the Aircraft

An aircraft struck by lightning becomes a conducting path for the return stroke current connecting the cloud and earth through the lightning channels. The aircraft can therefore be a part of the natural lightning discharge process. The aircraft skin becomes a path for the lightning current. However, the R, L, and C values of the aircraft are different from those of the lightning channel and vary with the conductivities of the materials used in the aircraft skin. The lightning discharge path via the aircraft is effectively represented using the DLCRM model. The calculated electrical elements R$_a$, L$_a$, and C$_a$ of the aircraft are lumped together and slotted in as one of the segments of the RLC transmission line model of the lightning return stroke. The respective aircraft R$_a$, L$_a$, and C$_a$ parameters are determined as outlined in the following paragraphs.

Aircraft resistance. The aircraft skin resistance is determined using the aircraft geometry and the materials used in the airframe skin wall and its thickness. The aircrafts used in this study are the airbus A380 and the F16 military aircraft. Thus, most of the electrical parameters of the fuselage materials are based on the geometry, the electrical conductivities of the skin and wall thickness of the two aircraft. The electrical conductivities of the materials used were either for the CFC airframe or for an aluminum metallic airframe. The resistance is calculated separately for both aluminum and CFC airframes using the respective electrical conductivities. The airbus A380 fuselage skin wall thickness taken from the airbus parameters is 2–4 mm. A skin wall of 4 mm thickness is used for both the airbus A380 and the F16 military aircraft.

The fuselage resistance (R$_{AF}$) for the two aircraft studied are calculated using,

Table 7.5 Electrical conductivities of various materials

Elements	Value of electrical conductivity (S/cm)
Silver	6.1×10^5
Copper	5.8×10^5
Aluminum	3.7×10^5
Stainless steel	1.4×10^4
Graphite	7.3×10^2

$$R_{AF} = \frac{L}{2\pi b t \sigma} \qquad (7.42)$$

where R_{AF} is the aircraft fuselage resistance, L is the aircraft fuselage length, b is the fuselage radius, t is the wall-skin thickness, and σ is the conductivity of the material. The airbus A380 has a fuselage radius of 4.205 m and a length of 72.72 m. The F16 military aircraft has a fuselage radius of 1 m and a length of 15.03 m. The conductivity of a skin wall of aluminum is 3.7×10^7 S/m and that for graphene composite used in aircraft is 73,000 S/m. Table 7.5 shows the conductivities of some common materials used in an aircraft body.

Applying the respective parameters in (4.50), the resistance values for an airbus A380 and F16 military aircraft fuselage made of metallic (aluminum) airframes are $R_{Al380} = 1.823 \times 10^{-5}$ Ω/m and $R_{AlF16} = 1.0544 \times 10^{-6}$ Ω/m respectively. Similarly, for an Airbus A380 made of graphite composite fiber, the per-unit resistance is $R_{CFC380} = 1.296 \times 10^{-4}$ Ω/m while that for F16 military aircraft, assumed to be made of graphite composite airframe, is calculated to be $R_{CFCF16} = 5.54 \times 10^{-4}$ Ω/m. Further, the wing resistance will form a part of the aircraft circuit model when the lightning channel (the bidirectional leaders) is attached to the wings. The resistive components of the wings are calculated separately. The wings comprise the ailerons, the flaps, and the engine skin wall that can be lumped together to approximate the overall wing resistance.

The wing resistance will be used as part of the aircraft-lightning bidirectional leader forming the DLCRM model of the aircraft wings. The bidirectional leader can begin at the wing tip and be swept through the wing to the radome to discharge to the ground. Alternatively, it can begin at the radome or tail and be swept through the fuselage to any of the wings and discharge to the ground. However, in the case of an aircraft caught between two oppositely charged cloud cells, the bi-directional leader will attach at one wing tip and the current would flow through the wing and mid fuselage; and exit through the other wing tip to connect a second leader to the other oppositely charged cloud cell.

The aircraft wing resistance is calculated using the wing skin-wall thickness. The wing design used for the A380 wing is similar to that of the Boeing 747. The aircraft skin wall thickness of 4 mm is used. The area calculation is based on the geometry of the wing design and its thickness. Similarly, the F16 military aircraft wing resistance is calculated using the F-18 (blue angel) wing geometry. The wing of the Airbus A380 is divided into two parts, A and B, and the dimension is determined from the

Airbus A380 data. The equation for the resistance of a sheet of metal (7.43) is used, which also applies to a sheet of CFC material. The resistance of a sheet of metal is calculated using,

$$R = \rho \frac{l}{Wt} \qquad (7.43)$$

where R is the resistance, ρ is the resistivity of the material, l is the length, W is the width, and t is the thickness of the sheet of material.

From the geometry of the wing and the equation of sheet resistance, the overall resistance is calculated with the two parts (A and B) making up the triangular sheets. The overall resistance is simply the sum of the two triangular sheets for the respective material used. Note further that (7.43) only gives the resistance of the upper sheet skin wall. The overall wing resistance includes the lower portion and upper portion of the wing skin walls making up the complete wing resistance. The overall wing airframe structure begins at the root (base) attached to the fuselage, the leading and trailing edges of the wing and the interconnection at the wing tip making up the complete wing skin wall structure. The total resistance is considered a parallel interconnection of the upper and the lower sheet metal/composites that make up the overall wing structure.

Aircraft capacitance. The aircraft capacitance is determined using the three-dimensional dipole methods as discussed in Chap. 2. For the CFC material, we ignore additional resistive and capacitive elements that exist inside the CFC material itself. The dipole modeling of electrostatic charges on an aircraft gives a succinct representation of the distribution of electrostatic charge build-up on aircraft surfaces. The method makes use of an elementary theory of electrostatic induction on the distribution of charges within an object that occurs as a reaction to the presence of a nearby charge. The analogy is applied to an aircraft as it enters a charged electric storm causing migration of polarized charges on the surface with positive electric charges gathered in the direction of the negative cloud center and the negative electric charges gathered towards the lower, positively charged earth surface. This is for a negative cloud to ground earth flash. In the case of a positive cloud to ground flash, the base of the cloud is positive and the ground plane below it becomes negatively charged. Thus, an aircraft flying below the region of such a charged thundercloud would have the negative charges spread out on the top surface or the aircraft while positive electric charges would spread out on the belly of the aircraft.

Aircraft build up electrostatic charges just by virtue of flying through the atmosphere, largely due to air friction. However, the breakdown electric fields due to the electrostatic charges occur as the aircraft enters a charged electric storm. Thus, the pre-breakdown charges and the capacitances are determined using the 3D dipole model technique outlined in Chap. 2.

The 3D electric dipole model incorporates the real geometrical dimensions of an aircraft with surface electric charge distribution represented by diploes of various separation distances positioned along the top and bottom surfaces of the radome, wings, fuselage, and the tail end of the aircraft. The cloud electric charge and its image electric charge (of opposite electric charge polarity to the source electric charge polarity) are taken into account as these two electric charges highly influence the overall electric field on the surface of an aircraft. The surface electric charge layer on the aircraft surface is modeled as an electric line charge with an electric dipole moment per unit area. The field of the charged electric dipoles on the top and bottom of an aircraft surface is obtained by representing the aircraft as a floating electrode isolated in space and charged to a specific voltage. The aircraft dipole model is shown in Chap. 2, Fig. 2.2. in three dimensions on the aircraft fuselage, the radome, the wings, and the stabilizers. Notice that a few electric dipoles are sufficient to compute the capacitance of an aircraft. However, to represent more accurately the aircraft geometry, more electric dipoles are required.

The cloud charge is determined from the capacitance and cloud voltage using the equation

$$Q_{CL} = C_{Sph} V_{CL} \tag{7.44}$$

where C_{Sph} is the cloud capacitance of a Gaussian spherical cloud charge center, V_{CL} is the cloud voltage, and Q_{CL} is the cloud charge.

The aircraft capacitance is determined from the aircraft altitude and the geometry of the aircraft. That is, the aircraft capacitance is calculated using the coefficient of potential of the dipole charges and their image charges. Figure 7.9 shows the placements of the aircraft dipole charges, and the cloud charge and their respective image charges. Note that Fig. 7.9 only shows a two-dimensional dipole positioned along the fuselage, radome, and vertical stabilizes. The 3D placements of the aircraft dipoles charges are shown in Fig. 2.4, which also indicates the dipoles for the wings, the engines, and the stabilizers.

Using the 3D dipole charges positioned on the surface of the aircraft and the altitude of the cloud charge, the following equation for the dipole position vectors is derived using the following expression:

$$d_{g,h} = \sqrt{(x_g - x_h)^2 + (y_g - y_h)^2 + (z_g - z_h)^2} \tag{7.45}$$

where $d_{g,h}$ is the separation distance between monopole g and monopole h that make up a dipole in three-dimensional space (x, y, z). Similarly, the distance between an electric charge and its image charge ($D_{g,h}$) is defined by:

$$D_{g,h} = \sqrt{(x_g - x_h)^2 + (y_g - y_h)^2 + (z_g + z_h)^2}. \tag{7.46}$$

The coefficient of potential ($p_{g,h}$) is calculated from (7.46) using the above position vector matrices.

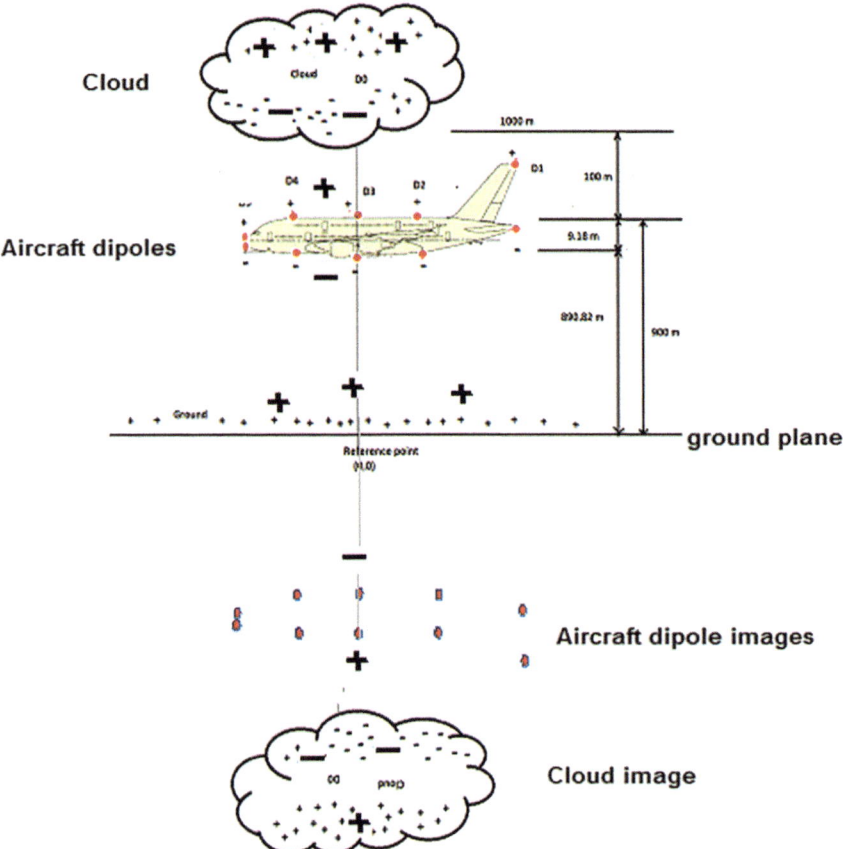

Fig. 7.9 The cloud charge and aircraft dipoles and their image charges

$$p_{g,h} = \frac{1}{4\pi\varepsilon_0} \cdot \ln\left(\frac{D_{g,h}}{d_{g,h}}\right). \tag{7.47}$$

Using this, the equation for the potential is as given in,

$$v = p_{g,h}q \tag{7.48}$$

where v is the voltage, $p_{g,h}$ is the coefficient of potential, and q is the electric charge.

If $p_{g,h}$ is a non-singular matrix, (7.48) can be inverted to get the electric charge q as given in,

$$q = p_{g,h}^{-1}v. \tag{7.49}$$

Thus, the capacitance of the cloud-aircraft ground system is simply the reciprocal of the coefficient of potential and is defined by,

$$C = p_{g,h}^{-1}. \tag{7.50}$$

Thus, the capacitance in Farad per unit length gives a good estimate of the aircraft capacitance.

Aircraft Inductance, L_a. Since an aircraft is like a floating conductor in free space, its inductance would be due to the self-inductance of the materials that make up the skin of the airframe. That is, the effects of nearby conductors are neglected assuming the current of the return stroke is only the displacement currents that flow through the capacitive elements. The equation of the internal inductance of the aircraft is calculated using the skin depth and the conductivity of the material at the frequency band of lightning flash. The skin depth is calculated at a high frequency such that the skin depth $\delta > 2r$ where r is the radius of the aircraft's cylindrical shell. A typical cyndrical shell is used for the range of lightning current frequencies up to 10 MHz. Thus, a frequency of 5 MHz is used in the calculation of the skin depth from (7.51). The cylindrical shell would yield the same inductance as a solid cylinder of the same radius at 5 MHz. The equation of the skin depth is given by,

$$\delta = \sqrt{\frac{1}{\pi f \mu \sigma}} \tag{7.51}$$

where δ is the skin depth of the material, f is the frequency, μ is the permeability of the material. Since aluminum (metallic) and CFC are both non-ferromagnetic materials, the relative permeability (μ_r) is approximated to 1 giving $\mu = \mu_0$, and σ is the conductivity of the material used. The conductivity for CFC is $73,000$ S/m and that of metallic (aluminum) is 3.77×10^7 S/m. For a CFC material, the $\delta_{CFC} = 8.33 \times 10^{-4}$ m while that for a metallic material such as aluminum, it is $\delta_{AL} = 3.66 \times 10^{-5}$ m. From the calculated values of the skin depth, a skin wall thickness of 4 mm is obtained for the aircraft skin. The return stroke current will not penetrate through the aircraft skin wall but flow along the aircraft body.

The internal or self-inductance of the aircraft can now be computed using the skin depth and the frequency, and it is given by,

$$L = \left(\frac{\mu_0 l}{4\pi}\right)\left(\frac{\delta}{b}\right) \tag{7.52}$$

where L is the inductance, l is the length of the fuselage, δ is the skin depth, and b is the radius of the fuselage. The inductance of an F16 military aircraft fuselage of 15.03 m length and 1 m radius, computed using the skin depth method, is $L_{F16AL} = 5.51 \times 10^{-11}$ H/m and $L_{F16CFC} = 1.252 \times 10^{-9}$ H/m for metallic (aluminum) and CFC materials, respectively.

7.7.4 The F16 Military Aircraft and Lightning Strike

The long DLCRM is used in modeling the lightning channel through the ionized air between a negatively charged cloud, aircraft, and to the ground as shown in the schematic Figs. 7.10 and 7.11. It is made up of several segments of a π-network, each representing the lightning channel from negative polarity of the cloud connecting the tip of the aircraft right wing through the mid-fuselage connecting to the ground via the tip of the left wing. A circuit model is shown in Fig. 7.11 with the strike point located at the tip of the right wing. Note that Fig. 7.11 shows only three π-segments for convenience. In the actual simulation model, several segments are used especially for the bidirectional lightning channels from cloud to the aircraft and from the aircraft to the ground. Further, the cloud capacitance is modeled as a spherical Gaussian surface with a diameter of 200 m. The simulation is carried out for a lightning initiated in

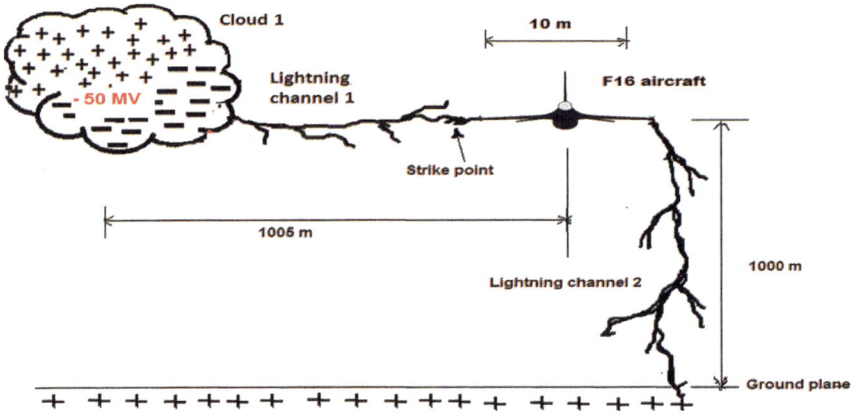

Fig. 7.10 An illustration of an F16 military aircraft struck by lightning

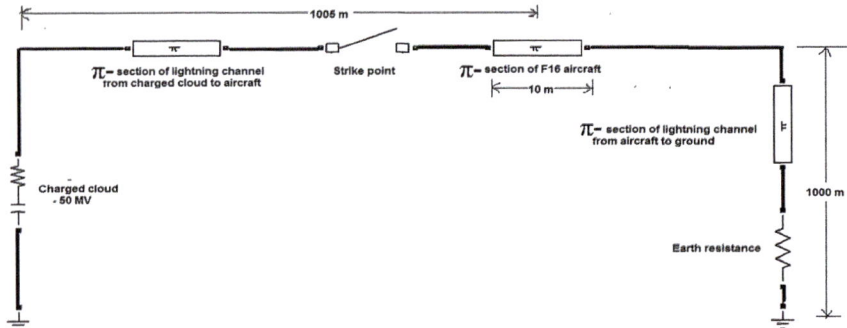

Fig. 7.11 A simplified DLCRM circuit model of the charged cloud and the F16 military aircraft showing only three π-segments for convenience

a thunderstorm at an altitude of 1000 m and the aircraft at the same altitude but separated by a distance of 1000 m from the center of the charged thundercloud as shown in Fig. 7.11.

A model of an F16 military aircraft is used with the specified dimensions. The π-segment parameters for the lightning channel are determined from the lightning channel parameters. The aircraft segment inductance is determined from (7.51) and (7.52). The aircraft capacitance is determined from the potential coefficients of the aircraft charges based on the 3D dipole model and the lightning cloud charges and their respective images. For an F16 military aircraft at 1000 m altitude, the capacitance is 7.4×10^{-11} F.

The aircraft resistance is calculated for a lightning flash swept for a distance along the wingspan as shown in Fig. 7.12. In this case study, a lightning flash of intensity of -50 MV strikes the F16 military aircraft at the tip of the right wing and then swept across the left wing via the mid-fuselage and attached to the ground. The total swept distance is 10 m according to the F16 military aircraft wingspan dimension.

The total resistance is calculated for both metallic and CFC airframes using the conductivities of both materials and a skin wall of 4 mm thickness. The resistances are 1.88×10^{-5} Ω and 0.0098 Ω for metallic and CFC airframe structures, respectively. Note that both resistances are lumped over the swept distance. The calculated inductances are 5.51×10^{-11} H and 1.252×10^{-9} H for metallic and CFC airframe structures, respectively. The values of the inductances are lumped over the entire aircraft. The earth resistance is 200 Ω while the cloud columnar resistance (the resistance of the vertical cloud structure) for an average cloud column is 165 Ω. For a large cloud cover, the cloud resistance is 353 Ω. However, the total columnar resistance

Fig. 7.12 F16 Geometry

reduces to 200 Ω close to mountains. Thus, an approximate columnar resistance of 200 Ω is used here.

The DLCRM network is simulated for a cloud-to-ground flash initiated from the negative cloud center, charged at −50 MV, through the aircraft fuselage to ground. From the strike point (at the rightwing tip), a bidirectional leader propagates from the aircraft structure linking the charged cloud to ground through the two wing tips of the aircraft. Figures 7.13 through to Fig. 7.16 show the voltage and current transients as well as their derivatives over the aircraft surfaces for both the metallic and CFC airframe structures. Table 7.6 gives a summary of the various first return stroke current parameters for an F16 military aircraft of metallic and CFC structures.

From Fig. 7.13, it can be seen that the current along the aircraft for both airframe structures reaches the high peaks of 31.263 kA and 33.343 kA for metallic and CFC airframes, respectively. These high current magnitudes are capable of causing severe direct effects such as melting or burning at the aircraft skin at the attachment points,

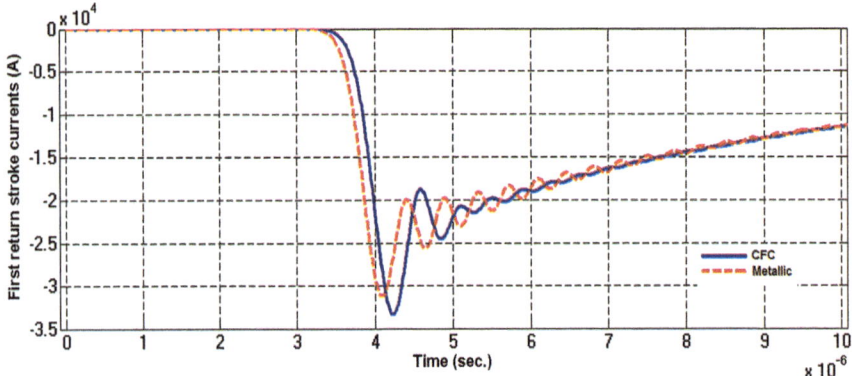

Fig. 7.13 First return stroke currents along an F16 military aircraft of CFC and metallic airframes

Table 7.6 First return current stroke parameters

Parameters	Metallic	CFC
Main Peak current (A)	31.263×10^3	33.343×10^3
Time to 10% of main peak (s)	3.5696×10^{-6}	3.6902×10^{-6}
Time to 90% of main peak (s)	3.9816×10^{-6}	4.1164×10^{-6}
Rate of rise (10–90% main peak) (A/s)	60.705×10^9	62.587×10^9
Time to main peak (s)	4.0876×10^{-6}	4.2241×10^{-6}
Main peak di/dt (A/s)	76.483×10^8	$78.94.23 \times 10^8$
Charge transfer (C)	−0.5561	−0.5561
Action integral (A².s)	4.0105×10^3	4.0096×10^3

namely the wing tips. Joule heating, which is proportional to the action integral of the lightning current, is capable of causing thin conductors to fuse explosively. The direct effects of joule heating on a CFC airframe can also result in melting and vaporization of the epoxy, leading to delamination damage of the carbon fiber. Further, magnetic forces arising from the high currents can crush or drive together or pull apart conductors. For CFC aircraft skin, the non-malleability and ductility of the material can cause burning, shredding, and arcing, which can pose a serious threat to aircraft safety.

It is observed from Fig. 7.13 that both currents possess fast fronts in sub-nano seconds reaching peaks within a period of 412 and 426.2 ns for metallic and CFC airframes, respectively. The transient front duration (time taken from 2 kA to peak value) is 568.7 ns for a metallic airframe and 593.5 ns for a CFC airframe, both falling below the 95% frequencies of the statistical tabulated values of 1.1 μs. The rate of rise from 10 to 90% of the peak values is 60.705×10^9 A/s (60.705 kA/μs) for a metallic airframe and 62.587×10^9 A/s (62.587 kA/μs) for a CFC airframe.

A further observation of the two current waveforms in Fig. 7.13 is the time delay to reach peak values. The current across the aircraft of a metallic airframe reached the peak value in 4.0876 μs and for the CFC airframe at 4.2241 μs. For the CFC airframe, the time delay indicates the dwell time for the swept stroke along the aircraft structure of the CFC exterior skins, rather than through substructures such as internal spars and ribs. For an aircraft of metallic airframe, the current is dissipated fast through the conductive structures; thus there is a less likely chance to induce any adverse effects. However, the slow current dissipation along the low conductivity CFC airframe is most likely to result in current penetration. That is, the slow decay currents along the CFC airframes can penetrate through coupling mechanisms such as the electromagnetic or resistive coupling. This can have adverse effects in the internal system of the aircraft. There are oscillations in the current along the metallic airframe due to its low resistance and inability to damp out the oscillation at the natural frequency of the lightning-aircraft DLCRM circuit.

In Fig. 7.14, it is observed that the current derivatives reach peak values of 78.888×10^9 A/s within 3.8671 μs for the metallic airframe, and 81.617×10^9 A/s within 3.9931 μs for the CFC airframe. The large current derivatives along the aircraft airframe can give rise to a time-changing magnetic field capable of inducing transient voltages ($v = L \frac{di}{dt}$) in the aircraft inductive elements of the wiring and other circuit components. Further, this can cause damages or interruptions to the avionics circuits of the communication, control, and command (CCC) systems.

The voltage transients shown in Fig. 7.15 reach peak amplitude of -7.6241 MV and -7.5997 MV for metallic and CFC airframe, respectively. The magnitudes of the voltage derivatives are 16.572×10^{11} V/s and 17.666×10^{11} V/s for metallic and CFC airframes, respectively, as shown in Fig. 7.16. Such a large lightning-induced voltage along the CFC airframe can give rise to changing electric fields capable of inducing high transient currents (CdV/dt) within the capacitive elements of the aircraft circuitries. The large rate of rise of voltage can result in high currents ($i = C \frac{dV}{dt}$) in the capacitive elements of the aircraft's electrical and avionics circuitries. Further, the large voltage time derivative can induce electric fields on the aircraft's

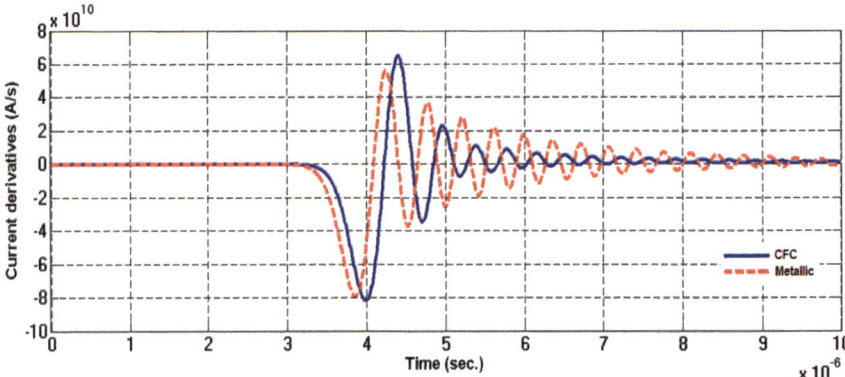

Fig. 7.14   The current derivative waveforms for an F16 military aircraft

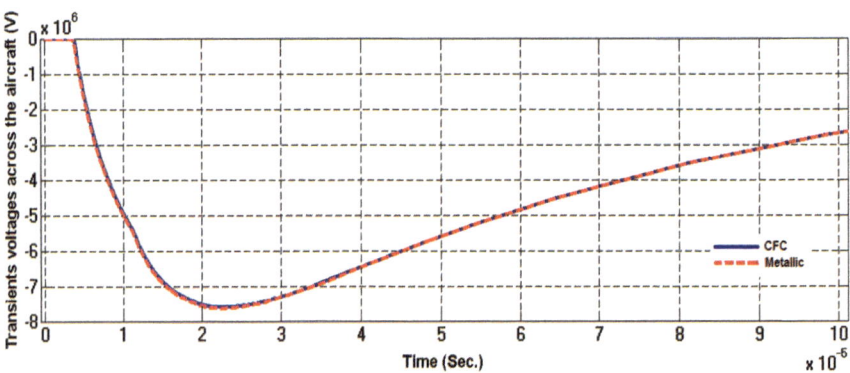

Fig. 7.15   Voltage transients across an F16 military aircraft

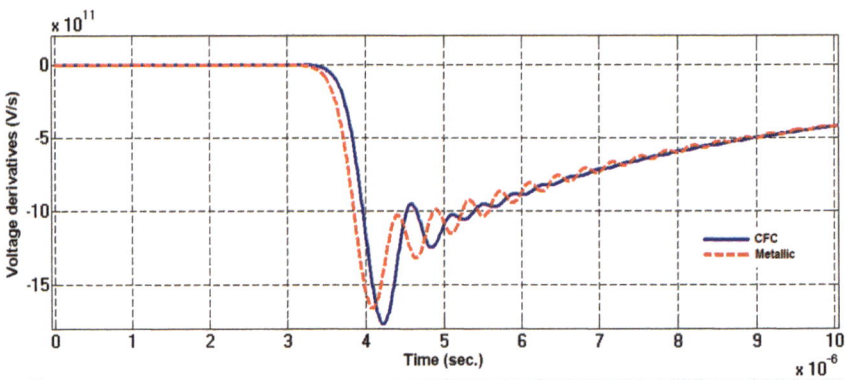

Fig. 7.16   Voltage derivatives for an F16 military aircraft

surface. The coupling of the fields can exist on the inner surface of the aircraft which can result in a resistive voltage drop.

The charge transfer through the F16 aircraft is -0.5561 C for both the metallic and CFC airframes. The action integral values observed are 4.0105×10^3 A^{2}s for a metallic airframe and 4.0096×10^3 A^{2}s for a CFC airframe. The severity of a CG flash of -50 MV flash on an aircraft shows currents of high magnitudes observed along the aircraft. Such high currents can be a major problem for modern aircraft with airframes made of electrically insulated carbon fiber/epoxy composites. The high current along the aircraft airframe can give rise to changing magnetic fields that can induce transient voltages in the wiring capable of causing damage or interruptions to the avionics systems. The non-metallic airframes can become entry points of lightning as they absorb lightning strikes instead of conducting and dissipating it.

Modern aircraft industries are employing non-metallic structures and highly digital and computerized control technologies in aircraft command and control systems that are susceptible to failures, instabilities, and damages if no extra protection addressing the severity of lightning is in place. Further, the threats of aging in carbon composite structures and the uptake of moisture in-flight coupled with the severity of lightning flash and the high-intensity burst of induced currents can result in the loss of mechanical strength of the airframe structure. Thus, aircraft industries need an improved definition of the threats that lightning poses in order to continue to drive the protection standards for the safety of aircraft.

7.7.5 The Airbus A380 Commercial Aircraft and Lightning Strike

The scenario analyzed in this case study is the simulation of the induced current and voltage waveforms for an airbus A380 of both metallic and CFC airframes. The aircraft is taken to be at the same 1000 m altitude as the electrically charged cloud, but at a distance of 300 m from the charged cloud center, as depicted in Fig. 7.17.

The aircraft initiates the lightning flash from the left wing. A bidirectional leader is initiated from the tip of the left wing (the stroke point) to the charged cloud and through mid-fuselage detaching from the radome to the ground, as shown in Fig. 7.17. Figure 7.18 shows the A380 π-segments of the aircraft for the metallic airframe structure.

The lightning flash swept stroke makes up the full length of the left wing (35.67 m), which includes the ailerons and the flaps resistance lumped together. The resistance calculated for the wing is the DC resistance and includes only the skin resistance. The fuselage segment comprises the length from the root of the wing to the radome making up a distance of 36.36 m. The total lightning flash swept stroke distance, swept from the left wing tip to the wing root, and the fuselage and radome, making a total distance of 72.03 m. The aircraft swept stroke distance has a total resistance of 2.147×10^{-5} Ω for the metallic airframe and 0.0111 Ω for the CFC airframe.

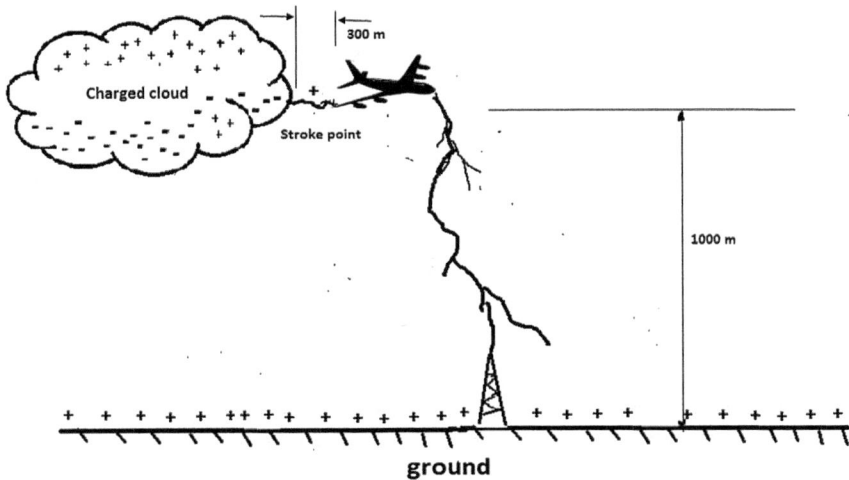

Fig. 7.17 A lightning trigger point at aircraft at an altitude of 1000 m as the charged cloud at a distance of 300 m from the charge center

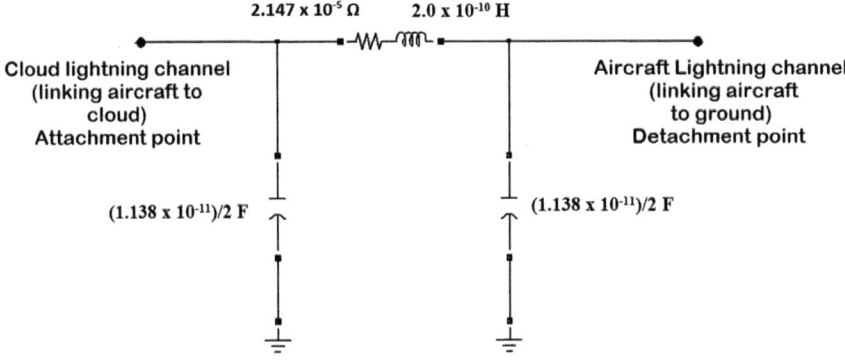

Fig. 7.18 The A380 airbus aircraft DLCRM π-segment for the metallic airframe

The inductance, calculated using the skin depth, is 2.0×10^{-10} H for the metallic airframe and 4.5×10^{-9} H for the CFC airframe. The capacitance is determined from the coefficients of potentials yielding a value of 1.138×10^{-11} F and is independent of the aircraft materials.

A lightning strike to an aircraft produces a specific structural current distribution of various intensities starting at the entry point to the detachment point structural current distribution. In this case study, the entry point is the right wing tip, and the return stroke current flows along the right wing through the root of the wing, along the front end of the fuselage to the radome tip (the detachment or exit point). The lightning-induced current distribution will depend on the lightning intensity. The thundercloud electric charge center potential is taken to be -50 MV.

In analyzing the first return stroke current along the metallic airframe, a simulation was done for currents at three different points as illustrated in Fig. 7.19. Ammeter *a* gives a measure of the first return stroke current burst at the entry point (the stroke point), ammeter *b* gives the first return stroke current along the airframe, and ammeter *c* measures the current exiting the aircraft (at the exit point). Figure 7.19 shows the current waveforms for the three ammeters for an A380 airbus of metallic airframe. In Fig. 7.20, the frequency spectrum of the entry point current is given.

The incident currents exhibit high peaks at the entry point due to the transition from a high surge impedance medium (lightning channels) to a low surge impedance medium of the aircraft structures. The discontinuities at the entry and exit points result in reflected current waves, which are observed in Fig. 7.19 for a metallic airframe. The oscillations are due to the reflected current waveforms at discontinuities being superimposed on the main return stroke current.

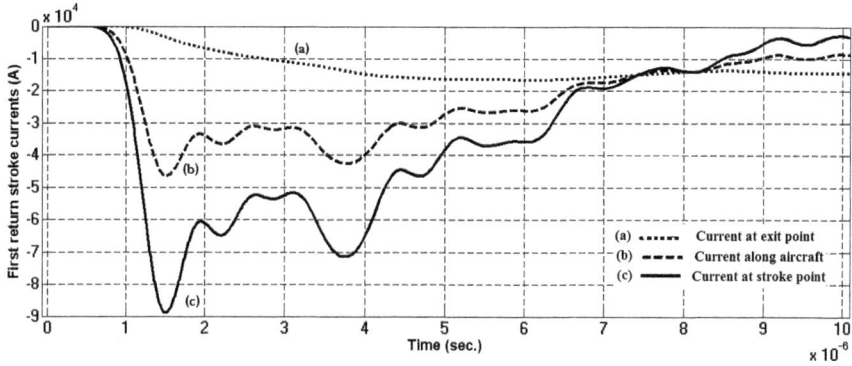

Fig. 7.19 Return stroke currents for an A380 airbus of metallic airframe at three different current measurement points: **a** at the exit point, **b** along the aircraft, and **c** at the stroke point

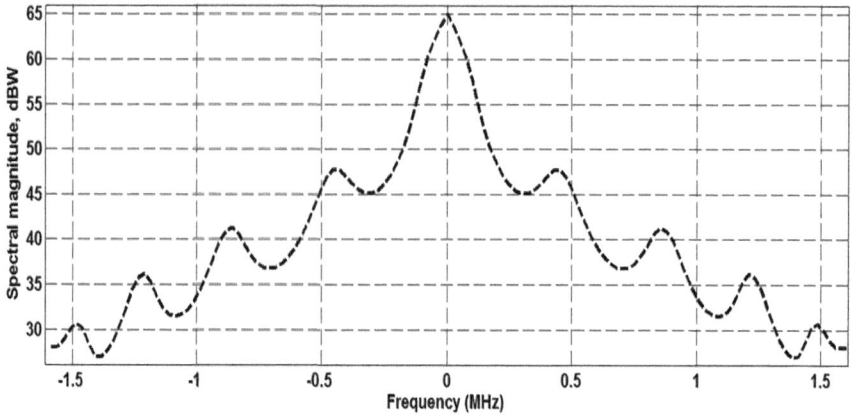

Fig. 7.20 Frequency spectrum of the current burst at the aircraft entry point for metallic airframe

7.8   Swept Stroke Mechanism

From the attachment points of a lightning flash leader arises the phenomenon called the swept stroke, the dragging along of the attachment point due to the movement of the aircraft. For an aircraft in flight, the attachment point on the surface may be swept, or dragged, over some distance from the initial attachment point along the aircraft structure. During the sweeping process, the lightning channel can reside along the surface. However, due to the motion of the aircraft in flight creating an unstable aerodynamic flow, the arc root can be swept along the surface and even jump establishing a new attachment point. The arc root may either dwell at the same spot and follow the fuselage displacement or continuously sweep over the fuselage for small distances. In both cases, the result is a large deformation of the lightning arc and an increase in the electric field in the air gap between the channel and the fuselage. The physical impact of lightning attachment includes vaporization, dielectric breakdown in the material, Joule heating, magnetic and acoustic forces, thermal and radiative fluxes.

There is an unstable air flow in the lightning arc along its surface. At the aircraft surface, it is estimated that the duration of a typical lightning arc flash is of the order of 100 ms. This effect of sweeping can be a threat to flight safety if the airframe is of a non-metallic structure. The large return stroke current will remain for a period of time prior to its dissipation through the airframe. This could couple into the aircraft avionics and circuits through electromagnetic and resistive coupling, thus posing a threat to aircraft safety.

7.9   Metallic Versus Carbon Fiber Composite Aircraft

The threat of lightning-induced voltages and currents is not a major problem for aircraft with aluminum and other conductive airframes. This is because the lightning-induced currents/voltages can easily dissipate into the atmosphere through the conducting surface compared to the non-conducting airframes. However, today's modern aircraft coming off the assembly lines are making extensive use of composite materials. The increasing use of carbon fiber composite (CFC) in aircraft structures has been propelled by the possibility of low running costs in terms of fuel consumption due to weight reduction and fast speed with less flight hours in traveling. Other properties of CFC airframes are the high tensile strength and corrosion resistance. In military aircraft, the use of CFC is advantageous for fast maneuvering, which is important in combat missions. Further, the need for carbon emission reduction is another factor in aircraft industries for the adaptation of CFC airframes. With the Intergovernmental Panel on Climate Change (IPCC) reporting that by the year 2050 up to 15% of total greenhouse gas emissions could be caused by aviation, it is important to appreciate how the past, current, and future use of advanced materials and design could help prevent this scenario.

   The CFC airframes present difficulties in lightning protection effectively to dissipate electrical charges and currents from the airframe structures due to its low electrical conductivity ($\sim 10^4$ S /cm) compared to aluminum ($\sim 10^7$ S/cm). Despite the improvement in using conductive fillers or microgrid or nanostrand conductive meshes integrated into the carbon composite materials and even conductive paints, it does not completely shield off electromagnetic interference (EMI) due to lightning radiated electromagnetic pulses (LEMPs).

   The severity of lightning threat is heightened further for fly-by-wire aircraft made up of composite materials and equipped with the latest digital state of the art technologies. Fly-by-wire aircraft also may be easily destabilized by LEMP sending the aircraft spinning out of control. Airframes of electrically insulated carbon fiber/epoxy composites can be damaged, particularly at the entry point of a direct lightning strike, since they absorb the lightning-induced voltage and currents instead of conducting and dissipating them. Thus, the magnitude of peak current and voltage is capable of inducing a higher electric field along the surface since the time transient is short, which can have severe effects on the command, communication, and control systems of the aircraft.

   The summary of cloud-to-ground flash first return stroke currents and voltage pulses on the aircraft for aircraft at various altitudes is given in Figs. 7.21 and 7.22. The results show that both aircraft surface return stroke currents and voltages increase with increase in cloud sizes and with increase in aircraft altitudes. For ground to cloud (GC) flashes, the first return stroke currents and the voltages reached higher values than for cloud to ground (CG) flashes.

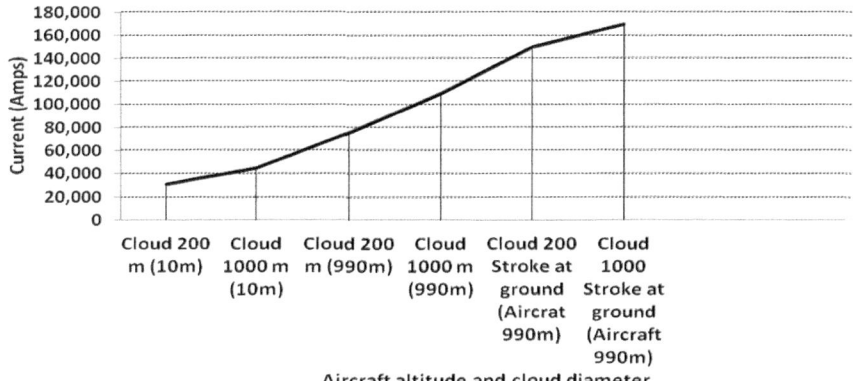

Fig. 7.21 Cloud to ground flash first return stroke currents at various altitudes (given in brackets) with varying charged cloud diameter

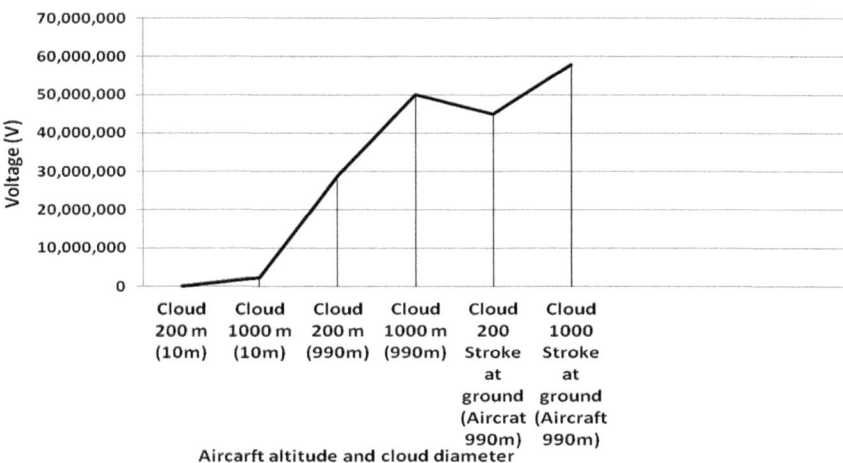

Fig. 7.22 Cloud to ground flash aircraft voltages at various altitudes (given in brackets) with varying cloud charges

7.10 Significance of Lightning Testing Standards and Certifications

7.10.1 Procedural Requirements

Lightning interaction with large power system apparatus (e.g. transformers and circuit breakers) and aircraft is still a complex phenomenon, made more complex with uncertainties surrounding climate change and its driving severe thunderstorms. Thus, it will be difficult to reproduce all its interactions and processes in a laboratory environment for power apparatus or aircraft body and lightning protection testing. The only way to demonstrate the resistance of structures and apparatuses to the lightning phenomenon is to model its direct and indirect effects, computer code them and perform computer simulation studies to design protection and shielding for various components and systems of components against these effects. The major task in the protection of electrical power systems, electronics, and aircraft against lightning threats is to identify systems and system components that are vulnerable to either direct and indirect effects.

The protection of mission critical systems against direct and indirect effects of lightning requires very stringent processes that have to comply with certification procedures set by various authorities. For an aircraft to be air worthy, for instance, it is the responsibility for aircraft manufacturers to provide the overall assurance for adequate lightning protections. This process requires certification plans for tests done on components or systems of components such as the airframes, power and electrical wirings and electronics, fuel systems and components, avionics and communication systems such as the radar, and other control and automation components. The certification plans for the tests are done either within the aircraft manufacturers' laboratories or the component suppliers' laboratories. In a nutshell, the protection of aircraft against lightning strike can be summarized in the following steps: Determining lightning attachment zone; building a lightning environment; finding out systems and components that are likely to be damaged by lightning; setting lightning protection standards for such identified systems and components; making protection design for those systems and components; confirming the rationality of the protection design by the use of laboratory or computer testbed tests.

7.10.2 Direct Lightning Effects Protection

The protection of aircraft against the direct effects of lightning strikes requires the protection of various systems or system of components that may be susceptible either to direct lightning attachment or to current flow between lightning attachment (entry and exit) points. For aircraft protection, components or subcomponents located in different sections of the aircraft are likely to experience different degrees of susceptibility to lightning. Some physical consequences of a lightning direct effect on an aircraft include: (i) dielectric puncture of skin covering electrically conducting elements, producing holes that can result in direct attachment of the lightning channel to the enclosed equipment, (ii) thermal convection from the ionized (electric plasma) lightning channel and the aircraft surface, (iii) exploding conductors due to lack of sufficient cross-sectional area to transfer the lightning current, (iv) resistive heating at the lightning arc contact point that can decompose the CFC resin, (v) thermal sparking at interface joints between two parts with insufficient cross-sectional areas to transfer the lightning current, (vi) voltage sparks due to induced voltages in the aircraft structure or wiring, and (vii) overpressure due to the explosion of the lightning channel, which leads to the propagation of a strong shock wave in the radial direction away from the arc. Thus, lightning zoning is a fundamental step in determining the appropriate lightning protection measures for aircraft in order to mitigate the threats due to such direct effects.

The major task for aircraft manufacturers and aerodynamicists when designing protection from direct effects of lightning is to identify points along the aircraft surface with the highest degree of lightning attachments. These are usually the extremities or the sharp points on the airframe structure . The extremities or sharp points are categorized into different zones. The zones will be dependent upon the aircraft's geometry, materials, and operational factors. Lightning attachment zones often vary from one aircraft type to another. However, the basic objective is to divide an aircraft body structure into zones of severity of potential lightning strike damage. The first step in locating the lightning strike zones is to determine the locations where lightning leaders may initially attach to an aircraft. Various methods are available to accomplish this task. They include such methods as similarity MEANING?, testing and analysis, the rolling sphere method, and the electromagnetic method. In this book, in Chap. 2, we have presented a highly accurate thundercloud electrostatics computation method that may be used to determine the most likely points of lightning strike for aircraft and large ground systems and structures.

7.10.3 Indirect Lightning Effects Protection

Protection for aircraft against indirect lightning effects relates to the level of transients that would be induced within the wiring circuitry and related components. There are two types of transient levels: the transient control level and the equipment transient design level. The verification of certifications is done for wiring and interconnecting components. It is necessary to demonstrate that the actual transient levels (ATLs) in the wiring and interconnecting components do not exceed the transient control levels (TCLs). That is, these tests are intended to measure the Actual Transient Levels (ATL) induced into aircraft electrical wiring as a result of lightning attachment to an aircraft to ensure that the ATL does not exceed the wiring TCLs. In order to measure ATLs, simulated lightning currents are injected, and the resultant currents and voltages on the wiring inside the aircraft are recorded.

Demonstrating that aircraft equipment will tolerate equipment transient design levels (ETDLs) is another requirement. This sets qualification test levels for the systems and equipment. The ETDL test defines the voltage and current amplitudes and waveforms that the systems and equipment must withstand without any adverse effects. The ETDLs for a specific system depend on the anticipated system and wiring installation locations within the aircraft, the expected shielding performance of the wire bundles and structure, and the importance of the system.

Demonstrating that interconnected and operating systems will tolerate the applicable ETDLs without component damage or functional upset is also required. Another acceptable method of a certification plan is through verification by similarity, which is demonstrated by detailed comparison of drawings, parts lists, operating parameters, and installation details. Basically, the certification plan must show that the system of components including airframe or structures is identical to a previously certified system from a lightning-protection test. For an indirect effects verification, it must be shown that the TCLs and ETDLs and margins will remain similar to those of previously identical components or systems of components.

7.10.4 Scaling Test Method

This method appears at the end of the development of the aircraft unlike the TCLs and EDTL, which are usually done at the early stage of the aircraft development. The scaling test method is simply performing a test on a scaled model of an aircraft under high voltage. A full-scale model can also be tested on with lightning current injections on the equipped aircraft to define the actual transient level (ATL) through measurement of transfer functions. A "worst case" lightning strike is simulated at each entry and exit point and measurement of current and/or voltage is conducted on the cables and wiring systems of the aircraft. However, with a full-scale model, the intensity of the current is less than that imposed by standardized documents due to the difficulty of generating current with a fast-rising front within the laboratories. Further, aircraft manufacturers are often reluctant to inject severe current pulses in a full-scale test (due to the high cost of possible damage to aircraft). Thus some normalization of a standardized waveform is applied. With these limitations and high costs, the development of exactly modeled and computer-coded lightning electrostatics and electrodynamics tests as presented in this book becomes a crucial challenge for the future.

7.10.5 Measurements on Aircraft Struck by Lightning in Flight

Several measurements were undertaken on board variously sized aircraft in flight within the vicinity of lightning and or within the thundercloud. A number of these are documented for four different aircraft, which were equipped with sensors and cameras to observe and analyze the process involved in the lightning phases. On one of the aircraft, the Transall C160 aircraft, which is a cargo and tactical transport aircraft used by the military, the breakdown electric field of 760 kV/m occurred at the ambient electric field of 75 kV/m, at an altitude of 4.6 km, with the electric charge accumulated being −0.84 C. The variation of the current to the change in the electric field showed a decrease in the electric field, which occurs upon the attachment of the aircraft leader connecting the cloud leader initiating the first return stroke current. The first return stroke current reached a peak of about 40 kA. The net negative charge increases on the aircraft due to the removal (neutralization) of positive electric charge by the propagating positive leader, and the field enhancement on the aircraft increases due to the increasing length of the overall conducting system of aircraft plus positive leader.

The increase in negative charge on the aircraft produces an increase in the electric field pointing towards the aircraft surface at all points on the surface. A few milliseconds after the initiation of the positive leader, the electric field value on the aircraft necessary for launching a negative leader is reached. The negative leader develops from an opposite extremity of the aircraft and propagates in a direction opposite to both the ambient electric field and the direction of extension of the positive leader. The negative leader development serves to reduce the negative charge on the aircraft leading to a reduction in the electric field pointing towards the aircraft surface.

7.10.6 Airport Lightning Protection

Protection of an airport requires threefold lightning protection systems protected. These are, firstly, the runway. The runway lights are necessary for an aircraft to make safe and correct landing during night time, when the pilot cannot see the runway without lightning. Damaging voltage surges may hit the runway lamps when lightning strikes nearby ground creating underground voltage surges, or through the electric energy supply system and the underground cables connecting the lamps. This requires a surge protection device for every single lamp. Runway lamps are placed on either side of the runway. Secondly, adequate lightning protection must be provided for the communication systems used for the control tower to aircraft communication system. This is similar to lightning protection systems used for rooftop communication and television receiver systems. Moreover, the instrument landing systems (ILS), or microwave landing systems (MLS), radar antenna and systems placed at the end of the runway as well as by the side of the runway to safely guide the aircraft to the runway when landing under all weather conditions, including under storm conditions and when there is no visibility. The third lightning protection system is for the lightning protection of the control tower and internal lightning protection systems for electronic circuits and equipment inside the airport buildings.

7.10.7 Lightning Engineering: The Present and the Future

The book has sought to collect together the present knowledge on lightning, the interaction of lightning with structures and systems, both ground and airborne, and lightning protection of structures and systems. In Fig. 7.23a, an image of a lightning flash during a severe storm is shown. Lightning activity is closely tied to climate

Fig. 7.23 **a** The lightning flash is closely tied to other severe weather conditions, such as tornadoes and hurricanes. Credit: NASA, USA. **b** Supercell thunderstorm. Credit: NSSL_NOAA, USA

(e.g. temperature) change and other severe storms such as tornadoes (normally a few hundred feet in diameter, formed during supercell thunderstorms) and hurricanes (typically hundreds of miles in diameter). Figure 7.23b shows a highly organized supercell thunderstorm (as opposed to small single-cell thunderstorms) which may last for 1 h, which is associated with large and violent tornadoes.

The major part of lightning activity takes place close to the equatorial line where the earth surface temperature is maximum. With climate change and increase in the earth surface temperature, lightning activity will change, and the electrical power of the lightning flash is expected to be more severe and destructive. Moreover, as observed in Chap. 1, lightning activity over smart cities with vast, complex electrical and electronic systems, has increased in frequency. This will raise new challenges and changes that will need to be introduced to address lightning protection of ground and airborne structures and systems. Hence, all the different kinds of lightning flashes (shown in Fig. 7.24a) will continue to demand more research, understanding, prediction, and the development of engineering systems to protect against the adverse effects of lightning. Although the maximum amount of lightning activity used to take place in the Congo, in Africa, in the 2000s, the lightning activity center has moved to Lake Maracaibo in Venezuela of South America. At Lake Maracaibo, the number of lightning strikes per km^2 per year is 235. Figure 7.24b shows lightning flashes over Lake Maracaibo.

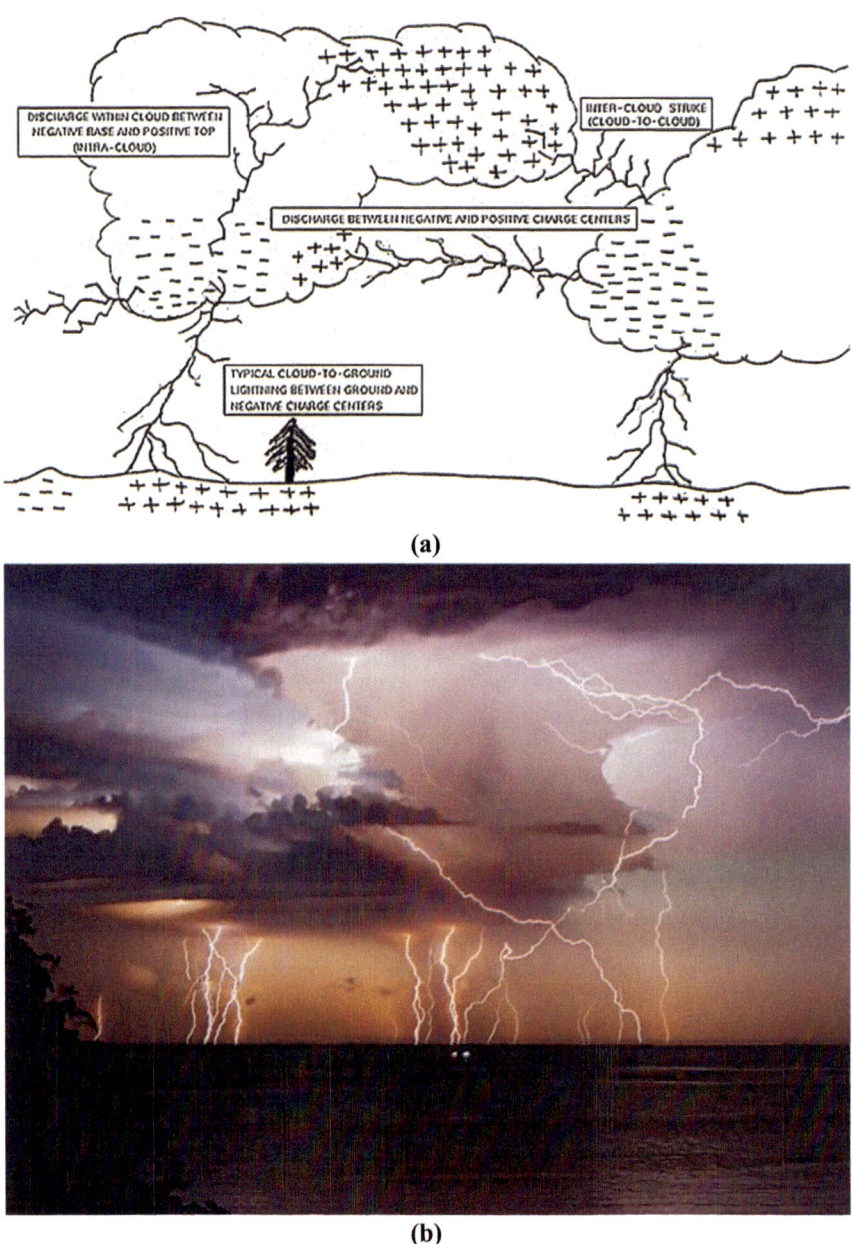

(a)

(b)

Fig. 7.24 **a** Cloud-to-Ground, Cloud-to-Cloud, Cloud-to-Air, Intracloud lightning flashes. **b** Intense lightning activity over Lake Maracaibo in Venezuela. Credit: NASA, USA

Bibliography

Abdelal, G.F., Murphy, A.: Nonlinear numerical modelling of lightning strike effect on composite panels with temperature dependent material properties. Compos. Struct. (2013). https://doi.org/10.1016/j.compstruct.2013.11.007

Airbus A380, Aircraft Characteristics Airport and Maintenance Planning, Airbus Technical Data and Support and Services, December (2014)

Anderson, J.G.: Lightning Performance of Transmission Lines. Transmission Line Reference Book 345 KV and Above, Electric Research Council and Electric Power Research Institute, Palo Alto (1982)

Anderson, J.G.: Transmission Line Reference Book 345 KV and above. Electric Research Council and Electric Power Research Institute, Palo Alto (1982)

Arevalo.L.: Numerical simulations of long spark and lightning attachments. A Ph.D. thesis submitted at Uppsala Universitet, Sweden (2011). https://ttu-ir.tdl.org/ttu-ir/bitstream/handle/2346/59453/31295004771985.pdf?sequence=1

Baba, Y., Rakov, V.A.: Electromagnetic models of the lightning return stroke. J. Geophys. Res. **112**, D04102 (2007). https://doi.org/10.1029/2006JD007222

Bane, S.P.M.: Spark ignition: Experimental and numerical investigation with application to aviation safety. A Ph.D. thesis submitted at California Institute of Technology, USA (2010). http://thesis.library.caltech.edu/5868/1/thesis_SBane.pdf

Baumgaertner, A.J. G., Lucas, G.M., Thayer, J. P., Mallios, S.A.: On the role of clouds in the fair weather part of the global electric circuit. Atmos. Chem. Phys. **14**, Open Access, 8599–8610 (August 2014). https://doi.org/10.5194/acp-14-8599-2014

Black, S.: Lightning strike protection strategies for composite aircraft. Composite World. http://www.compositesworld.com/articles/lightning-strike-protection-strategies-for-composite-aircraft

Blaj, M.A.: Protecting Electronic Equipment in Composite Structures Against Lightning. Ph.D. dissertation, Center for Telematics and Information Technology, University of Twente,, Netherlands (2015). http://doc.utwente.nl/96062/1/thesis_A_Blaj.pdf

Bitner, K.: Modeling of the lightning return stroke current at a tall structure using the derivative of the Heidler Function. Master's Thesis submitted at the Ryerson University, Canada (2002). Accessed from http://digital.library.ryerson.ca/islandora/object/RULA%3A1986/datastream/OBJ/download/Modeling_of_the_Lightning_Return_Stroke_Current_at_a_Tall_Structure_Using_the_Derivative_of_the_Heidler_Function.pdf

Blaj, M.A.: Protecting Electronic Equipment in Composite Structures Against Lightning. Ph.D. dissertation, Center for Telematics and Information Technology, University of Twente,, Netherlands (2015)

Cairo, F.: Airborne measurement program. In: The AGILE 10th Science Workshop on Lightning, Terrestrial Gamma-Ray Flashes, and Meteorology, Rome (April 2012). http://www.asdc.asi.it/10thagilemeeting/participants.php

Chippendale, R.D.: Modelling of the Thermal Chemical Damage Caused to Carbon Fibre Composites. A Ph.D. Thesis submitted at the University of Southampton, (July 2013)

Boer, R.K., Smeets, M.: Lightning Strike behavior of thick walled composite parts. National Aerospace Laboratory Report No: NLR-TP-2013-401 (2013). http://reports.nlr.nl:8080/xmlui/bitstream/handle/10921/917/TP-2013-401.pdf?sequence=1

Chishom, W.A.: Shielding Failure and Stroke Flashover. https://www.inmr.com/shielding-failures-stroke-flashovers-2/

Cobham Lightning Testing Service (2006, July 6), High voltage lightning testing of antenna, radome, and other structures. http://www.cobham.com/communications-and-connectivity/antenna-systems/specialist-technical-services-and-software/lightning-testing-services/high-voltage-lightning-testing/

Cooley, W.W., Shortess, D.L.: Lightning Simulation Test Techniques Evaluation. A Technical Report for US Department for Transportation, Federal Aviation Administration, Report No. DOT/FAA/CT-87/38 (1988). http://www.dtic.mil/dtic/tr/fulltext/u2/a200916.pdf

Cooray, V., Rakov, V.: Engineering lightning return stroke models incorporating current reflection from ground and finitely conducting ground effects. IEEE Trans. EMC **53**(3), 773–781 (2011)

Cooray, V., Becerra, M.: On the velocity of lightning upward connecting positive leaders. In: IX International Symposium on Lightning Protection, Foz Do Iguacu, Brazil (Nov 2007)

Dargi, M., Rupke, E., Wiles, K.: Certification of lightning protection for a full-authority digital engine control. NASA. Kennedy Space Center. In: International Aerospace and Ground Conference on Lightning and Static Electricity, vol. 2, pp.1–12, USA (1991). https://ntrs.nasa.gov/arc hive/nasa/casi.ntrs.nasa.gov/19910023403.pdf

Dawson, D.: Prseus update: hybrid wing body passenger cabin takes shape. Composite World (2015). http://www.compositesworld.com/articles/prseus-update-hybrid-wing-body-passenger-cabin-takes-shape

De Boer, A.I., et al.: In-flight lightning damage assessment system (ILDAS): initial in-flight lightning tests and improvement of the numerical methods. In: ICLOSE Conference, Oxford, U.K., September 6th-8th (2011)

De Boer, A.I., et al.: Inflight Lightning measurement system ILDAS: system architecture definition, based on analysis of first flight data. In: 23rd SFTE-EC Symposium, Amsterdam, Holland, 11–13th June (2011)

Deng, D.: Carbon fiber electronic interconnects. Ph.D. dissertation, Department of Mechanical Engineering, University of Maryland (2007). http://drum.lib.umd.edu/bitstream/handle/1903/6997/umi-umd-4508.pdf;jsessionid=390C062928AF333BD46AAC4DC30D228A?sequence=1

Deng, Y., Wang, Y., et al.: Improved electrogeometric model for EHV and UHV transmission lines developed through breakdown testing for long air gaps. Energ. J. **10**(10) (2017). https://doi.org/10.3390/en100030333

Deo, R.B., Jr. Starnes, J.H., Holzwarth, R.C.: Low-cost composite materials and structures for aircraft applications. A Paper presented at the RTO AVT Specialists' Meeting held in Loen, Norway, 7–11 May 2001, and published in RTO-MP-069(II). http://docshare01.docshare.tips/files/5085/50851653.pdf

Electro Magnetic Applications Inc.: Lightning certification of transport aircraft (2017). http://www.ema3d.com/lightning/. (2001)

Dong, Z., Yijun, Z., Luwen, C.: Climate lightning activity and its correlations with meteorological parameters in South China. In: XV International Conference on Atmospheric Electricity, pp. 1–14, June (2014)

Electrical knowhow: Course Lightning-2: Lightning Protection System Design and Calculations Design (2013). http://www.electrical-knowhow.com/2014/02/Introduction-to-Design-Calculations-of-Lightning-Protection-Systems.html

F-16 Fighting Falcon-Military Aircraft, (2015, February 18). Accessed from http://fas.org/man/dod-101/sys/ac/f-16.htm

FAS-Military Analyst Network (2000, March 05), F-16 Fighting Falcon (2000). https://fas.org/man/dod-101/sys/ac/f-16.htm

Fisher, F.A., Plumer, J.A., Perala, R.A.: Lightning protection of Aircraft, 2nd edn. Lightning Technologies Inc. (1999)

Fisher, F., Taeuber, R., Crouch, K.: Implications of a recent lightning strike to a NASA jet trainer. In: AIAA 26th Aerospace Sciences Meeting, AIAA Paper 88–0394 Reno USA (1988)

Fisher F., Plumer, J.A, Perala R.A.: .Aircraft lightning protection Handbook Report No. DOT/FAA/CT-89/22, U. S. Department of Transportation Handbook Federal Aviation Administration Technical Centre, Atlantic City, Sept. (1989)

Fisher, F.P., Perala, R.A.: Lightning protection of Aircraft, 2nd edn. Lightning Technologies Inc.(1999)

Fisher, J., Hoole, P.R.P., Pirapaharan, K., Hoole, S.R.H.: Applying a 3D dipole model for lightning electrodynamics of low-flying aircraft. IETE J. Res. (2015) https://doi.org/10.1080/03772063.2014.986543. (Jan. 2015)

Fisher, J., Hoole, P.R.P., Pirapaharan, K., Hoole S.R.H.: Parameters of cloud to cloud and intra-cloud lightning strikes to CFC and metallic aircraft structures. In: Proceedings International Symposium on Fundamentals of Engineering, IEEE Explore Digital Library, pp. 1–6, Jan (2017). https://doi.org/10.1109/ISFEE.2016.7803216

Fisher, J., Hoole, P.R.P., Pirapaharan, K., Hoole, S.R.H.: Pre-lightning strikes and aircraft electrostatics. In: International UNIMAS Science, Technology, Engineering Conference, Universiti of Malaysia, 24th-26th October (2016)

Fuller C (2016) Going the distance: Aircraft and lightning, how far is far enough? In: IEEE International Symposium on Electromagnetic Compatibility (EMC), vol. 2016, pp. 588–593

Gamerota, W.R., Elismé, J.O., Uman, M.A., Rakov, V.A.: Current waveforms for lightning simulations. IEEE Trans. Electromagn. Compat. 54(4), 880–888 (August, 2012)

Goodloe, C.C.: Lightning protection guidelines for aerospace vehicles. A report Written for NASA, Marshall Space Flight Center, Report No: NASA/TM -1999-209734, (1999). https://ntrs.nasa.gov/archive/nasa/casi.ntrs.nasa.gov/20000004589.pdf

Golding, W.L.: Lightning strikes on aircraft: How the airlines are coping. J. Aviat. Aerosp. Educ. Res. 15(1), 41–50 (2005)

Gomes, C., Cooray, V.: Electromagnetic transients in radio/microwave bands and surge protection devices. Prog. Electromagn. Res. 108, 101–130 (August, 2010)

Gouda, O.F., Den, A.Z.E., Amer, G.M.: Parameters affecting the back flashover across the overhead transmission line insulator caused by lightning. In: Proceedings of the 14th International Middle East Power Systems Conference (MEPCON'10), Cairo University, Egypt, December 19–21 (2010)

Gouda, O.F., Den, A.Z.E., Amer, G.M.: Parameters affecting the back flashover across the overhead transmission line insulator caused by lightning. In: Proceedings of the 14th International Middle East Power Systems Conference (MEPCON'10), Cairo University, Egypt, December 19–21 (2010)

Grisby, L.L: Electric Power Engineering Handbook: Power Systems, 2nd edn. Taylor & Francis Group, LLC, CRC Press (2006)

Guemes, J.A., Hernando, F.E.: Method for calculating the ground resistance of grounding grids using FEM. IEEE Trans. Power Deliv. 19(2), 595–600, April (2004)

Ha, M.S., Kwon, O.Y., Choi, H.S.: Improved electrical conductivity of CFRP by conductivity silver nano-particles coating for lightning protection. In: The 17th International on Composites Materials Proceedings, Edinburgh, Scotland (2009). http://www.iccm-central.org/Proceedings/ICCM17proceedings/Themes/Industry/AEROSPACE%20APPLICATIONS/INT%20-%20AEROSPACE%20APPLICATIONS/IA2.3%20Ha.pdf

Happ, F., Brüns, H.D., Mavraj, G., Gronwald, F.: Numerical computation of lightning transfer functions for layered, anisotropically conducting shielding structures by the method of moments. Adv. Radio Sci. 14, 107–114 (2016). http://www.adv-radio-sci.net/14/107/2016/ars-14-107-2016.pdf

Heidler, F., Flisowski, Z., Zischank, W., Bouquegneau, Ch., Mazzetti, C.: Parameters of lightning current given in IEC 62305-background, experience and Outlook. In: 29th International Conference on Lightning Protection, Uppsala, Sweden, 23rd-26th June (2008)

Heidler, F., Flisowski, Z., Zischank, W., Bouquegneau, Ch., Mazzetti, C.: Parameters of lightning current given in IEC 62305-background, experience and Outlook. In: 29th International Conference on Lightning Protection, Uppsala, Sweden, 23–26 June (2008). http://www.iclp-centre.org/pdf/Invited-Lecture-3.pdf. (25 September, 2011)

Hess R.: The electromagnetic environment. In: The Avionics Handbook. CRC Press LLG (2001)

Heidler F, Flisowski Z, Zischank W, Bouquegneau Ch. Mazzetti C.: Parameters of lightning current given in IEC 62305-background, experience and Outlook. In: 29th International Conference on Lightning Protection, Uppsala, Sweden, 23–26 June 2008. http://www.iclp-centre.org/pdf/Invited-Lecture-3.pdf

Hess R.: The electromagnetic environment. In: The Avionics Handbook. CRC Press LLG (2001)

Hoole, P.R.P.: Modelling the lightning earth flash return stroke for studying its effects on engineering systems. IEEE Trans. Magn. **29**(2), 732–740, March (1993)

Hoole, P.R.P., Pirapaharan, K., Hoole, S.R.H.: Handbook of Engineering Electromagnetics. WIT Press, UK (2013)

Hoole. P.R.P., Hoole, S.R.H.: Guided waves along an unmagnetized lightning plasma channel. IEEE Trans. Magn. **24**(6), 3165–3167 (1988)

Hoole, P.R.P., Hoole, S.R.H.: A distributed transmission line model of cloud-to-ground lightning return stroke: Model verification, return stroke velocity, unmeasured currents and radiated fields. Int. J. Phys. Sci. **6**(16), 3851–3866, 18 August, (2011)

IEEE Guide for Improving the Lightning Performance of Transmission Lines, IEEE Std. 1243-1997. The IEEE Working Group on Estimating the Lightning Performance of Overhead Transmission Lines (1997)

Jalali M.: Improving electromagnetic shielding with metallic nanoparticles. Doctoral Dissertation, Concordia University, June (2013)

Jones, C.H., Rowse, D., Odam, G.: Probabilities of catastrophe in lightning hazard assessments. Paper No 2001-01-877. In: International Conference on Lightning and Static Electricity, Seattle, USA (2001)

Kalanchiam, M., Chinnasamy, M.: Advantages of composite materials in aircraft structures. Int. J. Mech. Aerosp. Ind. Mechatron. Manuf. Eng. **6**(11), 2428–2432 (2012). http://waset.org/public ations/5121/advantages-of-composite-materials-in-aircraft-structures

Kendall, C., Black, E., Larsen, W.E., Rasch, N.O.: Aircraft generated electromagnetic interference on future electronic systems. Federal Aviation Administration Report No: DOT/FAA/CT-83/49, December (1983)

King, D., Inderwildi, O., Carey, C.: Advanced aerospace materials: past, present and future. A document on Aviation and the Environment of the Oxford University's Smith School of Enterprise and the Environment. http://www.chriscarey.co.uk/a&e_materials.pdf

Lago, F.: Lightning in aeronautics. J. Phys. Conf. Ser. **550** (2014) 012001 (2014). https://doi.org/10.1088/1742-6596/550/1/012001/pdf

Lalonde, D., Kitaygorsky, J., Tse, W., Brault, S., Kohler, J., Weber, C.: Computational electromagnetic modelling and experimental validation of fuel tank lightning currents for a transport category aircraft. In: International Conference on Lightning and Static Electricity, Toulouse, France (2015). http://www.ema3d.com/downloads/ICOLSE_2015_MODELING_EXPERIMENTAL_VALIDATION_OF_FUEL_TANK_LIGHTNING_CURRENTS_FOR_AIRCRAFT.pdf

Laroche, P., Blanchet, P., Delannoy, A., Issac, F.: Experimental studies on lightning strikes to aircraft. J. Aerosp. Lab. Light. Hazards Aircr. (5) ALO5–06, 1–10 (Dec. 2012). http://www.aerospacelab-journal.org/sites/www.aerospacelab-journal.org/files/AL05-06_0.pdf

Maradei, F.: Numerical simulations for EMC compliance and standard development: a new trend in European standards. In: Emerging Technologies in EMC Workshop, Buenos Aires, Argentina (April, 2011). https://www.inti.gob.ar/electronicaeinformatica/emc/pdf/1-Maradei.pdf

McDonald, J.D.: Electric Power Engineering Handbook: Electric Power Substations Engineering, 3rd edn. Taylor & Francis Group LLG, CRC Press (2012)

Michimoto, K., Uda, H.: A prevention method of lightning strikes to aircrafts. In: XV International Conference on Atmospheric Electricity, 15th-20th June (2014)

Miranda, F.J., Jr. Pinto, O., Saba, M.M.F.: Occurrence of characteristic pulses in positive ground lightning in Brazil. In: 19th International Lightning Conference and 1st International Conference on Lightning Meteorology, Tucson, U.S.A, April (2006)

Mrazova, M.: Advanced composite materials of the future in aerospace industry. INCAS Bull. **5**(3), 139–150 (2013). https://doi.org/10.13111/2066-8201.2013.5.3.14

Naccarato, K.P., Campos, D.R., Meireles, V.H.P.: Lightning urban effect over major large cities in Brazil. In: XV International Conference on Atmospheric Electricity, Norman, Oklahoma, USA, 15–20 June (2014)

Nag, A., Rakov, V.A.: A transmission-line–type-model for lightning return strokes with branches. In: XV International Conference on Atmospheric Electricity, Norman, Oklahoma, USA, 15th-20th June 2014. Accessed from http://www.nssl.noaa.gov/users/mansell/icae2014/preprints/Nag_200. pdf

NASA Glenn Research Centre, Wing area (2017). https://www.grc.nasa.gov/WWW/k-12/Virtua lAero/BottleRocket/airplane/area.slide.html

NOAA, 'Lightning Safety': US Department of Commerce and National Oceanic and Atmospheric Administration, and Office of the National Weather Service (1995). http://www.srh.noaa.gov/jet stream/lightning/lightning_safety.html

NPFA, Standard for the Installation of Lightning Protection System, NPFA-780 (2020)

Parmantier, J.P., Issac, F., Gobin, V.: Indirect effects of lightning on aircraft and rotorcraft. J. Aerosp. Lab. Light. Hazards Airc. Launchers, AL05 10, (5), 1–27 (2012). http://www.aerospacelab-jou rnal.org/sites/www.aerospacelab-journal.org/files/AL05-10_0.pdf

O'Loughlin, J.B., Skinner, S.R.: General aviation lightning strike report and protection level study. Technical Report for US Department for Transportation, Federal Aviation Administration, Report No. DOT/FAA/AR-04/13 (2013). http://www.tc.faa.gov/its/worldpac/techrpt/ar04-13.pdf

Peters, M., Leyens, C.: Materials science and engineering. J. Aerosp. Space Mater. 3 (2009). (Encyclopedia of Life Support Systems (EOLSS)). http://www.eolss.net/sample-chapters/c05/e6-36-05-03.pdf

Prabhakaran, R.: Lightning strike on metal and composite aircraft and their mitigation. In: International Conference on Intelligent Design and Analysis of Engineering Products, Systems, and Computation, Coimbatore, India, pp. 208–221, July (2010)

Price, C.: Thunderstorm, lightning and climate change. In: 29th Internation Conference on Lightning Protection Conference, Uppsala, Sweden, June (2008)

Ranjith R, Myong, R.S., Lee, S.: Computational investigation of lightning effects on aircraft components. Int. J. Aeronaut. Space Sci. 15(1), 44–53, March (2014)

Rakov, V.A.: Lightning phenomenology and parameters important for lightning protection. In: IX International Symposium on Lightning Protection, 26th–30th November, Foz do Iguacu, Brazil (2007)

Rakov V.A.: Lightning Discharge and Fundamentals of Lightning Protection. J. Light. Res. 4(Suppl. 1: M2), 3–11, June (2012)

SAE ARP 5412 Committee Report, Aircraft Lightning Environment and Related Test Waveforms. April (2012)

Rakov, V.A., Uman, M.A.: Review and evaluation of lightning return stroke models including some aspects of their application. IEEE Trans. Electromagn. Compat. 40(4), 403–426 (Nov. 1998). https://pdfs.semanticscholar.org/59a8/de4e2f6a6b08c83ef3cd9ee333f3bc1ddbee.pdf

Schoene, J.D.: Direct and nearby lightning strike interaction with test power distribution lines. Ph.D. Dissertation, Department of Electrical and Computer Engineering, University of Florida (2007)

Smith, F.: The use of CFC in aerospace: past, present, and future challenges. A presentation document by Avalon Consultancy Services Ltd. (2013). https://avaloncsl.files.wordpress.com/2013/01/avalon-the-use-of-composites-in-aerospace-s.pdf

Space News–Aeronautical Engineering: Wing construction and their components (2013). http://lea rnsmartengg.com/wing-contructiontheir-components/

Qie, X., Jian, R., Laroche, P.: Triggering lightning experiments: an effective approach to the research of lightning physics. J. Aerosp. Lab. (5) ALO5–05, 1–12 (December 2012). http://www.aerosp acelab-journal.org/sites/www.aerospacelab-journal.org/files/AL05-complete-issue.pdf

Teng, X., Liu, G., Yu, Z., Zhuang, X., Zhao, Y.: Research on initial lightning attachment zone for aircraft. In: 3rd International Conference on Electric and Electronics, Hong Kong (December 2013)

Thirukumaran, S., Harikrishnan, R., Hoole, P.R.P., Jeivan, K., Pirapaharan, K., Hoole, S.R.H.: A new electric dipole model for lightning-aircraft electrodynamics. J. Comput. Electr. Eng. (COMPEL) 33, 540–555 (2014)

Thirukumaran, S., Hoole, P.R.P., Harikrishnan, R., Kanesan, J., Pirapaharan, K., Hoole, S.R.H.: Aircraft-lightning electrodynamics using the transmission line model part I: review of the transmission line model. Progress Electromagn. Res. M **31**, 85–101 (May, 2013). https://doi.org/10.2528/PIERM12110303. http://www.jpier.org/pierm/pier.php?paper=12110303

Thirukumaran, S., Hoole, P.R.P., Harikrishnan, R., Jeevan, K., Hoole, S.R.H.: Electrostatic Discharges (ESD): Rate of Rise of Currents and Radiated Electric Fields. In: Progress In Electromagnetics Research Symposium, (March 2012)

Tooren, M., Krakers, L.: Multi-disciplinary design of aircraft fuselage structures. In: 45th AIAA Aerospace Sciences Meeting and Exhibit, Aerospace Sciences Meetings USA (2007). https://doi.org/10.2514/6.2007-767

Uman, M.A.: A comparison of natural lightning and the long laboratory spark with application to lightning testing. A report prepare for Department of Federal Aviation Administration, Naval Aviation Facilities Experimental Centre, Report No: NA-69-27, August (1970)

Upadhya, A.R.: Manufacturing and certification of composite primary structures for civil and military aircrafts. A presentation at ICAS Biennial Workshop on Advanced Materials and Manufacturing–Certification & Operational Challenges, Stockholm, Sweden (December 2011). http://www.icas.org/media/pdf/Workshops/2011/ICAS%20Workshop%20presentation%2006%20Upadhya.pdf

West, L.: Lightning induced waveforms 4 and 5A in composite airframes, the inability of copper braid to shield it, and a new layered copper braid and high-mu foil shield. Interaction Note 608 Revision A, (April 2011). http://ece-research.unm.edu/summa/notes/In/IN616.pdf

Willet, J., Le Vine, D.M., Idone, V P.: Lightning return-stroke current waveforms aloft, from measured field change, current, and channel geometry. J. Geophys. Res. Atmos. (October, 2006). https://ntrs.nasa.gov/archive/nasa/casi.ntrs.nasa.gov/20070018241.pdf

Zhang, Y., Ma, M., Weitao, L., Tao, S.: Review on climate characteristics of lightning activity. Acta Meteor. Sinica **24**(2), 137–149 (2010)

Zwemmer, R., et al.: In-flight lightning damage assessment system (ILDAS): Results of the concept tests. In: ICLOSE Conference, Pittsfield, U.S.A., September 15th–17th (2009)

Appendix
STAT2ARC2EMP: A Computer-Based High-Voltage Testbed for Electrostatic and Transient Current Threats to Ground and Airborne Structures and Equipment: For Arcs and Lightning Flashes

Abstract In Chaps. 2, 5, 6, and 7 of this book, for the first time a unified approach, termed STAT2ARC2EMP, to handling the electrostatics, arc or lightning currents and radiated electromagnetic pulses associated with high voltage breakdown in electric power systems or lightning generated electrostatic stress, surge currents, and near to far electromagnetic pulses (LEMP). These are critical for break down prediction, high-voltage equipment and structure analysis, design, and protection of both high-voltage apparatuses and microelectronic systems and devices. STAT2ARC2EMP stands for statics-to-arc/lightning-to-electromagnetic pulse radiation. The initial production of high electrostatic field stress due to thunderclouds or high-voltage apparatuses in a digital substation with many microelectronic systems is modeled and determined by using a dipole electric charge simulation method. This is a very useful and necessary computational tool for many high voltage-related and pre-lightning strike analyses and designs of equipment and systems. In the next stage, the STAT2ARC2EMP routine uses the electric fields calculated to determine the electric network parameters, and in particular, the capacitances that store electrostatic energy and discharge it destructively once the path of the initial electric discharges and leader is completed. The second stage uses an electric circuit technique to determine the transient electric voltages and currents that are generated. In lightning, the large transient current is called the return stroke current. The third stage of the STAT2ARC2EMP techniques uses the electric currents in the arc or lightning channel to calculate the electromagnetic pulses radiated. An analytical solution for the fields radiated from a finite length of an antenna is used to determine the radiated electromagnetic fields. The results are summarized for all three stages. This is a unified approach to high voltage breakdown and lightning phenomena which without using a strong scientific, self-contained approach to study accurately the high-voltage break down phenomena from electrostatic fields to transient high currents to radiation of electromagnetic pulses that pose indirect threats.

© The Editor(s) (if applicable) and The Author(s), under exclusive license to Springer Nature Switzerland AG 2022
P. Hoole and S. Hoole, *Lightning Engineering: Physics, Computer-based Test-bed, Protection of Ground and Airborne Systems*,
https://doi.org/10.1007/978-3-030-94728-6

289

Introduction

As the aerospace industry expands into both manned and unmanned commercial and military vehicles, preventing electric field enhanced aircraftinitiated lightning strikesand protections againstserious damage and accidents becomes a major concern to the aerospace industry. When an aircraft flies into the environment of an electrified cloud, it enters into an enhanced electric field region surrounding the cloud, which in most cases has a large negative charge center in its lower region. The electric fields will induce an electric dipolecharge over the body of the aircraft, with positive electric charges on the top surface of the aircraft and negative electric charges over the underbelly of the aircraft, resulting in an electric dipolecharge structure. These can be sufficiently enhanced to result in electric discharges; for instance, resulting in positive leaders emanating from the radome of the aircraft. With this, at another extremity of the aircraft a negative leader may develop from electrostatic discharges occurring at another electric field enhanced part of the aircraft body. The negative leader will move towards the ground or another nearby thundercloud. It is important to determine the electric field enhanced areas of the aircraft in order to design preemptive measures to reduce lightning strikerisks, even to design and to maintain the aircraft to reduce electric field enhancements in these high-risk areas. Knowledge of the electric charges induced on the aircraft body and the electric field distribution is also essential to decide on the safe placement of sensitive microelectronic systems associated with aircraft measurement and navigational systems.

Although direct measurements of electric currents and electric fields have been made by flying instrumented aircraftinto thunderclouds and into thunderclouds and their vicinity, these are limited in scope because of the immense expense involved in carrying out these experiments, they being confined to specific thunderclouds which were investigated, as well as the limited number of events observed (e.g. a total of about 50 lightning strikes to a CV580 aircraft). Only about three types of aircraft have been used for these experiments. Measurements are made only in around five different locations on the aircraft. Seeing the need to develop a well-attested computational testbed for aircraft design, testing and protective measures on any type of aircraft at different locations and inclinations with reference to the thundercloud, work is being done to develop an electromagnetic testbed which is in the process of extension to several important aspects of aircraft-thundercloud interaction, both before and after a lightning strike. An electric dipole-based method was proposed and its results compared with laboratory-based measurements. It has been used to model the aircraft in great detail and further tests were carried out for both cloud to ground and cloud to cloudlightning strikes in which an aircraft becomes part of the lightning flash channel. Computer-based simulation studies are also being carried out on the electrostatic environment of the thundercloud and the electric charges and electric fields produced on the aircraft surface before the aircraft is struck by lightning.

In this book, specific observations were made on the prestrike electric field which is perpendicular to the aircraft body, the electric field which is parallel to the aircraft body that generates electric currents and the specific regions in which the induced electric charges are large resulting in electric fields that exceed the breakdown electric field strength of the air surrounding the aircraft. The reliable simulation and the development of simulation for certification continue to be areas of intense research and development.

Electric Field Induced on the Aircraft and Its Components

Consider the thundercloudcharge center as a sphere with charge Q and the aircraft at a distance R below the electric charge center. This electric charge will also be induced on the underside of the aircraft, with the sign reversed, thus resulting in a dipole electric charge produced on the aircraft by the thundercloudcharge center.

Consider now the horizontal electric field E_h along the surface of the aircraft. It will produce an electric current given by $J = \sigma E_h$, where σ is the conductivity of the aircraft body. For an aluminum body aircraft the conductivity is large, whereas for carbon compositeaircraft, the value is low, and the value of this current remains small. Very close to the cloud, with a large thundercloudelectric charge (e.g. 20 °C), significant surface current can flow along the aircraftor aircraft equipment (where shielding is poor as in the case of aircraft with a composite body), and may give rise to an electric field build up over a few milliseconds or so before theelectric breakdown commences on the aircraft body. However, in most cases where the aircraft is far away from the thundercloud (say, 500 m), it is the vertical electric field E_v that induces the electric charge on the aircraft surface leading to the breakdown and subsequent sharp, transient of currents.

Identifying Regions of Large Electric Fields and Induced Electric Charges

The entry of an aircraft into an ambient electric field can be regarded as a sudden introduction of a conductor into an electric field which intensifies the local electric fields. This enhances the local electric field buildup around the aircraft. The electric field enhancement will reach maxima along the aircraft extremities that are oriented towards the ambient fields. Typically, an ambient field of 100 kV/m at the radome could be enhanced to 1 MV/m; similarly at tail tips.

The electric field at the wing tips could rise to 400 kV/m, and to 200 kV/m at the tips of the turbo-engines. The charging of the aircraft produces a potential gradient between it and its surroundings. The potential gradient builds up to a sufficient level that corona discharge results. The corona discharges occur at the extremities of the aircraft and initiate a bidirectional leader that connects the cloud charge electrically to ground. Hence, there are two distinct phases to lightning-aircraft interaction. The first is the development of streamers and the leader which develops at the field enhanced parts of the aircraft. The second phase involves the high currents produced by the first and subsequent return strokes. The second phase therefore induces the high energy transient current pulse, subsequent re-strikes and the long duration of the slow currents.

In order to simulate the electric charges induced on the aircraft body, a large number of electric dipoles are placed on it, with more dipoles placed on the aircraft components that are more susceptible to electric field enhancement. The smoother surfaces, such as the aircraft wing surfaces, are assigned fewer electric dipoles. At high altitudes, electric field breakdown may occur at an electric field of about 400 kV/m. This value is much lower than the breakdown electric field of about 3000 kV/m at sea level. The charge calculation makes use of the coefficients of potential of the electric dipole charges and their mirror images on the ground to a selected point on the surface of the aircraft. This requires the distances of the dipoles and their images to a selected point on the surface of the aircraft.

Arcs and Lightning Flashes: Transient Phase

The lightning flash electric gas discharge channel may be modeled by a trasmission line, with its parameters determined from fundamental gas discharge physics and electrosttic theory. Moreover, the electric power lines struck by lightning, the lightning conductor and an aircraft may all be modeled using the electric circuit papramenters.

The near and intermediate radiated electromagnetic pulses may also be determined if the additional terms representing them are added on to the far away electric and magnetic fields. These radiated electromagnetic field equations form the third and the last module of the STA2ARC2EMP technique to determine the LEMP or arc generated electromagnetic fields that may be picked up by the smart antennas installed in a digital substation and used to detect corona and partial discharges for early warning systems in a power substation.

The Electrostatic Discharge (ESD) is a well-known threat to power apparatuses, aerospace vehicles, and digital electronic systems. As old power substations are converted into digital substations, the monitoring and control of the system apparatus depends increasingly on digital electronics sitting under high-voltage busbars and lines. An aircraft flying or parked under a thundercloud is also subjected to large electrostatic threats.

In Chaps. 2, 5, 6, and 7 of this book, a versatile and accurate method to determine the electrostatic field and electric charges produced by high-voltage sources on apparatuses and electronic circuits positioned under them. In aircraft, the electrostatic threat has increased the use of non-metallic, composite materials for the aircraft body. Moreover, the severe lightning flashesto aircraft also commence with ESD on the aircraft body that results in the initiation of positive leaders that grow towards the thundercloud from one part of the aircraft and a negative leader that is launched towards the ground or another cloud.

We discussed the induced electric charges due to the vertical electric field component of the thundercloud charge center, as well as the electric currents induced on the surface of the aircraft body or equipment by the horizontal component of the thundercloud generated electric field. Moreover, from electrostatic fields computed prior to the initiation of corona or the initial leader, we show that in addition to the most commonly identified part of the aircraft from which leaders are initiated, namely the radome, the main wing tips, the curved surface of the mid-wing and the stabilizer tips also experience highly enhanced electric fields that may lead to the generation of electric breakdown.

The thundercloud charge center induced electric fields that are perpendicular to the aircraft body is of greater importance than the electric fields parallel to the aircraft body with regard to the electric charges induced on the aircraft body. However, when the aircraft is very close to the thundercloud or inside the thundercloud, the parallel, or tangential, electric field may produce significant current flow and electric charge build up with time. Among the aircraft regions most susceptible to electric breakdown and enhanced electric fields are the radome, the stabilizer tip, and the wing tip. But other parts including the middle of the wing and fuselage also experience significantly large electric fields.

The STAT2ARC2EMP technique is a comprehensive, unique, computer-based testbed to perform analysis and design studies on electrostatic fields (STAT), producing arcs and lightning flashes (ARC), which in turn radiate electromagnetic pulses (EMP). It may be used for testing aircraft and ground-based high-voltage power systems equipment and substations. It has the versatility to simulate scenarios hard to capture in high-voltage laboratories, as well as to explore the future threats to airborne and ground systems due to climate change-related severity of electric charges generated in, for instance, the thunderclouds. The STAT phase allows a design engineer to explore threat reduction to not only high-voltage apparatuses but also to microelectronic boards designed to function in electrically harsh environments. The ARC phase allows one to explore lightning and arc generated voltage swings and current surges in power systems and on aircraft. The EMP phase of the technique allows one to compute near, intermediate, and far electric and magnetic fields generated by arcs and lightning flashes.

Index

A

Absorber, 34–36
Acoustic wave, thunder, 168
Action integral, 9, 20, 31, 167, 233, 234, 242, 243, 267, 269
Airbus A380, 66, 69–72, 74, 76, 77, 79, 231, 258–260, 269
Aircraft, 1, 3, 4, 7, 9, 14, 21, 22, 27, 32–34, 46, 51–53, 59–74, 77–79, 102, 158, 163, 171, 194, 195, 197–200, 204, 231, 233–235, 252–255, 257–267, 269–278, 288–291
Aircraft capacitance, 260, 261, 263, 265
Aircraft inductance, 263
Aircraft resistance, 258, 265
Airport, 3–5, 62, 209, 278
Airport lightning protection, 278, 288
Air termination, 38, 89, 94, 95, 97, 98, 101, 105, 112, 115, 130, 134, 136–144, 148, 149
Air-to-Air communication, 4
Animal, 32, 43
Antenna, 29, 89–91, 94, 122, 127, 129, 130, 140, 148, 183–185, 209, 211, 212, 223–227, 278, 287, 290
Apollo, 4
Array antenna, 222, 224–226
Arrester, 4, 34, 36, 88, 89, 97, 107, 125–129, 132, 245
Artificial Neural Networks (ANN), 223–227
Attractive radius expression, 251

B

Back flashover, 34, 36, 232, 235, 237, 238, 240, 241, 244, 245
Basic impulse insulation level, 252
Basic Impulse Level (BIL), 38, 236, 240, 252
Basic protection, 89
Bearings, 246
BIL, *see* Basic Impulse insulation Level
Biological effect, 28
Blades, 43, 101
Blue jets, 43, 45
Boats, 85, 94–96
Bonding, 38, 89, 91–93, 101, 103, 105, 113, 114, 121–124, 199
Boundaries, 26, 40, 56, 112, 118–121, 124, 126–129, 162
Branch, 3, 7, 22, 58, 158, 244, 248, 249
Breakdown, 3, 53, 57, 58, 63, 65, 67, 69–72, 74, 76–78, 236, 248, 249, 260, 272, 277, 287, 289–291
Buildings, 7–9, 22, 23, 27, 38, 42, 43, 45, 51, 53, 58, 86, 88–94, 96, 97, 101, 103–105, 108–112, 114–121, 127–134, 136–145, 147, 152, 153, 185, 198, 203, 233, 248, 275, 278

C

Cable connection, 3
Cable shielding, 153
Capacitor, 11, 92, 169
Carbon composite, 198, 269, 273, 289
Carbon fibre, 267, 269, 272, 273
CFO, *see* Critical Impulse Fashover

Channels, 1, 5, 12–16, 18–23, 26, 45, 58,
    59, 64, 158, 160, 162–165, 168, 169,
    171–175, 177, 178, 180–182, 184,
    185, 194, 197, 203, 210, 249, 254,
    255, 257, 258, 272
Channel terminations, 15
Charge, 4, 11, 13–15, 17, 20–22, 26, 30, 33,
    34, 43, 51, 53–59, 63–65, 67–71, 73,
    76–78, 160, 166, 169, 172–174, 181,
    182, 191, 194, 213, 233, 235, 236,
    242, 243, 248, 249, 251, 257,
    260–262, 265, 266, 269, 270, 273,
    274, 277, 278, 288–290
Circuit breakers, 35, 86, 92, 99, 100, 133,
    157, 249, 251, 274
City, 2, 3, 7, 9, 42, 45, 62, 241, 247
Climate change, 8, 9, 27, 43, 45, 46, 231,
    246, 272, 274, 279, 291
Cloud discharges, 43
Cloud to air, 51, 158
Cloud to cloud, 51, 288
Cloud-to-Ground (CG) lightning, 7, 61,
    134, 195, 249
Collisions, 13
Column, 16, 26, 58, 177
Common mode, 85, 91, 114
Communication systems, 1, 33, 36, 46, 151,
    224, 225, 242, 275, 278
Computer-based testbed, 22
Computer testbed, 275
Conductivity, 1, 12, 13, 19, 20, 25, 26, 29,
    32, 153, 168, 171–173, 182, 192,
    250, 257–259, 263, 265, 267, 273,
    289
Connecting leaders, 172
Continuing current, 20, 64, 159, 165, 169,
    181, 233
Control tower, 4, 278
Coordinated protection, 126
Corona, 13, 19, 26, 42, 56, 57, 59, 63, 65,
    171, 235–237, 289–291
Corrosion, 95, 101, 121, 123, 137, 147,
    152, 272
Critical electric field, 57
Cross-sectional areas, 275
Cumulus cloud, 53, 54
Current delay, 227, 267
Currents, 1–4, 9–14, 16, 18–27, 29–32, 34,
    36, 38, 43, 45, 46, 52, 53, 55, 57–59,
    61–64, 66, 72, 74, 85–87, 89–91, 93,
    97, 99, 101, 105, 107, 114–116, 118,
    121–123, 125, 127–129, 131–133,
    143, 149–151, 153, 154, 157–159,

    162–173, 177–182, 184, 185, 190,
    192–198, 209, 211–213, 220, 225,
    227, 231–235, 238, 240–245,
    248–255, 258, 259, 263, 266–273,
    275–277, 287, 288, 290, 291
Current waveforms, 163, 170, 171, 182,
    197, 235, 240, 241, 243, 267, 271
Curve Fitting Model (CFM), 19, 168, 170

D
Damage, 1–4, 7, 9, 27, 28, 30, 33, 36, 42,
    43, 52, 87, 88, 90, 92, 94, 99, 105,
    106, 108–114, 116, 117, 125, 127,
    133, 138, 140, 143, 198, 199, 233,
    234, 237, 245, 250, 251, 267, 269,
    276, 277, 288
Dart leader, 10, 14, 58, 64, 164, 249
Dart-stepped leader, 57
Data lines, 98, 120, 125, 127
DEHN, 88, 94–100, 102–104, 107, 115,
    117, 130, 132, 137, 147, 149
Diodes, 86
Direct lightning strikes, 2, 28, 34, 36, 38,
    87, 90, 101, 105, 110–113, 116, 117,
    129, 131, 134, 138, 140, 150, 250
Discharge current, 101, 128, 133
Dispersion, 25, 26, 137, 164, 173–176, 179,
    210
Distributed LCR Model (DLCRM), 18, 22,
    163, 164, 168, 171–173, 178–182,
    192–194, 197, 198, 240, 252, 253,
    255, 258, 259, 264, 266, 267, 270
Distribution lines, 232, 238, 239, 245
Distribution Static Compensator
    (D-STATCOM), 37
Distribution systems, 90, 231
Diverter, 34
DM, see Differential Mode
Down conductor, 38, 89, 90, 101, 103, 105,
    112, 115, 130, 134, 136–138,
    141–149
Downward leader, 14, 38, 63, 134, 172,
    240, 248, 251

E
Earth electrodes, 85, 95, 100, 101, 121,
    138, 145–148
Earthing, 2, 4, 36, 85, 86, 95, 100, 101,
    103, 105, 113, 120–123, 131, 137,
    138, 147, 148, 153–155
Earthing conductor, 2, 95, 98, 121, 122, 129

Earth resistance, 30, 113, 122, 145–147, 177, 181, 255, 258, 265
Earth termination, 99–101, 113
Earth termination system, 38, 89, 95, 98, 105, 134, 136, 137, 145
Electrical systems, 87, 89, 97, 105, 147, 242
Electrical wires, 94
Electric breakdown, 3, 11, 16, 51, 65, 289, 291
Electric charge, 1, 3, 7–9, 12, 14–16, 18, 20–22, 24, 26, 30, 33, 45, 51–53, 55–59, 63–65, 67–74, 76–79, 159, 171, 181, 182, 185, 194–196, 203, 231, 233, 248, 251, 252, 255, 260–262, 270, 277
Electric dipole, 26, 33, 65, 68, 261, 288, 290
Electric discharge, 33, 43, 45, 51, 54, 55, 57, 58, 64, 72, 232, 249, 287, 288
Electric fields, 4, 21, 23, 24, 33, 45, 51, 55, 59, 63–67, 69–74, 76–79, 100, 160, 171, 172, 183, 192–195, 197, 224, 225, 248, 260, 267
Electrodes, 13, 51, 86, 89, 90, 100, 101, 122, 137, 138, 146, 147, 172, 173, 181, 261
Electrodynamic effect, 28, 29
Electrodynamics, 4, 31, 53, 67, 148, 158, 231, 234, 235, 277
Electro-geometrical method, 38
Electromagnetic Compatibility (EMC), 118, 120, 148, 154, 194
Electromagnetic field, 19, 20, 23, 24, 45, 64, 87, 107, 117, 118, 120, 126, 131, 149, 150, 163–165, 167, 170, 171, 174, 178, 179, 181, 184, 185, 192, 195, 231, 234, 252, 255
Electromagnetic waves, 1, 19, 22, 24, 25, 45, 157–160, 162, 164, 168, 179, 252, 253, 255
Electron density, 13, 25, 26
Electronic systems, 1, 2, 8, 10, 88, 93, 106, 111, 112, 114, 116, 118, 120, 148–150, 157, 163, 164, 198, 199, 234, 279, 290
Electrostatic charge, 53, 57, 59, 63, 65, 79, 260
Electrostatic Discharge (ESD), 27, 33, 65, 79, 227, 288, 290
Electrostatic field, 22, 51, 53, 64, 65, 72, 74, 79, 172, 196, 287, 290, 291
Elves, 43, 45

Empirical Model (EM), 16, 168, 170, 209, 218
Enhanced protection, 90
Environmental shielding, 247
Equipotential bonding, 36, 38, 95, 98–101, 103, 113, 116, 118–132, 140, 147, 152–154
Exploding wire, 34
Explosion, 2, 14, 16, 29, 111, 116, 134, 224, 275
External lightning protection system, 105, 115, 134, 139–142, 144, 145, 148
External power supply, 95
External zone, 116, 127

F
F16 military aircraft, 72–75, 77–79, 231, 258, 259, 263–266, 268
Fair weather, 10, 54
Filters, 99
Finite difference, 172, 173, 176–178
Finite Difference Time-Domain method (FDTD), 19
Finite element, 22, 143
Finite element methods, 22, 143
First strokes, 5, 23, 57, 165–167
Flash, 1–3, 5, 7–10, 12, 14–16, 18–23, 27, 28, 30, 34, 36, 42, 43, 45, 46, 51–55, 57–59, 61, 63–65, 68–72, 74, 78, 79, 101, 106, 108–110, 116, 157–159, 163–167, 180, 181, 183, 184, 192–199, 203, 209, 210, 220, 222, 224–227, 232, 238–240, 245, 246, 248–251, 254, 260, 263, 265, 266, 269, 272–274, 278–280, 287, 288, 290, 291
Flash density, 42, 94, 110, 111, 238, 247, 249
Flat ground, 54
Fork lightning, 5, 14
Forrest fire, 7, 27, 43, 54
Foundation earth, 100, 101, 122, 138, 145–147
Frequency converter, 85, 99
Frequency dependent, 146, 147, 209
Fuses, 86, 91, 92, 97, 100, 133, 267

G
Gas discharge tube, 154
Grid size, 173
Ground conditions, 3
Ground conductivity, 32

Ground flashes, 14, 15
Grounding components, 245
Grounding high-frequency, 36
Grounding lightning protection, 38, 46
Grounding resistance, 237
Grounding systems, 35, 89–91, 245, 250
Ground Potential Rise (GPR), 91, 92
Group wave, 175

H

Hazardous area, 2, 138
High-frequency groundings, 36
High-voltage stations, 5
High-voltage testing, 231
Historic buildings, 85, 103
House, 27, 52, 67, 79, 85, 87–91, 93, 106
Human life, 27, 43, 106, 108, 110, 113, 163
Humid sulphurous atmosphere treatment,
     148

I

Identification, 10
Indirect effects, 2, 9, 10, 29, 38, 46, 51, 53,
     103, 106, 232, 234, 245, 274, 276
Indirect lightning strikes, 113, 127, 132
Indirect strikes, 2, 28, 29, 113, 132
Induced over voltages, 2
Inductance, 9, 118, 158, 160, 163, 168,
     172–175, 181, 233, 238, 257, 263,
     265, 270
Inductive electric charge, 56
Inductors, 92, 169
Infrastructure, 38, 245
Inspection, 113, 119, 142
Insulation breakdown, 65
Insulator, 232, 235–237, 240, 244, 248, 250
Integral technique, 22, 192
Intercepted lightning, 59, 61
Internal lightning protection, 96, 114–116,
     132, 278
International Electrotechnical Commission
     (IEC), 106, 115
Internet of Things, 3, 10, 231
Intracloud discharges, 42
Intracloud flashes, 14, 15, 158, 280
Ionization, 12, 14, 25, 45, 53, 69, 74, 168
Ionized gas, 16, 25, 26, 173
Ionized region, 13
IT system, 27, 85, 86, 128, 129

J
Joints, 140, 141, 275

K
K-changes, 21

L
Launch vehicle, 4
Leader inception criterion, 9
Leader stroke, 1, 3, 6, 7, 10–12, 15, 16, 27,
     64, 65, 169
Leader velocity, 14
LEMP protection, 114, 118, 119
Lengths, 18, 19, 21, 22, 31, 32, 38, 45, 58,
     64, 88, 91, 100, 109, 130, 139–141,
     143, 146, 147, 153, 154, 157, 171,
     173, 175, 178, 181, 184, 185, 189,
     190, 192, 212, 225, 236–238,
     253–255, 257–260, 263, 269, 277,
     287
Lightning channel, 1, 3, 5, 16, 18–22, 25,
     26, 28, 29, 34, 42, 64, 71, 149, 158,
     160–162, 168, 171–174, 178–180,
     182, 185, 194, 195, 197, 200, 235,
     240, 246, 253–255, 257–259, 264,
     265, 271, 272, 275, 287
Lightning conductor, 86, 171, 290
Lightning currents, 8, 9, 22, 28–30, 36, 58,
     64, 89–91, 94, 95, 101, 103,
     105–107, 114, 115, 117, 120,
     124–132, 134–138, 140, 143–145,
     147–150, 153, 157, 165, 170, 183,
     196, 198, 199, 213, 218, 232–235,
     240, 244, 245, 258, 263, 267,
     275–277, 287
Lightning damage, 1, 2, 105, 106
Lightning detection, 238
Lightning discharges, 9, 51, 53, 55, 57, 63,
     116, 134, 165, 194, 258
Lightning electric fields, 45, 270
Lightning electromagnetic fields, 87, 116,
     118, 131
Lightning Electromagnetic Pulse (LEMP),
     2, 8, 10, 16, 19–23, 29, 38, 87, 88,
     97, 103, 105, 106, 114, 116–118,
     149–151, 157, 158, 164, 165, 170,
     183, 185, 192, 195, 198, 199, 203,
     209, 210, 213, 220, 222, 224–227,
     229, 234, 273, 287, 290
Lightning equipotential bonding, 101, 105,
     112, 115, 116, 120–122, 125–127,
     142

Lightning frequency, 109, 110, 246
Lightning-induced voltages, 1, 10, 267, 272, 273
Lightning leaders, 12–16, 19, 26, 61, 95, 158, 162, 164, 168, 169, 210, 245, 246, 276
Lightning parameters, 42, 119, 180, 242
Lightning protection, 2, 4, 7, 10, 27, 30, 31, 33, 35, 36, 38, 46, 51, 72, 74, 85–90, 95–98, 101, 102, 105–108, 110, 112–121, 124–126, 129–135, 137, 138, 141–144, 148, 157, 198, 199, 231, 234, 273–275, 278, 279
Lightning Protection Level (LPL), 107, 108, 112, 116, 135
Lightning Protection System (LPS), 88, 90, 100, 101, 103, 105–108, 112–114, 116, 124, 134, 136, 137
Lightning Protection Zone (LPZ), 102, 106, 107, 116–118, 121, 150, 154
Lightning standards, 33, 46, 106, 107, 114, 116
Lightning strikes, 2–5, 8, 20, 22, 23, 27–30, 32–34, 37, 38, 43, 46, 52–54, 60–64, 67, 69, 72, 77–80, 87, 90, 92, 94, 97, 105, 108, 110–112, 115, 116, 121, 134, 135, 139–141, 143–145, 147, 151, 153, 154, 194, 198, 199, 232, 234, 235, 238, 240, 245–250, 258, 264, 269, 270, 275–279, 288
Lightning surge protection, x
Lightning surges, 2, 87, 89–91, 97, 99, 106–108, 114, 159, 169
Lightning transients, 27, 43, 57, 108, 184, 200, 232, 234, 249, 276
Lightning warning systems, 102, 112
Localisation, 209, 210, 217, 227, 229
Low-voltage networks, 27
Lumped Circuit Model (LCM), 19, 164, 168, 170, 172, 198, 200

M
Magnetic fields, 16, 20, 21, 29, 64, 66, 149, 152, 157, 160, 163, 165, 167, 168, 171–173, 183–185, 192–195, 197, 232, 253, 269
Magnetic shield, 152
Mast, 36–38, 94, 95, 138
Maxwell's equations, 160, 168, 172, 183, 252, 255
M change, 21, 165
M-components, 16, 42

Measurements, 7, 14, 16, 19, 21, 23, 24, 33, 42, 72, 78, 113, 157, 163–167, 170, 179, 181, 182, 192, 195, 224, 231, 235, 242, 277, 288
Mesh method, 136
Metal, 33, 74, 80, 90, 95–97, 99, 101, 107, 112, 115, 118, 120, 121, 124, 125, 127, 130, 131, 137, 139–141, 144–148, 152, 253, 260
Metrological Measuring System, 40
Metal Oxide Varistor (MOV), 93
Mission critical systems, 27, 52, 274

N
Nacelle, 101, 102
NASA, 4, 54, 66, 279, 280
Nearby lightning strike, 112, 151, 152
Negative lightning, 43, 57, 58
Negative strokes, 150, 165
Networks and interactive Service, 100
NOAA, 52, 159, 279
Non-convection graupel-ice mechanism, 56
Non-inductive convection mechanism, 56
Non-inductive electric charging, 56
Normal mode, 85, 91
NSSL, 159, 279

O
Offshore oil platforms, 7
Optical fiber, 127, 130
Overhead power lines, 174
Overvoltages, 29, 30, 34, 38, 85, 114, 116, 133

P
Parabolic antenna, 140
Partial lightning current, 115–117, 124, 128, 130, 131
Peak current, 9, 14, 16, 30, 37, 164–166, 170–172, 182, 193, 233, 234, 241–244, 266, 273
Perceptron, 225–227
Personal protection, 27
Petrol tank, 2
PhotoVoltaic systems (PV), 88, 96, 98
Plasma, 1, 13, 14, 16, 18, 19, 25, 26, 162, 165, 168, 275
Positive flashes, 12, 22, 58, 165, 167, 194
Positive lightning, 12, 42, 194
Potential gradient, 32, 59, 106, 289

Power lines, 2, 5, 7, 27, 53, 87, 90, 91, 112,
    113, 198, 235, 245, 246, 248, 290
Power transmission lines, 232
Pre-lightning, 51, 53, 60, 64, 79, 287
Preliminary breakdown, 57, 58
Probability, 33, 61, 62, 67, 69, 72, 74,
    76–78, 90, 94, 105, 111–113, 115,
    135, 199, 246
Propagation velocity, 19
Protective angle method, 135, 136, 139, 141
Protective devices, 86–88, 99, 112, 114,
    116, 125–129, 131–133

R
Radar, 4, 33, 275, 278
Radio wave, 209, 211, 212, 217, 218, 229
Radius, 13, 19, 37, 38, 42, 101, 107,
    134–136, 146, 151, 171–174, 181,
    182, 192, 235–238, 257–259, 263
Radome, 3, 7, 33, 59, 60, 64, 66, 69–72, 74,
    76, 77, 79, 200, 259–261, 269, 270,
    288, 289, 291
Rate of rise, 20, 23, 24, 34, 163, 165–167,
    182, 193, 194, 196, 197, 231, 241,
    242, 244, 245, 266, 267
Residential buildings, 116
Residual Current Device (RCD), 85, 86, 99
Resistance, 9, 11, 18, 19, 30, 31, 34, 36, 62,
    86, 103, 113, 122, 123, 137, 138,
    142, 143, 145–147, 154, 160, 163,
    168, 172, 173, 175, 177, 179–181,
    199, 233, 235, 237, 238, 250,
    257–260, 265–267, 269, 272, 274
Return-stroke, 1, 3, 6, 7, 10–23, 26, 28, 29,
    34, 42, 54, 57–59, 62–64, 149, 151,
    157–160, 162–174, 178–185,
    192–196
    action integral, 20, 167
Return-stroke current waveform, 240
Return stroke model, 18, 19, 21, 157, 163,
    165, 194, 234, 252
Return stroke velocity, 10, 18, 23, 165, 167,
    172, 178, 179, 180
Ring conductors, 147
Risk, 9, 32, 33, 77, 79, 90, 105, 106,
    108–111, 113, 116–118, 123, 128,
    133, 135, 147, 153, 238, 248, 288
Risk analysis, 105, 119
Risk assessment, 105, 108, 110, 116, 119,
    125, 133
Risk management, 9, 106, 108
Rocket, 4, 6, 14, 163–165, 195

Rocket-triggered lightning, 6, 165
Rocky soil, 147
Rolling Sphere Method (RSM), 37,
    134–136, 138, 140, 141, 276
Route selection, 247

S
Safety, 1, 3, 8, 46, 79, 108, 143, 150, 267,
    269, 272
Safety system, 143
Salt mist treatment, 148
Sand soil, 181
Scaling test, 277
Screening, 56, 58
Sea, 43, 57, 62, 94, 95, 181, 236, 290
Separation distance, 65, 67, 73, 74, 78, 105,
    115, 116, 130, 138, 141–144, 147,
    237, 248, 249, 260, 261
Shield earthing, 153, 154
Shielding, 10, 36, 37, 46, 52, 101, 105,
    112–114, 118, 119, 125–127, 131,
    149–154, 234, 245–247, 250, 274,
    276, 289
Shielding failure, 34, 232, 235, 243–246,
    250
Shield wire, 35–37, 151, 232, 235–237,
    239, 240, 244, 245, 249–251
Shield wire surge impedance, 236
Smart city, 2, 3, 10, 46, 224, 231, 279
Soil resistivity, 238, 247, 249, 250
Space charge, 56
Spark gaps, 34, 86, 121, 131, 154
SPD, *see* Surge Protection Device
SPD location, 114, 133
Sprites, 43, 45
Standardized lighting currents, 110
Static wire, 36
Steeple, 103, 139, 142
Stepped leader, 14, 57, 58, 63, 64, 69, 164,
    169, 249–251
Striking distance, 4, 14, 37, 134, 135, 139
Submicrosecond, 241–245
Subsequent return stroke, 14, 16, 20, 34,
    57–59, 149, 163, 164, 181, 233, 249,
    290
Substation, 5, 34, 36, 37, 51, 53, 67, 110,
    148, 227, 231, 234, 235, 240–246,
    249–251, 287, 290, 291
Surge, 1, 2, 18, 27, 34–36, 46, 85–87,
    90–93, 95, 97, 99, 100, 103, 106,
    108, 111, 112, 114, 116, 117,
    125–129, 131–133, 154, 233, 251,
    278

Surge arresters, 93, 97, 106, 129, 131–133, 154, 250, 251
Surge current rating, 91
Surge impedance, 18, 38, 235–237, 251, 252, 271
Surge protection, 90, 92–94, 99–101, 105, 106, 114, 131, 132, 142
Surge Protection Device (SPD), 92, 99, 101, 105, 107, 112, 114, 118, 124, 128, 130, 132, 133, 155, 241, 244, 278
Surge Protection Measures (SPM), 114
Swept stroke, 34, 64, 70, 72, 267, 269, 272
Switches, 89, 115, 157, 200

T
Tank flammable material, 2
Telecommunication, 2, 7, 8, 87, 131, 132, 158, 174, 234
Telecommunication lines, 106, 113, 127, 129, 131
Telecommunication systems, 38, 88, 121, 131, 163, 231
Telecommunication towers, 5
Temperature, 2, 9, 12, 13, 20, 25–27, 30, 31, 41–43, 56, 58, 62, 140, 144, 148, 199, 211, 231–233, 257, 279
Test, 26, 33, 67, 69, 123, 148, 158, 178, 197, 209, 211, 212, 217, 218, 274–277, 288
Thermal effect, 28, 29, 164
Thunder, 7, 16, 28, 29, 63, 79, 110, 111, 158, 163, 164, 168, 209, 224, 229, 239
Thundercloud, 3, 4, 6, 8–16, 27, 33, 43, 45, 51–60, 62–67, 72, 74, 79, 158, 159, 169, 181, 182, 199, 203, 231, 260, 265, 270, 276, 277, 287–291
Thundercloud electric field, 4, 6, 12, 15, 33, 45, 51, 52, 55, 56, 64, 67, 270
Thunderstorm days, 238
Thunderstorm life cycle, 40
TN system, 86
Touch voltages, 95, 112, 116, 123
Tower ground resistance, 237
Tower height, 247, 249
Tower surge impedance, 235
Tower, wind turbine, 248

Transformer, 34, 36, 91, 95, 102, 110, 124, 125, 148, 157, 249–251, 274
Transmission Line Modelling (TLM), 252
Transmissions lines, 19, 34, 35, 158–160, 162, 163, 168, 171, 173, 175–181, 232, 235, 240, 245–258
Tree damage, 42
Triggered lightning, 6, 14, 59, 61, 63, 164, 165
Triggered lightning current, 14, 63
Troposphere, 9, 27, 41–43, 52
TT system, 85, 86
Turbine blades, 43, 101
Type A earth electrodes, 146
Type B earth electrodes, 147

U
Unshielded cable, 131
Upward flash, 15, 16, 18, 43, 51, 101
Upward leader, 14, 38, 63, 134, 172, 249
Upward lightning flashes, 248

V
Velocity, 10, 13, 14, 16, 18–20, 23–25, 45, 151, 163, 164, 166, 167, 170, 173–175, 178–181, 197, 252
Vertical ground rods, 36
Visual effect, 28

W
Wave equation, 176
Waveform, 34, 91, 99, 107, 116, 158, 163, 169, 170, 182, 185, 224, 226, 227, 232, 244, 249, 268, 269, 276, 277
Wavelength, 150–152, 178, 179, 225
Wave number, 24, 160, 162, 179, 225
Waveshape, 241, 242, 244
Wave velocity, 25, 174, 210
Wind turbines, 43, 85, 101–103, 139, 197, 248
Wing, 4, 7, 33, 59, 65, 66, 69–74, 76–79, 199, 259–261, 264–267, 269, 270, 289–291
Wire, exploding, 34
Wireless, 93, 151, 225, 234

Printed by Books on Demand, Germany